Active Rheology Control of Cementitious Materials

This book reports on cutting-edge research within the new field of active rheology control of cementitious materials, presenting new ideas developed within the ERC Advanced Grant Project, SmartCast (hosted at Ghent University), which extend the possibilities of admixtures and additions beyond current options.

The research presented here develops a new method of actively controlling the rheology of fresh concrete during casting operations by incorporating specially designed responsive components. This results in real-time changes to the rheological behaviour of the cementitious material, allowing the user to intervene actively after the cementitious material has left the mixing phase. This newly gained agility contributes to increased processing speed and placement reliability in the case of traditional casting methods and can also facilitate advanced 3D concrete printing. The different routes followed to achieve this active rheology control are explained within.

The book suits researchers and innovative practitioners and is the first comprehensive text to present these new findings.

Modern Concrete Technology Series

A series of books presenting the state-of-the-art in concrete technology.

Series Editor
Geert De Schutter
Department of Structural Engineering and Building Materials
Faculty of Engineering and Architecture
Ghent University
Ghent
Belgium

16. **Fibre Reinforced Cementitious Composites – 2nd edition**
    A. Bentur and S. Mindess

17. **Sustainability of Concrete**
    P. C. Aïtcin and S. Mindess

18. **Concrete Surface Engineering**
    B Bissonnette, L Courard and A Garbacz

19. **Textile Reinforced Concrete**
    A. Peled, A Bentur and B Mobasher

20. **Durability of Concrete: Design and Construction**
    M.G. Alexander, A. Bentur and S Mindess

21. **Concrete Permeability and Durability Performance: From Theory to Field Applications**
    Roberto J. Torrent, Rui D. Neves and Kei-ichi Imamoto

22. **Shotcrete: Materials, Performance and Use**
    M. Jolin and D.R. Morgan

23. **Active Rheology Control of Cementitious Materials**
    Geert De Schutter and Karel Lesage

For more information about this series, please visit: https://www.routledge.com/series-title/book-series/MCT

Active Rheology Control of Cementitious Materials

Geert De Schutter and Karel Lesage

CRC Press
Taylor & Francis Group
Boca Raton London New York

CRC Press is an imprint of the
Taylor & Francis Group, an **informa** business

Cover image: Geert De Schutter and Karel Lesage

First edition published 2023
by CRC Press
4 Park Square, Milton Park, Abingdon, Oxon, OX14 4RN

and by CRC Press
6000 Broken Sound Parkway NW, Suite 300, Boca Raton, FL 33487-2742

British Library Cataloguing-in-Publication Data
A catalogue record for this book is available from the British Library

Library of Congress Cataloging-in-Publication Data
Names: De Schutter, Geert, editor. | Lesage, Karel, editor.
Title: Active rheology control of cementitious materials / [edited by]
Geert De Schutter and Karel Lesage.
Description: First edition. | Boca Raton : CRC Press, [2023] |
Series: Modern concrete technology | Includes bibliographical references and index.
Identifiers: LCCN 2022043171 | ISBN 9781032266909 (hbk) |
ISBN 9781032266916 (pbk) | ISBN 9781003289463 (ebk)
Subjects: LCSH: Cement–Additives. | Concrete–Viscosity. |
Rheology. | Viscous flow. | Adaptive control systems.
Classification: LCC TA434 .A33 2023 | DDC 620.1/3–dc23/eng/20221130
LC record available at https://lccn.loc.gov/2022043171

ISBN: 978-1-032-26690-9 (hbk)
ISBN: 978-1-032-26691-6 (pbk)
ISBN: 978-1-003-28946-3 (ebk)

DOI: 10.1201/9781003289463

Typeset in Sabon
by Newgen Publishing UK

Contents

Preface   vii
Author and contributor biographies   ix
Funding information   xiii

1 Introduction to active rheology control of cementitious
   materials   1
   GEERT DE SCHUTTER AND KAREL LESAGE

2 Rheology and processing of cementitious materials   37
   KAREL LESAGE, KHADIJA EL-CHEIKH, AND GEERT DE SCHUTTER

3 Active rheology control of cementitious materials with
   responsive mineral particles   77
   DENGWU JIAO, MERT YUCEL YARDIMCI, KAREL LESAGE,
   AND GEERT DE SCHUTTER

4 Active rheology control of cementitious materials with
   responsive polymers   137
   METWALLY EZZAT, RICHARD HOOGENBOOM, KAREL LESAGE,
   AND GEERT DE SCHUTTER

5 Application of active rheology control to pumping of
   cementitious materials   171
   ROBIN DE SCHRYVER, MERT YUCEL YARDIMCI, KAREL LESAGE,
   AND GEERT DE SCHUTTER

6 Application of active rheology control to concrete
   formwork leakage   189
   CHIZYA CHIBULU, MERT YUCEL YARDIMCI, KAREL LESAGE,
   AND GEERT DE SCHUTTER

7   Application of active rheology control to 3D printing of
    cementitious materials   221
    DENGWU JIAO, KAREL LESAGE, AND GEERT DE SCHUTTER

8   Numerical simulation of active rheology control of
    cementitious materials   233
    ROBIN DE SCHRYVER, KAREL LESAGE, AND GEERT DE SCHUTTER

9   Overview and remaining challenges of active rheology
    control of cementitious materials   269
    KAREL LESAGE AND GEERT DE SCHUTTER

    Index   277

Preface

Although the Romans mastered concrete construction in an amazing way (considering the scientific and technological knowledge in that era), we now have about a century of experience with modern concrete construction. However, new developments in powder and admixture technology have given a boost to the materials science of cementitious materials only in the past three decades. Modern concrete is a high-tech material, which is a remarkable example of applied nanotechnology combining many disciplines like chemistry, physics and mechanics. While traditional concrete needs active compaction (vibration), its performance in fresh state has now been outmastered by highly technological mixtures based on tailor-made blends combined with specially designed admixtures, showing rheological properties that make the fresh concrete self-compacting (SCC, also called self-consolidating concrete in North-America). The importance of the introduction of this revolutionary type of concrete, with significant ecological benefits (reduced energy consumption, less noise, less vibration, etc.), is illustrated by the fact that the first European project on self-compacting concrete was the first construction-related project ever to reach the final round of the prestigious European Descartes Prize competition.

In spite of its many advantages, more than three decades after its introduction SCC has still gained only a low market share. The excellent rheological properties of SCC have not been sufficiently translated into new production opportunities, as traditional production processes cannot exploit the full potential of the new material. Encountered problems concern high pumping pressures, leaking of formworks and excessive formwork pressures. Current approaches only rely on the passive behaviour of the fresh cementitious material, influenced by practical and environmental conditions, with no means to actively intervene once the material has left the mixer. For many applications, with contradictory rheological requirements for the different processing steps (pumping, formwork casting, 3D printing, etc.), one mixture cannot optimally serve the purpose of all steps if no adjustments can be made while processing.

Within the SmartCast project, funded as an Advanced Research Grant by the European Research Council (ERC), a ground-breaking innovative concrete casting concept has been scientifically studied. Inspired by the application of responsive polymers in other fields, and combined with responsive mineral particles or powders, the rheology of fresh cementitious materials can be adjusted in an active way, post-mixing. While the principal goal was to fundamentally upgrade classical formwork-based production methods, the obtained results are also very promising for the further development of 3D concrete printing techniques. The SmartCast concept is introduced in this text book, which can be considered as a concise scientific report of the project as completed in September 2022, after 6 years of research (one year longer than foreseen because of the coronavirus disease (COVID) pandemic).

We are very thankful for the dedication and contribution of all SmartCast researchers, for their excellent research during the previous years, leading to a successful lab-scale proof of concept. Their help in writing the different book chapters is also truly applauded. We gratefully acknowledge the support given by the European Research Council, including the academic freedom we were given to chase our scientific dreams. The generous extension of the project end date because of the pandemic was very helpful and welcome. We are also very happy with the support given by Ghent University. Specials thanks are addressed to the co-workers of the UGent EU-Team who really facilitate the life of researchers working on major European projects. We were also very privileged that we could perform the main experimental work at the Magnel-Vandepitte Laboratory, Department of Structural Engineering and Building Materials, Faculty of Engineering and Architecture, as well as the main chemical work and synthesis at the Supramolecular Chemistry Group, Department of Organic and Macromolecular Chemistry, Faculty of Sciences. Thank-you to all co-workers of both groups who contributed to the success of SmartCast!

Now that the final report of SmartCast is written, showing the scientific progress and breakthrough, a lot of further work still needs to be done before the concept can be practically applied on a larger scale in concrete industry. We hope that further research and development actions will be initiated by many groups worldwide, in academia and industry in combined effort. We truly hope that the Open Access publication of this book will provide a good basis for others to continue the efforts, in parallel to further actions planned in our own group in Ghent. Concrete is a splendid and advanced building material deserving motivated and dedicated researchers from industry and academia to even further improve it in service of mankind, as reflected in the motto of our Department of Structural Engineering and Building Materials: 'Building[for]Humanity'.

Geert De Schutter
Karel Lesage
Ghent, September 2022

Author and contributor biographies

Chizya Chibulu is a PhD candidate in the Department of Structural Engineering and Building Materials at Ghent University in Belgium. Her current research mainly focuses on rheology of cementitious materials, application of magneto-rheology to cementitious materials, formwork pressure and formwork leakage. She received a Bachelor of Science in Civil Engineering in 2013 from the University of Cape Town, South Africa. The following year, she joined the Concrete Materials and Structural Integrity Research Unit (CoMSIRU) as a master's student and went on to obtain a Master of Science in Civil Engineering, specializing in Structural Engineering and Structural Materials in 2016. Her research topics then included concrete repair, durability and the use of high volumes of supplementary cementitious materials in concrete.

Robin De Schryver is a postdoctoral researcher at the Magnel-Vandepitte Laboratory in the department of Structural Engineering and Building Materials of the Faculty of Engineering and Architecture at Ghent University. After graduating magna cum laude for his degree of Master of Science in Civil Engineering in 2016, he pursued an academic path obtaining his associated PhD degree in June 2022. During his PhD journey of exploring the fundamentals of non-Newtonian flows and rheology in view of active rheology control, he gained deep expertise in experimental quantification and numerical simulation of concrete and cementitious suspensions or non-Newtonian fluids in general and related data processing. Meanwhile, he published several articles on the fundamental behaviour description of pipe flows and the numerical simulation thereof. As while growing up in the region of Ghent (Belgium) with a passion for science, programming and mathematics, one of his life philosophies is that a model is only an approximation of a considered physical problem constrained by its mathematical simplicity and assumptions; the emphasis of his research was put on several aspects of numerical validation.

Geert De Schutter is a full professor of 'Concrete Technology' and head of the Department of Structural Engineering and Building Materials, Faculty of Engineering and Architecture, Ghent University, Belgium. He obtained a Master of Science (1990) and PhD degree (1996) in Civil Engineering, both from Ghent University. He is a fellow of RILEM and ACI, and a member of Fib. He is a laureate of several national and international awards, among which are the Vreedenburgh Award 1998, the RILEM Robert L'hermite Award 2001 and the ACI Anderson Medal in 2014. He was appointed as a guest professor at several universities worldwide, including Oita University (Japan), University of Cergy-Pontoise (France), University of Nantes (France) and Tongji University Shanghai (China). In 2012 he was also appointed as Francqui Chair at the University of Liège (Belgium). His research interests include rheology and processing (including 3D printing), hydration and microstructure development, and durability of cementitious materials. In 2016, he was awarded an Advanced Grant of the European Research Council. Prof. G. De Schutter is (co-)author of more than 700 papers in peer-reviewed journals and conference proceedings, and of a few books, including *Self-Compacting Concrete* and *Damage to Concrete Structures*, published by Taylor & Francis Group. He is currently the editor of the Book Series 'Modern Concrete Technology' published by Taylor & Francis.

Khadija El-Cheikh was born in Tripoli, Lebanon, in 1987. She graduated from the Faculty of Engineering in Lebanon with a Bachelor's degree in Civil Engineering. After her Bachelor, she received a full scholarship from the University of Joseph Fourier in Grenoble, France, where she obtained her Master's degree in Geomechanics, Civil Engineering and Risks. In 2015, Khadija received her PhD degree from IMT Lille Douai University, France. Her thesis focused on the study of the formation of the interface layer between fresh concrete and rough formwork. Focusing on concrete rheology, Khadija spent three years as a postdoctoral researcher in Ghent University, Belgium, under the supervision of Prof. Geert De Schutter. Her research interests lie primarily in understanding the rheological behaviour of the cement-based materials. She is now working as a project leader in the Concrete Technology Laboratory at Belgian Building Research Institute (BBRI) in Belgium.

Metwally Ezzat was born in 1989 in Sharkia, Egypt. He obtained his Bachelor of Science degree in Chemistry in 2011 from Assiut University, Egypt. In 2015, Metwally received a full scholarship from the National Central University, Taiwan, to further his Master of Science degree in Biomedical Engineering under the supervision of Prof. Chun-Jen Huang. During his master's degree, Metwally developed novel phospholipid polymeric micelles for drug delivery applications. In addition, he studied the superhydrophilicity of zwitterionic polymers for antifog/antifrost coating applications. In 2022, Metwally received his PhD degree in Chemistry from Ghent University under the supervision of Prof. Richard Hoogenboom and Prof. Geert De Schutter. Currently he is a postdoctoral researcher at Ghent University.

Richard Hoogenboom is heading the Supramolecular Chemistry (SC) group at Ghent University that was founded based on his appointment as an associate professor in 2010 and he was promoted to full professor in 2014. He obtained a PhD degree from Eindhoven University of Technology under the supervision of Prof. Ulrich S. Schubert and performed postdoctoral research with Prof. Martin Möller (RWTH Aachen) and Prof. Roeland Nolte (Radboud University Nijmegen). Prof. Hoogenboom has published more than 500 refereed scientific articles that received 22,500+ citations (h-index 71) and he is listed as an inventor on 21 patent families. He is currently the editor-in-chief of the *European Polymer Journal* and an associate editor for *Australian Journal of Chemistry*. Prof. Hoogenboom is the recipient of the inaugural Royal Society of Chemistry (RSC) Polymer Chemistry award (2015), the PI IUPAC young investigator award (2016), the ACS Macromolecules/ Biomacromolecules Young Investigator award (2017) and the ACS Carl S. Marvel Creative Polymer Chemistry Award (2021) and is an elected fellow of the RSC, the young academy of Europe and the POLY division of the ACS. Since January 2018, Prof. Hoogenboom is also the cofounder of Avroxa BVBA that commercializes poly(2-oxazoline)s as Ultroxa®.

Dengwu Jiao is a postdoctoral researcher in the Magnel-Vandepitte Laboratory at the Department of Structural Engineering and Building Materials, Ghent University, Belgium. He obtained his Master of Engineering from China Building Materials Academy (Beijing, China) in August 2017, and obtained the degree of Doctor of Civil Engineering from Ghent University (Ghent, Belgium) in January 2021 and the Doctoral Degree in Engineering from Hunan University (Changsha, China) in April 2021. He has authored around 40 publications in reputative journals and conference proceedings, including 17 first-authored papers indexed by SCI. He is a member of RILEM, active in Technical Committee PCC 'Pumping of concrete'. His research interest includes concrete rheology, mixture design and 3D concrete printing of cement-based materials.

Karel Lesage is a part-time professor at the Magnel-Vandepitte Laboratory in the Department of Structural Engineering and Building Materials at Ghent University in Belgium. He obtained his Master of Science from KU Leuven (2009) and his PhD degree from both KU Leuven and Ghent University (2014). Until 2023 he was a postdoctoral researcher in the team of Prof. Geert De Schutter. During that time he co-authored several patents and multiple papers in peer-reviewed journals. His main research focuses on the rheology of highly concentrated suspensions with a particular interest in experimental control of particle interactions due to shear, surface chemistry and chemical admixtures. He also took part in Rilem Technical Committees 209-RFC, 266-MRP, 276-DFC, PCC and PFC.

Mert Yucel Yardimci obtained his Bachelor's degree in Civil Engineering from Firat University and master's and doctorate degrees from Dokuz Eylul University in Türkiye. He held research and teaching positions at different universities in Türkiye. He also worked as a postdoctoral researcher at Israel Institute of Technology-Technion between May 2008 and May 2009. While working as an associate professor at Istanbul Okan University, he joined Prof. Geert De Schutter's research team at Magnel-Vandepitte Laboratory for Concrete Research at Ghent University, Belgium, in January 2019, and he has been working as a postdoctoral researcher in an ERC Advanced Grant Project- SmartCast, Smart casting of concrete structures by active control of rheology, since then. His scientific expertise includes the rheology of cementitious materials, alkali-activated materials, textile-reinforced composites and ultra-high-performance cementitious composites. He has been active in RILEM technical committees and the Turkish Chamber of Civil Engineers.

Funding information

European Research Council
Established by the European Commission

This report is a deliverable of the European Research Council (ERC) Advanced Grant project 'SmartCast'. This project has received funding from the ERC under the European Union's Horizon 2020 research and innovation programme (grant agreement No. 693755).

UNIVERSITEIT
GENT

The research has been performed at the Magnel-Vandepitte Laboratory, Department of Structural Engineering and Building Materials, Faculty of Engineering and Architecture, Ghent University, Belgium, in due cooperation with the Supramolecular Chemistry Group, Department of Organic and Macromolecular Chemistry, Faculty of Sciences, Ghent University, Belgium.

DISCLAIMER

This book reports cutting-edge research results, and is to be considered as a scientific research report only. The authors and/or the related institutions or organizations will in no event be liable for any loss or damage including without limitation, indirect or consequential loss or damage, or any loss or damage whatsoever arising from, or in connection with, the use of information given in this report.

OPEN ACCESS SOURCES

Parts of the chapters in this book are based on previous publications of the same group of authors, published as internal documents or in Open Access journal papers:

- G. De Schutter (2015) Smart casting of concrete structures by active control of rheology (SmartCast), Advanced Grant Research proposal as submitted to the European Research Council (ERC)
- G. De Schutter, K. Lesage (2018) Active control of properties of concrete – A (p)review, *Materials and Structures*, 51:123, Open Access, CC BY 4.0, published by Springer, doi.org/10.1617/s11527-018-1256-2
- D. Jiao, K. El Cheikh, C. Shi, K. Lesage, G. De Schutter (2019) Structural build-up of cementitious paste with nano-Fe_3O_4 under time-varying magnetic fields, *Cement and Concrete Research*, 124, 105857, CC BY 4.0, published by Elsevier, doi 10.1016/j.cemconres.2019.105857
- R. De Schryver, G. De Schutter (2020) Insights in thixotropic concrete pumping by a Poiseuille flow extension, *Applied Rheology*, 30, 77–101, Open Access, CC BY 4.0, published by De Gruyter, doi.org/10.1515/arh-2020-0103
- D. Jiao, K. Lesage, M.Y. Yardimci, K. El Cheikh, C. Shi, G. De Schutter (2020) Rheological properties of cement paste with nano-Fe_3O_4 under magnetic field: Flow curve and nanoparticles agglomeration, *Materials*, 13, 5164, Open Access, CC BY 4.0, published by MDPI, doi:10.3390/ma13225164
- D. Jiao, K. Lesage, M.Y. Yardimci, K. El Cheikh, C. Shi, G. De Schutter (2021) Quantitative assessment of the influence of external magnetic field on clustering of nano-Fe_3O_4 particles in cementitious paste, *Cement and Concrete Research*, 142, 2021, 106345, Open Access, CC BY 4.0, published by Elsevier, doi.org/10.1016/j.cemconres.2020.106345
- D. Jiao, K. Lesage, M.Y. Yardimci, K. El Cheikh, C. Shi, G. De Schutter (2021) Rheological behavior of cement paste with nano-Fe_3O_4 under magnetic field: Magneto-rheological responses and conceptual calculations, *Cement and Concrete Composites*, 120, 104035, Open Access, CC BY 4.0, published by Elsevier, doi.org/10.1016/j.cemconcomp.2021.104035

- D. Jiao, K. Lesage, M.Y. Yardimci, K. El Cheikh, C. Shi, G. De Schutter (2021) Structural evolution of cement paste with nano-Fe_3O_4 under magnetic field – Effect of concentration and particle size of nano-Fe_3O_4, *Cement and Concrete Composites*, 120, 104036, Open Access, CC BY 4.0, published by Elsevier, doi.org/10.1016/j.cemconcomp.2021.104036
- D. Jiao, K. Lesage, M.Y. Yardimci, C. Shi, G. De Schutter (2021) Possibilities of fly ash as responsive additive in magneto-rheology control of cementitious materials, *Construction and Building Materials*, 296, 123656, Open Access, CC BY 4.0, published by Elsevier, doi.org/10.1016/j.conbuildmat.2021.123656.
- R. De Schryver, K. El Cheikh, K. Lesage, M.Y. Yardimci, G. De Schutter (2021) Numerical reliability study based on rheological input for Bingham paste pumping using a finite volume approach in OpenFOAM, *Materials*, 14, 5011, Open Access, CC BY 4.0, published by MDPI, doi.org/10.3390/ma14175011.
- D. Jiao, R. De Schryver, C. Shi, G. De Schutter (2021) Thixotropic structural build-up of cement-based materials: A state-of-the-art review, *Cement and Concrete Composites*, 122, 104152, Open Access, CC BY 4.0, published by Elsevier, doi.org/10.1016/j.cemconcomp.2021.104152
- D. Jiao, K. Lesage, M.Y. Yardimci, C. Shi, G. De Schutter (2021) Flow behavior of cementitious-like suspension with nano-Fe_3O_4 particles under external magnetic field, *Materials and Structures*, 54, 209, Open Access, CC BY 4.0, published by Springer, doi.org/10.1617/s11527-021-01801-y
- C. Chibulu, M.Y. Yardimci, D. Jiao, R. De Schryver, K. Lesage, G. De Schutter (2021) Active stiffening control by magnetically induced blocking in confined flow of fly ash pastes, *Construction and Building Materials*, 313, 125485, Open Access, CC BY 4.0, published by Elsevier, doi.org/10.1016/j.conbuildmat.2021.125485
- D. Jiao, C. Shi, G. De Schutter (2022) Magneto-rheology control in 3D concrete printing: A rheological attempt, *Materials Letters*, 131374, Open Access, CC BY 4.0, published by Elsevier, doi.org/10.1016/j.matlet.2021.131374
- D. Jiao, C. Shi, G. De Schutter (2022) Magneto-responsive structural build-up of highly flowable cementitious paste in the presence of PCE superplasticizer, *Construction and Building Materials*, 327, 126925, Open Access, CC BY 4.0, published by Elsevier, https://doi.org/10.1016/j.conbuildmat.2022.126925.
- D. Jiao, M.Y. Yardimci, R. De Schryver, K. Lesage, C. Shi, G. De Schutter (2022) Effect of solid volume fraction on the magnetorheological response of nano-Fe3O4 incorporated cementitious paste, *ASCE J. Mat. Civ. Eng.*, 04022280-9, Open Access, CC BY 4.0, published by ASCE, doi:10.1061/(ASCE)MT.1943-5533.0004440.

- D. Jiao, C. Shi, G. De Schutter (2022), Estimation of magnetic force between micron-sized fly ash particles in cementitious suspension, *ASCE Journal of Materials in Civil Engineering*, Open Access, CC BY 4.0, published by ASCE .
- C. Chibulu, D. Jiao, M.Y. Yardimci, G. De Schutter (submitted 2022) Magneto-rheology control of cement paste containing Fe_3O_4 nanoparticles in view of reducing or blocking formwork leaching.
- G. De Schutter, M. Ezzat, K. Lesage, R. Hoogenboom (submitted 2022) Responsive superplasticizers for active rheology control of cementitious materials.
- M. Ezzat, K. Lesage, R. Hoogenboom, G. De Schutter (submitted 2022) 'Smart superplasticizers based on redox-responsive polymers for rheology control of cementitious materials'.
- M. Ezzat, K. Lesage, R. Hoogenboom, G. De Schutter (submitted 2022) Magnetic-responsive polymer composites for rheology control of cementitious materials.

DOCTORAL DISSERTATIONS

The SmartCast project resulted in following doctoral dissertations, which have also been the basis for several chapters in this book:

- Dengwu Jiao (2021) *Active rheology control of cementitious materials using magnetic field*, promoted by Prof. Geert De Schutter, Ghent University, Belgium, and Prof. Caijun Shi, Hunan University, PR China.
- Metwally Muhammad Ezzat (2022) *Concrete superplasticizers: From fundamental studies to the development of smart superplasticizers for rheology control*, promoted by Prof. Richard Hoogenboom and Prof. Geert De Schutter, Ghent University, Belgium.
- Robin De Schryver (2022) *Active rheology control of cementitious materials: Numerical and experimental pumping investigation*, promoted by Prof. Geert De Schutter, Ghent University, Belgium.
- Chizya Chibulu (in preparation) *Application of active rheology control (ARC) to concrete formwork leakage* (tentative title), promoted by Prof. Geert De Schutter, Ghent University, Belgium.

Chapter 1

Introduction to active rheology control of cementitious materials

Geert De Schutter and Karel Lesage

CONTENTS

1.1	Introduction: background and driving forces	2
	1.1.1 Context	2
	1.1.2 State of the art and step beyond	3
	1.1.3 Ground-breaking objectives	7
	1.1.4 Impact	8
1.2	Active rheology control and active stiffening control	9
1.3	Active control already available for cementitious materials	11
	1.3.1 Mechanical intervention	11
	1.3.2 Chemical intervention	13
	1.3.3 Thermal intervention	13
	1.3.4 Hygral intervention	14
	1.3.5 Intervention by pressure	15
	1.3.6 Magnetic intervention	15
	1.3.7 Electric intervention	17
	1.3.8 Intervention with microwaves	18
1.4	Active control in other fields	19
	1.4.1 Examples of applications	19
	1.4.2 Type of polymers	20
	1.4.3 Type of signals	21
	1.4.4 Type of responses	21
1.5	Active rheology control in other fields	22
1.6	Challenges for active control in concrete	25
	1.6.1 High pH	25
	1.6.2 Low signal conductivity	26
	1.6.3 Low signal transmission	26
1.7	Summary	27
	References	27

DOI: 10.1201/9781003289463-1

1.1 INTRODUCTION: BACKGROUND AND DRIVING FORCES

1.1.1 Context

Traditional concrete has a limited fluidity and needs to be actively compacted. Its performance has now been scientifically surpassed by highly technological mixtures based on multiple blends of powders combined with specially designed admixtures, showing rheological properties that make the fresh material self-compacting [1]. In spite of its many advantages (reduced energy consumption, less noise, less vibration, etc.), more than three decades after its introduction, self-compacting concrete (SCC) still only gained a very low market share. While in many countries the share of SCC in precast industry has been rising to levels of 20 to 30%, the market share in ready-mix industry remains at below 3 to 5% in most European countries. The excellent rheological properties of SCC have not been sufficiently translated into new production opportunities, as the traditional production processes do not fully match the potential of the new material. Problems occur with high pumping pressures, leaking of formworks and excessive formwork pressures. As local traditions often prevail over new scientific concepts and technologies, current production processes do not take full profit of the rheological potential of fresh cementitious materials, and are still very largely labour-driven and very sensitive to the human factor.

Within the European Research Council (ERC) Advanced Grant project 'SmartCast', awarded to Prof. Geert De Schutter, and running between 2016 and 2022, a new concrete casting concept was developed, which could lift the concrete industry (both precast and on-site) to a higher quality level and transform it into a modern and highly technological industry. This new concept, called Active Rheology Control (ARC) and also including Active Stiffening Control (ASC), enables an adjustment of the flowability of the fresh cementitious material while processing, and thus post-mixing. The SmartCast concept is based on two different routes. A first route is based on newly developed switchable superplasticizers, applying the concept of responsive polymers. A second route is based on available mineral particles that show a response to an external trigger signal, for example, magnetic fields. Combining the new admixture technology and/or the triggerable mineral particles with an external trigger signal enables active control of the desired rheological behaviour of the fresh cementitious material. While the principal goal of the project was to fundamentally upgrade classical formwork-based production methods, including pumping, the obtained control concepts can also be applied to other production methods, for example, three-dimensional (3D) concrete printing. This book reports the new ARC and ASC concepts as developed within the ERC Advanced Grant Project SmartCast.

Some parts of the book chapters are based on earlier papers in international journals with peer review, published in Open Access. In order to compile the innovative SmartCast concepts and obtained results and insights into one volume, this Open Access book is written as a final report of the ERC Advanced Grant Project SmartCast. It is the wish of all authors that this book will enable researchers all over the world to work on active rheology and stiffening control of cementitious materials, starting from the successes and failures obtained within the SmartCast project. We look forward to welcoming improved approaches and seeing smart answers and solutions to the remaining challenges. The SmartCast researchers feel like happy kids that just learned to walk. Running a marathon together is still a dream, to which many research teams worldwide will hopefully contribute.

1.1.2   State of the art and step beyond

SCC as it is known today was developed about three decades ago. Earlier 'look-alikes' do exist, though these are not defined as 'self-compacting', e.g. underwater concrete or highly flowable concrete. A crucial element was the invention of plasticizers in the 1970s. However, already in the early 1900s, some developments were made towards casting methods, which are presently considered to be only possible with self-compacting concrete [2]. One example is the system called 'Non Plus', applied in Germany, the Netherlands and Belgium in the 1910s and the 1920s. This system, with liquid concrete 'poured' into the formwork 'without any further compaction', was developed in Germany around 1906, and was applied for house construction, in spite of the heavy competition of the more traditional approach relying on masonry [3]. However, due to problems related to inappropriate concrete rheological properties and to the complex and expensive formworks needed, the 'Non Plus' system gradually faded away. It was impossible in the beginning of the 1900s to overcome the above-mentioned problems because of the lack of modern components such as plasticizers, powders, and viscosity modifying agents, governing concrete rheology. Furthermore, the example of the 'Non Plus' systems clearly illustrates that insufficient knowledge about, for example, the formwork pressure can lead to the abandonment of an otherwise promising new technique. This emphasizes the need to further study all aspects related to the introduction of SCC in an advanced and increasingly automated production of concrete structures.

Another example, also from the early 1900s, is the concept developed by Edison to cast concrete houses in one operation [4]. Although the idea was very innovative and relevant, similar to the 'Non Plus' system it proved to be very difficult to bring the idea into good practice. It was impossible in the beginning of the 1900s to overcome the problems related to formwork filling and pressure because of the lack of modern components such as plasticizers, powders and viscosity modifying agents to regulate concrete rheology.

With the introduction of SCC, the materials knowledge needed to automate casting operations as foreseen in the early 1900s became available [1]. At an international level, coordinated actions within world-wide technical and scientific organizations like RILEM (*International Union of Laboratories and Experts in Construction Materials, Systems and Structures*) and ACI (*American Concrete Institute*) have resulted in the compilation of state-of-the-art documents on many aspects of SCC ranging from the selection of materials and mix design, over hydration and microstructure development, to mechanical properties and durability [5, 6]. The development of SCC has also initiated a renewed attention and more in-depth analysis and application of fundamental rheological research in concrete science [7].

As stated by Shah in his '*future research needs*' paper [8], '*research in the rheology of SCC is a key to increasing SCC versatility in construction and utility in various construction types*'. This research goes beyond basic rheological parameters such as viscosity and yield stress, and needs to include complex behaviour such as thixotropy '*because of its benefits, such as in prevention of segregation during pumping, reduction of formwork pressure, timely green strength and shape stability in slip-form construction ...*'. The currently most advanced thixotropy model, the *Particle Flow Interaction Theory* defined as an extension of the *modified Hattori-Izumi Theory* [9], is very complex and difficult to implement in CFD codes. Simplified approaches exist [10], but with limited accuracy for complex situations. World-wide research efforts have been initiated to advance in this area. As an example, the effect of superplasticizers on (cement and powder) particle agglomeration, relevant to understanding thixotropy, has been studied fundamentally by Lesage in his doctoral research [11]. The SmartCast project, on which this book is reporting, had the ambition to advance beyond current approaches and understanding, developing methodologies to actively control rheology and stiffening. This includes aspects of controllable thixotropy by controlled clustering of particles, as well as controlled plasticizing by controlling adsorption or other functionalities of superplasticizers. While the newly developed concepts can in principle be applied to active control of rheological behaviour during processing of the fresh cementitious materials in general, SmartCast specifically considered pumping and formwork casting as relevant processing steps, as illustrated in Figure 1.1.

Pumping of traditional concrete was studied fundamentally more than a decade ago [12, 13]. Although SCC is apparently more flowable than traditional concrete, recent studies have shown that the pumping behaviour of SCC is more complex, involving more significant shear deformation in the material, along with the effect of the lubrication layer near the pipe surface. Considering potential shear-thickening behaviour of powder-type SCC, high

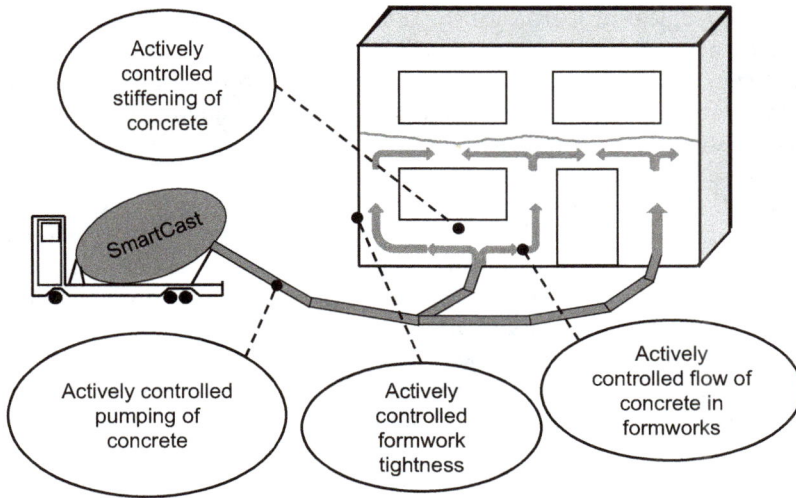

Figure 1.1 Schematic overview of the ERC Advanced Grant Project SmartCast.

pumping rates can lead to much higher pumping pressures than in case of traditional concrete. Furthermore, the additional role of thixotropy during pumping has been illustrated by Feys [14, 15]. At this moment, a good understanding of pumping of cementitious materials is available, although questions remain concerning the pumping behaviour of specific types of concrete in which the formation of a lubrication layer is not induced, or in which internal friction and Coulomb effects become prevalent (e.g. in case of ultra-high performance concrete). During current pumping operations, the only option is to control the pumping pressure or the discharge rate while operating the pump. Short interruptions of the pumping process can lead to major difficulties in resuming pumping operations due to the some-times tremendous effect of internal structural build-up or thixotropy [16]. Current pumping operations do not allow for the active control of the con-crete rheology during pumping. Once the concrete enters the pump, the operator can only passively consider the evolution of the rheological prop-erties of the cementitious material, and has no means to adjust the material properties. A first step in improving this situation has been introduced by Choi et al. [17] in a first pilot test, applying an externally imposed electro-magnetic field in order to improve the formation of the lubrication layer. This technique has been inspired by a recently developed formwork removal technique based on electro-osmosis [18, 19]. As Choi et al. used carbon steel pipes, the magnetic field was not able to penetrate the bulk of the pumped concrete, but somehow seemed to improve the formation of the lubrication layer, possibly by polarization effects. In any case, they still passively relied on the rheological properties of the pumped concrete. SmartCast aimed to introduce a ground-breaking approach by developing an innovative concrete mix design containing responsive polymer admixtures and/or triggerable mineral particles interacting with applied electromagnetic fields, enabling the active adjustment of the lubrication layer and/or the bulk concrete while pumping.

Currently, formworks are typically filled layer-wise from the top. Top-down casting operations do exist in concrete industry, mainly in precast industry although some examples also exist for on-site casting, e.g. casting of mega-columns in Shanghai Tower, China. Top-down casting can induce problems with blocking, segregation, entrapped air and poor-quality surface. A more optimal filling process is based on pumping operations filling the formwork from the bottom-up, thus avoiding air inclusions during casting. The current level of understanding enables a fairly accurate simulation of the concrete flow in formworks [10]. However, due to difficulties involved with filling bottom-up, this formwork filling concept has so far only been applied in some pilot cases, e.g. LNG terminal in Osaka, Japan, and a concrete villa in Gistel, Belgium. The highest level of bottom-up filling operations is currently available in precast industry, e.g. for the production of modular elements in factory conditions [20]. However, also in this situation, the system is passively relying on the rheological response of the fresh concrete to the set pumping and flow conditions. No active control of concrete rheology is considered at all during the casting operations. The major step forward of actively controlling the rheology of fresh concrete during casting operations was introduced in SmartCast, by modifying the concrete with responsive polymers and/or triggerable mineral particles.

In general, contractors refer to formwork pressure and leakage problems as one of the main reasons not to use SCC. In practice, formwork tightness requirements are defined in normative documents, e.g. limiting the extent of honeycombs. However, no fundamental study is available on the relationship between formwork design and detailing (e.g. formwork joints) on the one hand, and concrete rheology and pressure on the other hand. Some relevant state-of-the art knowledge is available concerning formwork pressures during top-down casting [21], but the situation in case of bottom-up pumping is considered in a too simplistic way in the literature, typically assuming formwork pressures equal to hydrostatic pressures. In reality, the bottom-up pumping process can induce very high pumping pressures exceeding hydrostatic pressures in case of high viscosity, dense reinforcement and high filling rates [22]. Furthermore, in current practice, due to lack of robustness of SCC, small changes in concrete composition can lead to increased flowability and reduced segregation resistance, increasing the risk of formwork leakage. SmartCast aimed to address this knowledge gap. More specifically, formwork leakage was studied fundamentally in relation to the rheology of the cementitious material, and providing options for active control.

Once the fresh concrete is in its final position, the flowability of the material becomes obsolete, and even unwanted as it contributes to the high formwork pressure. A fast stiffening process can reduce the formwork risks. In current state-of-the-art, the setting process due to the hydration of the cement or binder is gradually converting the cementitious material from a liquid into a solid material. This setting process is dependent on the physical and chemical properties of the material, and is influenced by environmental conditions, mainly temperature. Thixotropic properties of the cementitious material can also contribute to a rapid stiffening and can be enhanced by the addition of special (clay) powders to enable slipform casting [23]. However, thixotropy is a reversible process, and due to unwanted vibrations (e.g. running trucks near the freshly cast concrete), the structure could be broken down, making the material more flowable again. A recent development in the area of slipform casting can be found in ongoing research concerning 'Smart dynamic casting' [24]. SmartCast introduced an innovative concept, based on responsive admixtures and mineral powders, enabling active triggering of setting or stiffening once the material is in its final position, immediately reducing formwork pressures and leakage.

1.1.3   Ground-breaking objectives

The available state of the art at the start of the SmartCast project formed the basis for the optimization of casting operations. SmartCast however advanced beyond the state-of-the-art knowledge and beyond classical casting operations. Currently, the rheological properties of the concrete are defined by mix design and mixing procedure, without any further active adjustment being possible during the casting. The ground-breaking idea to achieve the main objective of ARC and ASC is to modify concrete by adding admixtures responsive to externally activated electromagnetic signals. The automation of casting operations involving ARC of optimized SCC containing specially developed admixtures and/or triggerable mineral powders could lift the quality in construction industry to higher levels, and could contribute to future-proof 3D printing operations.

The overall goal is extremely challenging, involving many specific technological difficulties. Active control techniques will be developed consisting of a matching interaction between applied electromagnetic fields and responsive admixed polymers or mineral powders. The selection of a compatible polymer having the appropriate response is technologically extremely challenging. The alkaline concrete environment with relatively low electrical conductivity poses boundary conditions to the eventual activation method. This method needs to act at different locations (pipe flow, placing and stiffening) and might serve several purposes. The active triggering of stiffening preferably needs to induce irreversible stiffening, making continuous control beyond this signal obsolete. This is fundamentally different from the active control of lubrication layer and rheology, for which continuous adjustment during the entire pumping and casting process should remain possible.

Gaining the capability to actively control concrete rheology and stiffening during processing is a unique and challenging goal, which could revolutionize formwork-based concrete industry. In a next step, after successful completion of SmartCast, active control features can be implemented in the design of a 3D printing system. Active control of fresh cementitious materials could mean a major contribution to successful 3D concrete printing in future, whereas passively relying on autonomous stiffening or structural build-up will always remain an obstacle in current approaches.

1.1.4 Impact

The SmartCast project has provided the basis for the active controlling of both the rheological and stiffening properties of cementitious materials. This is a total paradigm shift in comparison with current processing techniques that passively rely on the evolving properties of the fresh concrete. In the short term, achieving active control while pumping can have an immediate impact on the concrete industry, as it can be applied to pump trucks without interfering with the elements to be cast. In the longer term, concrete casting with active control of flow and stiffening could significantly change the concrete industry and push quality levels to significantly higher standards. The successful completion of the SmartCast project opened up a new scientific area, which in itself could contribute to a green (environment-friendly) and reliable concrete industry with improved quality levels due to more intensive automation and the gaining of control over the setting process. For society, it could mean more reliable construction, with less damage cases and less failures, while at the same time better preserving the environment (reduced carbon footprint due to replacement of Portland clinker by alternative powders, reduced energy consumption and reduced noise and vibration levels as external vibration becomes obsolete, reduced exposure of technicians to safety and health risks as already experienced in precast factories where SCC has been introduced).

The market potential linked to the newly developed concepts of ARC and ASC is very high, and of significant importance to Europe. According to the International Labour Organisation (ILO, cited in [25]), the construction output by value is mainly concentrated in the developed countries, with Europe accounting for 30% of global output, followed by the USA (21%) and Japan (20%).

'Innovation in construction is essential', as also advocated by Akintoye et al. [26], 'It is therefore asserted that a paradigm shift in thinking is now *required [...]. The higher the levels of innovation garnered, the greater the likelihood that this will in turn generate economic growth'.* By increasing the automation and control levels during production, reducing the influence of human error, the new concept will help overcome some important problems and challenges the construction industry is still fraught with, including *failure to meet market demands, skills shortage and poor-quality products* [25]. A proper introduction of the new concept will require an appropriate knowledge management, duly considering the *'knowledge paradox'* [27], finding an equilibrium between open information and intellectual properties. An adequate collaborative innovation procurement strategy will have to be developed, overcoming the current *'fully-fledged blame structure'* in construction industry [25]. It is the personal belief of the authors that with a cultural change in 'people' (education, training, respect and positive behaviour) and 'quality' (automation, control and focus on the end product), construction industry can move forward to enter a new era. It is a pleasant and comforting thought that the completed SmartCast project might contribute to this move, be it in a humble and limited way.

1.2   ACTIVE RHEOLOGY CONTROL AND ACTIVE STIFFENING CONTROL

The quality and performance of concrete, both in fresh and hardened state, largely depend on mix design, processing and casting. Obtaining the right concrete for the right job is primarily a matter of using the right type, quality and dosage of the constituent materials, further influenced by practical conditions and circumstances in the concrete factory or on the construction site. A proper mix design for a given application is based on scientific and technological knowledge concerning the properties of concrete [28], combined with long-term individual and collective practical experience. In addition to the mix design, the concrete properties are influenced by the mixing process and the applied casting method. The mixing equipment and protocol, going along with differences in dispersion of the constituent materials, can significantly influence the properties of the fresh material [29, 30]. Pumping operations also influence the material properties [31], while further influencing processing steps include (lack of) vibration and (lack of) curing. When processing and casting is performed following relevant standards or state-of-the-art prescriptions, it is expected to achieve the intended concrete properties as considered in the mix design.

However, many variations, intentionally or not, play an additional role, sometimes making it a hard challenge to achieve intended concrete properties in fresh and hardened state. As the properties of the constituent materials vary in between deliveries, maintaining constant concrete quality in spite of the changing constituent properties is not always straightforward. This is illustrated in literature [32] for the case of the effect of variable foaming potential of admixtures on the rheological behaviour of paste. Even with identical constituent materials, small variations in mixing and processing can have noticeable effects. A well-known aspect is the influence of variable moisture content and water addition during mixing. In some cases, e.g. some self-compacting concrete mixtures, small variations in water content can invoke unacceptable variations in workability, making the concrete non-robust [33, 34]. As illustrated in literature [35], inappropriate rheological properties of the fresh concrete can cause unwanted phenomena like segregation, lack of pumpability, increased formwork pressure, insufficient quality of the surface finish... Undesired extra water addition, excessive air inclusion, insufficient vibration during casting, etc., will all have a negative impact on the final mechanical and durability performance of the concrete structure.

Quality control consequently is an important aspect in concrete industry, involving quality control of the constituent materials, quality control of the mixing process, quality control during execution and quality control (compliance testing) of the final properties. While the quality control actions can be very active in the sense that a lot of people spend a lot of energy in getting things done in the right way, we typically passively rely on the behaviour of the concrete. Once the concrete is mixed and flowing in the pumping pipes or formworks, we have no active methods to change the behaviour of the concrete, e.g. requiring it to flow faster or become stiffer. It is true that pumping pressures could be actively changed, or vibration methods could be actively modified, but we nevertheless are stuck with a material with a pre-defined rheological response to the external changes. We have no means to modify the material behaviour while processing is ongoing, or when the material is in final position. In other words, we cannot actively change the constitutive behaviour of the concrete, either in fresh or in hardened state. We can only change external parameters (e.g. pumping pressure) to invoke an effect (e.g. increase in discharge rate) obeying the constitutive material behaviour.

In other fields, active control methods to change material properties 'on demand' do exist. As will be further explained in this chapter, special fluids are available responding to an external electric or magnetic trigger signal, changing the behaviour of the material from more fluid-like to more solid-like (or vice versa) on demand [36]. These rheofluids can, e.g., be used in smart structures, enabling active intervention for damping of vibrations. Rheological properties like yield stress and viscosity can be actively controlled in a reversible way by applying an external electric or magnetic field. Additional examples of active control of material properties or material actions in other fields will be mentioned further on in this chapter, showing very innovative and challenging concepts that will be inspiring for cementitious materials.

Going beyond passively relying on the behaviour of the fresh cementitious materials after mixing, ARC and ASC, as developed within the ERC Advanced Grant Project SmartCast, provide options for active intervention and adjustment of the flowability of the fresh cementitious material in post-mixing stage. The concept of active control involves an external action or signal to which the material responds by changing its behaviour. The signal can be given 'post-processing', which means after mixing, during the fresh or hardened state. The concept of active control includes the incorporation of special components or advanced (nano- or micro-) materials into the cementitious material that show a desired response to a given specific signal.

In the literature, the word 'control' is often used with respect to the flow of fresh concrete (e.g. [37]). Mostly the term 'control' in these contexts refers to efficient changes in mixture composition by adding (pre-processing) new or modified components, like, e.g., new generation admixtures, in view of obtaining (controlling) desired material properties. Once the cementitious material is mixed, no further intervention is possible to adjust the properties. With active control, we refer to an external signal that can be applied at any time during or after processing of the material, triggering a desired response in the material. The active control can be reversible, or in some cases irreversible.

1.3 ACTIVE CONTROL ALREADY AVAILABLE FOR CEMENTITIOUS MATERIALS

In current concrete practice, active interventions are manifold, although mostly not in the sense of actively modifying material behaviour by triggering a response of mixed-in special components. Nevertheless, an overview of current active intervention methods (as also schematically summarized in Table 1.1) is a good starting point to learn from, and to push the limit of our current abilities to control fresh and hardened concrete properties.

1.3.1 Mechanical intervention

The direct contact between (aggregate) particles leads to high friction, obstructing the flow. In fresh state, an obvious active mechanical intervention is given by the vibration process, using poker vibration or other techniques. By means of a poker, a vibration with a certain frequency and amplitude is (in this case internally) imposed on the concrete. This 'signal' has a certain range of action, triggering the (solid) particles in the cementitious material to respond (vibrate). As a consequence, the yield stress drops to zero within the affected zone [38], and the material liquefies and fills the formwork in a better way (i.e. compacts). While a poker requires an invasive action ('internal vibration'), other vibration methods are available operating from

Table 1.1 Overview of current practices with active intervention to control properties of concrete

Type of intervention	'Signal'	'Responsive' constituents	Effect
Mechanical	Vibration (frequency, amplitude, range)	Solid particles (vibration)	Elimination of yield stress (rheology control)
Chemical	Chemical trigger, e.g. retarder, accelerator, CO_2 ...	Cement particles, molecules in pore solution ...	Retardation or acceleration of hydration reaction ... (setting and hardening control)
Thermal	Temperature field	All constituents (through their thermal properties), cement particles (activation energy)	Control of temperature gradients, control of reaction rate
Hygral	Moisture	Pore system (capillary pressure, diffusion ...), Cement particles (reactivity)	Control of shrinkage (cracking), control of hydration process
Pressure	Reduced or increased pressure, or ultrasound	Air	Control of compaction, contribution to accelerated curing, dispersion
Magnetic	Magnetic field	Magnetizable materials (MNPs, steel strips or fibres...), water	Set-on-demand, micro-vibration, effect of magnetized water on hydration, healing on demand ...
Electric	Electric current or potential	Dipolar molecules, ions	Control of reaction rate (via heating due to resistivity), formwork release, improved pumping (via polarization)
Microwave	High frequency electric signal	Dipolar elements	Control of reaction rate (via heating due to internal dipolar vibration)

the surface of the concrete, e.g. formwork vibrators. External vibrators on pumping pipes are sometimes used at regular distances in case of pumping of concrete for tall towers, several hundreds of meters upwards.

Vibration of fresh concrete can thus be seen as the active application of an external signal (mechanical vibration with a certain frequency and amplitude), triggering a response from some constituents in the bulk of the concrete (in this case vibration of solid particles), changing the rheological properties of the fresh concrete (reducing the yield stress to zero). In this way, it could be said that ARC of fresh concrete is already performed in concrete industry since the 1930s due to the introduction of poker vibrators. However, ARC can go much further, as will be illustrated further.

1.3.2   Chemical intervention

An active chemical intervention is often applied in case of sprayed concrete, introducing accelerating admixtures in the spraying head. The added accelerating admixture triggers the cement particles, speeding up the hydration process in a chemical way, leading to faster setting (stiffening) and hardening of the sprayed concrete. Similar but even more complex chemical interventions can be applied when at the end of a working day the concrete within a long pumping pipe (e.g. for the construction of a tall tower) is removed for pipe cleaning. The concrete flowing back in a truck mixer can be retarded for a long night by adding a chemical retarder. The next morning, while resuming pumping operations, a chemical accelerator can be added, counteracting the effect of the retarder. This kind of active combination of retarders and accelerators has e.g. been studied by Lesage [11].

The active intervention in this case is giving a chemical signal to the cement particles, leading to a control of the hydration rate. After addition of the accelerator the triggered response of the cement particles is irreversible. However, the triggered response of the added retarder can be reversed by later addition of an accelerator. This kind of active intervention is limited to certain moments in the process, e.g. when the concrete is flowing through the spray head in case of sprayed concrete, or when the concrete is in the truck mixer in the case of re-use of concrete in a pumping pipe. The control is no longer possible at other timings, and it is also not possible to adjust the control levels once the admixtures have been added.

Another example of active chemical intervention is the use of carbon dioxide (CO_2) for the accelerated curing of cementitious materials, often in combination with the application of a higher temperature. This active chemical intervention has already been reported in 1972 [39], but has been given renewed attention in recent years because of environmental issues, as illustrated by Monkman and MacDonald [40], studying the addition of CO_2 gas during mixing of the concrete for the production of blocks, and by Nielsen et al. [41], studying carbonate-bonded construction materials from alkaline residues.

1.3.3   Thermal intervention

The effect of temperature on the hydration process of cement is well known, and can be calculated by means of the Arrhenius equation, with the activation energy as most important parameter [42]. This knowledge is applied in industry by means of heat curing, e.g. steam cycles in precast industry [43, 44]. This involves an active intervention, applying a thermal signal to the hardening concrete in view of accelerating the hydration process and achieving faster setting and strength development (at the cost of reducing somewhat the final strength [45]. The applied thermal field is triggering the cement particles (and/or supplementary cementitious materials) that are responding according to their activation energy, resulting in an acceleration of the hydration process. Steam curing (and heat curing more generally) can be considered as a type of active setting and hardening control.

Another type of active control involving a thermal field is given by cooling measures to mitigate thermal cracking in hardening massive concrete elements [46]. In embedded cooling pipes, a cooling agent is circulated through the massive element, imposing lower temperatures inside the hardening concrete [47, 48]. All constituents of the concrete respond to the applied thermal signal (lower temperature in embedded cooling pipes) following the physical laws of thermal diffusion, as described by Fourier's equation [49]. Important parameters in this process are the specific heat and the thermal diffusivity of the hardening concrete, as well as the heat production due to the exothermal hydration process [50], also responsive to temperature following the previously mentioned Arrhenius equation.

1.3.4   Hygral intervention

Freshly cast concrete needs to be protected against premature drying, in view of mitigating shrinkage cracking and creating optimal conditions for the hydration process. For on-site casting, protection against premature drying during the first days is for many elements achieved in a passive way by the formworks. After striking of formworks, or for those elements not protected by formworks (e.g. slabs), curing can be performed in an active way by water spraying or other techniques. The active intervention providing sufficient moisture near the concrete surface is interacting with the pore system, based on capillary forces and diffusion processes. As a result, the concrete is maintaining sufficient moisture conditions, mitigating or reducing early age shrinkage cracking, and improving the hydration process (ACI Committee 208 2016).

Modern concepts of internal curing by means of porous aggregate particles or super absorbent polymers (SAP) [51] cannot be considered as active interventions, as the water contained in those materials is only released internally as a result of the self-desiccation linked to the hydration process. In future, however, active intervention could be envisioned, by making SAPs or similar products to release their water content on demand by providing an external trigger. The triggered release of agents contained in hydrogel-type polymers is a known concept in other fields already now, as will be illustrated further.

1.3.5   Intervention by pressure

Concrete also includes air, showing a potential effect on the rheological properties depending on the size of air bubbles and the applied shear stresses [52]. During processing, air can influence the rheological behaviour, and vice versa, the processing can influence the air system of the concrete [31]. During mixing, pressures in the mixing pan can be intentionally reduced, resulting in vacuum mixing [53]. As a result, the air void system of the fresh (and later on hardened) concrete is influenced, resulting in reduced porosity and increased mechanical and durability performance. Modifying the air content in fresh concrete also influences the rheological behaviour [52]. Other researchers made use of frequently changing pressure waves by means of an inline ultrasound application [54]. A process referred to as cavitational shear disperses cement grains better and thereby reduces the setting time and admixture use. This pre-cast method allows to speed up stiffening and is preferentially applied to neat cement paste, before mixing it as concrete. Vacuum conditions can also be applied during casting, resulting in vacuum compaction of the fresh concrete [55]. While curing, air pressures could be increased, possibly combined with increased temperature and/or increased CO_2-content (see also higher), in view of accelerated hardening of the concrete element [56]. External pressure control can thus trigger changes in air content and pore structure, in view of achieving improved material performance.

1.3.6   Magnetic intervention

Magnetic fields can be relatively easily applied in industry, and can be an element of control during a production process. A magnetic field could be applied to magnetize water before its addition to cementitious materials. A magnetic field changes the physical properties of water, including, e.g., specific heat and evaporation rate [57]. Some researchers claim that the use of magnetized water in cementitious materials has an effect on fresh and hardened properties [58]. They observed that the use of magnetized water in concrete resulted in a very significant increase in concrete strength, possibly due to the higher surface area of magnetized water. Su et al. [59] also observed a strength increase when using magnetized water in mortar and concrete containing fly ash. Nevertheless, the use of magnetized water is still open to significant debate, as it is not clear how water could maintain its magnetization long enough after removal of the magnetic field. It has to be further studied whether the reported improvement of properties really is the effect of previous magnetization of water, or rather the effect of other experimental influences that have not been clearly identified.

Soto-Bernal et al. [60] studied the effect of maintained static magnetic fields applied during hardening on the performance of cement paste. They concluded that more calcium-silicate-hydrate (C-S-H) is produced, with denser morphology and reduced porosity. It is conjectured that the magnetic field is influencing the hydration process by means of a restructuration process at molecular scale, resulting in microstructural improvement and increased mechanical performance.

The application of magnetic fields to control rheology of cementitious materials containing magnetic particles in real time has been studied by Nair and Ferron [61]. They considered two types of carbonyl iron powder, added to Portland cement paste (water cement ratio 0.4) in dosages up to 10% by mass of cement (corresponding to 4% in volume). Magnetic field with strengths up to 1.0 T has been applied while performing rheological tests on fresh samples. It is experimentally shown that the rheological behaviour of the paste containing magnetic particles can be actively altered by applying the magnetic field. The effect depends on the field strength as well as on the dosage of magnetic particles. On the other hand, the authors conclude that the magnetic field does not seem to have a noticeable effect on the morphology and formation of early age hydration product, and does not influence the final compressive strength. This seems to be in contradiction with the previously mentioned conclusions obtained by Soto-Bernal et al. [60], showing that further research is needed to elucidate the effect of magnetic fields on cement hydration and microstructure development. Disregarding the discussion on the effect of magnetic field on hydration, the research of Nair and Ferron [61] is a major step forward towards ARC of cementitious materials.

Abavisani et al. [62] also considered magnetic fields to study the possibility to align steel fibres (chips) according to a certain orientation. They conclude that the application of a magnetic field on the fresh steel chip-reinforced concrete facilitates the compaction process, and leads to an increased compressive strength depending on the test direction. Following the optimum direction, the compressive strength increases by more than 17%. Particularly interesting with regard to rheology control is their observation that the vibration of the steel chips in the alternating magnetic field can be considered as a way to obtain *self-compactor elements* for concrete.

The same group of authors further studied the effect of alternating magnetic fields and alternating current on the behaviour of reinforced concrete columns [63], considering magnetic fields up to 0.5 T with frequency of 50 Hz, and alternating current (AC) current up to 36 A. The alternating magnetic field facilitated the compaction of the fresh concrete in the reinforced columns due to the triggered vibration of the reinforcing bars.

Applying the magnetic field on the hardened concrete seems to lead to contradictory results, with sometimes an increase [62] and sometimes a decrease [63] in performance. The authors state that the application of magneto-electric fields can be the basis for *real-time compressive behaviour controlling* of reinforced concrete elements, even enabling a control of failure mode. However, their claims are based on contradictory results, and are not fundamentally supported by a clear mechanism explaining the potential influence of the (electro-)magnetic field on the material and structural properties. While the principle of active control of concrete properties by magnetic signals has been set forward, it is not fully clear which mechanism could be the basis for the explanation of the reported effect of magnetization on hydration and microstructure development. This will need further research attention, with more rigorous test protocols in order to fully exclude potential influences of other parameters.

On the other hand, it seems clear that mixed-in magnetic particles or magnetizable materials like steel chips as well as preplaced steel rebars are responding to the magnetic signal, with effects on rheology, workability and even mechanical performance. This opens a window for new developments and applications, possibly involving new types of magnetizable materials [64] or new fields of application like on-demand healing [65, 66]. In some applications, the effect of the magnetic field triggers its response via another physical mechanism like (micro-)vibration (making the link with mechanical intervention) or temperature increase (making the link with thermal intervention).

1.3.7 Electric intervention

An electric control signal (AC current) has been mentioned in previous paragraph, where it was combined with the research on the magnetic control of the mechanical performance of hardened reinforced concrete columns. Other literature results are available on the application of electric fields on fresh or hardening concrete. As an example, Kovtun et al. [67] applied direct current for the curing of alkali-activated fly ash concrete. The current serves as an indirect way of heating, while the higher temperature achieved leads to the accelerated hardening of the material. In their research, Kovtun et al. used a setup in which the direct current, provided through a controllable unit with a maximum current of 12 A and voltage up to 260V, was continuously controlled in order to reach a curing temperature of 60°C in the concrete, with an initial heating rate of 20°C in order to avoid too fast heating, similar to what is done when applying steam cycles. While the external signal is electric, the physically controlling mechanism behind the accelerated hardening is the thermal energy. Nevertheless, the response of the alkali-activated material to the electric trigger depends on the activator used, through its influence on the resistivity of the concrete. A similar principle has been followed by Zhang et al. [68, 69] as described in their patents concerning the control of setting time of alkali-activated concrete through the application of an electric field. The electric field only serves as a practical means to control the temperature in the cementitious material, depending on its resistivity.

A more innovative application of an electrical field can be found in the papers of Goudjil et al. [18, 19]. They impose an electric potential on a steel formwork panel, up to a level of 3V during 10 minutes, in view of polarizing the concrete layer close to the formwork surface. The applied electric signal triggers a response in the material, resulting in a reduced adhesion between the hardening concrete and the formwork panel. The induced electro-osmotic effect thus facilitates the formwork release once the concrete has sufficient strength to be self-supporting. In this way, the application of demoulding oils can be avoided.

Inspired by this electrically controlled formwork removal technique, Choi et al. [17] applied a similar polarization technique to improve the flow of concrete in pumping pipes. In a first pilot test, they externally applied an electromagnetic field around steel pumping pipes in order to polarize the material near the pipe surface and improve the formation of the lubrication layer. Although the electromagnetic field applied by Choi et al. indeed seemed to improve the pumping behaviour of the concrete, as experimentally verified by an ultrasound velocity profiling technique, it is important to notice that they did not add any specifically responding material like magnetic particles to the concrete. They merely relied on the response of the classical constituents present in the concrete. Although they mentioned the increased mobility of magnetized water and the results of the pilot test were positive, the fundamental physico-chemical mechanisms behind the success of this technique need further study and elucidation. Understanding this in detail might open the door to the development of new materials that could be mixed-in, strengthening its response triggered by the polarization signal and potentially enabling to better control the response to an intended level.

Another example of electric intervention is the technique of cathodic protection by means of applied current in view of protecting an existing structure from reinforcement corrosion [70]. This technique is becoming more and more popular in concrete practice, and is a nice illustration of currently available active control mechanisms. The fundamental principle of its functioning can be understood by considering a Pourbaix diagram, explaining passive and active states of reinforcing steel in concrete [71].

1.3.8   Intervention with microwaves

As an extension of electric signals, microwaves can also be mentioned in this overview. Microwaves are electromagnetic waves with high frequencies

(300 MHz to 300 GHz) and wavelengths ranging from one millimetre to one meter. A dielectric material placed in a high-frequency electric field will heat up due to internal friction, as the induced vibration of the dipolar molecules is hindered by molecular attraction forces. As already reported in 1987 by Xuequan et al. [72], the promoted cement hydration can be attributed to the elevated temperature as a result of the exposure of the hardening concrete to microwaves. However, an additional effect seems to be the desiccation of free water prior to setting, resulting in decreased porosity. As a result, microwave curing of concrete yields very high early strength, while also providing slightly higher strength at 7 and 28 days [72, 73].

1.4   ACTIVE CONTROL IN OTHER FIELDS

In other fields, functional materials get a lot of research attention. For a multitude of potential applications, controllable materials are developed, responding to specific trigger signals, and resulting in an aimed-for effect. In this section, an (non-exhaustive) overview is given about some applications, type of materials, type of signals and type of responses. Some specific attention is given to different options to control rheology. The applications and control methods used in other fields can be inspiring in view of reaching active control of properties of concrete.

1.4.1   Examples of applications

A major field of research studying stimuli-responsive materials is given by *drug delivery on demand* [74–82]. Complex three-dimensional polymer structures have been developed that carry drugs or special agents. The polymers are stimuli-responsive, which means that as a result of some trigger signal they change their structure or undergo other modifications, resulting in the release of the drug at the intended location in the body. These drug delivery systems have great potential in the fight against cancer and other major diseases. In this context, the development of biosensors can also be mentioned, as they can be integrated within drug delivery systems [83].

Stimuli-responsive polymers are also used for *soft robotics* and *micro-actuators*, when the functional polymer can transform chemical energy into a mechanical action like folding or bending [84–88]. Typically, electro-responsive hydrogels are considered for this field of application, as the control is repeatable and reversible. In advanced applications, the stimuli-responsive polymers can be used for the development and control of *self-propelled nanomotors* [89].

In this non-exhaustive overview of applications of functional materials, reference can also be made to *adhesion on demand* [90], *smart windows* [91] and even the possibility to write *information* into triggered hydrogel by very precise control of the electric signal given to self-assembling molecules [92].

1.4.2 Type of polymers

The majority of current applications relying on stimuli-responsive control are based on *hydrogel*, generally defined as *three-dimensional network structures consisting of polymeric chains joined by tie points or joints and swollen in water up to thermodynamic equilibrium* [93]. Hydrogels have already been studied for a long time, with early work in 1930s and 1940s [93]. It is a very broad class of materials that can be synthesized in numerous ways in view of obtaining a wide range of properties and behaviour [94]. The concept of hydrogels is based on crosslinking of hydrophilic polymers into a spatial complex structure that is able to swell or de-swell depending on the thermodynamic equilibrium within the surrounding medium. In case of ionic hydrogels, the ionic interactions bring additional forces that influence the swelling equilibrium. A major parameter is the pH-value of the surrounding medium, which in current fields of application is typically much lower than in cementitious materials. Nevertheless, hydrogels can be applied in cementitious materials, as illustrated by the well-known application of SAP [51].

Although gels offer great possibilities, they have limitations concerning multi-responsivity and multi-functionality. An alternative approach is the use of *vitrimers*, covalently crosslinked polymer networks. Vitrimers can be modified in different ways, and even be combined with carbon nanotubes, resulting in many different functionalities, in response to even six different stimuli [95].

A specific group of materials is given by *core-shell type particles* with monodisperse polymeric cores and uniform shells with independent control of properties of core and shell [96]. Core-shell microspheres dispersed in oils are a nice example of electrorheological fluids that can be used as e.g. dampers, brakes and shock absorbers.

Stimuli-responsive polymers intended for the (reversible) control of shape are also called *shape memory polymers* (SMP). The mechanism can be based on different controllable transitions, such as reversible molecule crosslinking or anisotropic-isotropic transition. They can be reinforced by different materials, e.g. carbon nanotubes, carbon nanofibres, exfoliated nanoclay, etc., to improve the mechanical performance of the resulting material, called *shape memory polymer composites* (SMPC) [97]. They could also include Fe_3O_4-particles or other magnetizable materials in order to make them magneto-responsive [98, 99].

Magnetic nanoparticles (MNP) can also be incorporated into *hyperbranched polymers* to make them magneto-responsive. Hyperbranched polymers show good possibilities to make them stimuli-responsive in general, offering good options to include a wide choice of terminal functional groups and branches [100]. The concept and design of stimuli-responsive polymers in general is getting a lot of research attention nowadays. New design strategies [101] and new synthesis and polymerization methods [102] are important aspects in obtaining new stimuli-responsive materials that could also be applicable in cementitious environments, overcoming specific challenges (see 4) that currently pose problems to most of the existing responsive materials.

1.4.3   Type of signals

Most common signals concern temperature and pH (or pH difference), or the application of an electric, magnetic or electromagnetic field. Many other signals, however, have shown to be applicable in view of inducing a triggered response in functional materials, like, e.g., pressure [103], redox-signal [74, 83, 104–107], CO_2 [104, 108, 109], light [106, 110], ultrasound [77], UV [111–113]...

Most stimuli-responsive polymers respond to one or more trigger signals. In case of dual or multi-responsive materials, the desired action depending on the given combination of trigger signals is an important point of attention. In case of two signals, as an example, it could be that the desired response is only activated in the presence of both stimuli, while another option could be the activation in case of any single stimulus. Different other logic combinations can be chosen, as illustrated in [75].

The applicability of the trigger signal depends on the type of material, and further depends on practical considerations. While for some applications photonic signals are convenient and environmentally friendly, the responsive process (often chemical) typically does not show a high efficiency [114]. For opaque materials like cementitious materials, light signals are not a good option, unless only a response at the surface is needed. Signals based on change in pH or on electrochemical responses typically have a higher efficiency. Electric or redox signals further depend on the resistivity or conductivity of the material, and are thus not applicable in all circumstances. Heating and cooling seem to be easy trigger systems in principle, but are not very practical in many cases.

1.4.4   Type of responses

Roy et al. [115] give an interesting overview of stimuli-responsive materials, with clear illustrations of their effects. Numerous types of responses resulting in different effects can be listed. In general, control can be reversible or non-reversible, depending on the type of triggered response and the type of signal. A non-exhaustive list of different types of triggered responses is given hereafter:

- Triggering crosslinking resulting in gel formation, often by means of thermally induced sol-gel transition [116]

- Cleavage of polymer molecules, resulting in collapse of a hydrogel or a hyperbranched polymer, or influencing micellization [77, 110, 113]
- Controlling swelling/de-swelling and the release of an agent, due to changes in osmotic pressure by asymmetric distribution of ions in electric field [84], or by change of electronic charge surface potential [80]
- Changing the permeability of a responsive membrane, by closing or opening 'chemical gates', e.g. by pH control [115]
- Sorption control of water vapour of, e.g., electro-responsive polypyrole film [117]
- Control of solubility, often by changing temperature [118]
- Changing the nature of a complex molecule from hydrophilic to hydrophobic, also enabling control over the micellization process [116]
- Changing the self-assembly, e.g. CO_2 and redox-stimulated transformation of entangled worms to rod-like micelles [104]
- Changing the orientation of the polymers, e.g. by the application of an electric and/or magnetic field [119, 120]
- Controlling entanglement of polymer chains [121]
- Controlling aggregation/dispersion of particles in a medium [111]
- Control of trans-cis isomerization of polymers [112]
- Shape-morphing [122]
- Oscillatory shape change, e.g. oscillatory bending [122, 123]
- Crystallization/melting transition [97]
- Crystal anisotropic-isotropic transition [97]
- ...

The listed responses can induce changes in numerous material properties, or can lead to specific actions (like, e.g., drug release). Some of the listed responses could have good potential for rheology control. This is further discussed in the following paragraph.

1.5 ACTIVE RHEOLOGY CONTROL IN OTHER FIELDS

The rheological behaviour of a material is governed by fundamental physical properties, characterized by specific parameters, further depending on the type of material [124]. For solid materials, the ordering of crystals (in case of crystalline materials) is an important aspect, or the absence of ordering (in case of amorphous or glassy materials). The deformation behaviour, e.g. characterized by elastic properties, largely depends on the internal mutual interaction forces between crystals and/or amorphous phases. On the other end of the material spectrum, the deformation behaviour of simple liquids (like water) is mainly governed by the atoms or molecules, responding to Van der Waals forces and thermal agitation, and is, e.g., characterized by the viscosity which is the only relevant parameter in case of a Newtonian fluid.

When particles are introduced into a liquid matrix, the behaviour can be different depending on the size and nature of the particles, and the resulting inter-particle forces. In suspensions, depending on the density of the particles and of the suspending liquid, particles could segregate downwards or upwards, as governed by the interplay between gravitational forces and buoyancy forces. However, particles also interact, depending on their concentration in the suspending liquid, counteracting settlement forces. As a result, the viscosity of the suspension will be different from the viscosity of the suspending medium. For low particle concentrations, up to 30% of the maximum concentration, the viscosity increase is limited. However, at larger concentrations, the viscosity can increase significantly, as described in the Krieger–Dougherty equation or similar models [7]. Furthermore, orientation or alignment of particles can significantly influence the viscosity compared to the case with random orientation of particles.

In case of small particles, in the range of μm and below, colloidal interactions become dominant, mainly Van der Waals forces combined with Brownian motion, and the particles tend to form clusters or aggregates. This aggregation phenomenon could be counteracted by electrostatic repulsion or by steric hindrance, as typically done by superplasticizers in cementitious materials [125]. Depending on the internal structure formation by clustering of particles, the rheological response can be diverse and complex, involving concepts like yield stress (static and dynamic), thixotropy, shear-thickening or shear-thinning, etc.

When the particles in the liquid are polymers, droplets or air bubbles, the overall rheological behaviour of the resulting fluid can become even more complex, with additional governing parameters. Although advanced mathematical models are available [126], based on fundamental principles, the experimental characterization of the complex fluids is not an easy task. Measuring rheological properties is not straightforward [127], and rheological constitutive laws are mostly empirical with limited validity. In many research papers, small amplitude oscillatory shear (SAOS) is applied, resulting in the determination of the so-called storage modulus G' (representing the energy storage or elastic property of the fluid) and the loss modulus G" (representing the viscous dissipation in the fluid) [128]. In case of changing rheological properties of systems with smart functional materials, oscillatory rheometry is an interesting experimental method to study the reversible or non-reversible transition from more liquid-like to more solid-like behaviour.

Considering the governing physics and the resulting characteristic parameters of complex fluids, changing or controlling the rheological behaviour can involve different mechanisms or approaches. A non-exhaustive overview is given hereafter.

A well-known concept to change the rheological properties of a fluid is given by electrorheological (ER) fluids [36, 129, 130] and magnetorheological

(MR) fluids [131, 132]. ER response was first reported by Winslow already in 1949 [133]. In these suspensions, electro- or magneto-responsive particles align along the field lines of the applied electric or magnetic field, forming chain- or column-like structures, changing the rheological properties of the material from liquid-like to solid-like. The transition is reversible, and provides a means to actively control the rheological response of the material, with potential application in many fields of engineering.

In terms of rheological properties, it is reported that the yield stress (τ_y) in ER fluids depends on the electric field strength (E_0) as follows:

$$\tau_0 \propto E_0{}^a \tag{1.1}$$

with parameter a typically in the range of 1.5 to 2.0 [36]. Instead of applying Bingham-type models, a modified constitutive equation, called the Cho–Choi–Jhon (CCJ) model, is often referred to in case of ER or MR fluids [36, 130, 132]. Due to the formation of an internal structure under electric or magnetic field, the loss modulus G" typically shows an initial plateau as a function of shear stress, while the storage modulus G' is higher than the loss modulus G". In electro- or magneto-activated state, the suspension dominantly shows an elastic behaviour. However, when the shear stress exceeds a critical value, the storage modulus G' shows a sudden decrease, physically governed by the rupture of the internal structure and the resulting onset of material flow [36, 130, 132]. SAOS can elucidate in this way the internal structure formation in electric or magnetic fields, providing data in view of accurate ER or MR control.

Besides internal structure formation triggered by electric or magnetic field, other mechanisms can be considered for rheology control. One example is based on micro-vibration or micro-actuation. It was previously mentioned that alternating magnetic field applied to cementitious materials containing steel chips can make the chips work as micro-vibrators [62]. The same principle could be applied to alternating field acting on electro-responsive polymers or polymer composites, which could take the role of micro-vibrators [85–87].

Coming back to some fundamental parameters governing the rheological behaviour of suspensions and colloids, different triggered responses as listed before could also be useful in view of controlling the rheological behaviour. Hydrogels can be helpful in sol/gel transitions, release of agents potentially modifying viscosity of the suspending medium, swelling/de-swelling mechanisms, controlling micellization, shape morphing, etc. The wide range of potential control mechanisms of hydrogels also offer many opportunities to control the rheology of complex materials containing the hydrogels. It is to be mentioned, however, that rheology control by means of hydrogels has not been the main point of attention of hydrogel research up to now, while the focus was mainly on other intended applications, like, e.g., drug delivery.

Any controlled response influencing inter-particle forces, particle clustering, viscosity of the suspending liquid, chemistry of the suspending liquid, etc., offers a potential mechanism for rheology control. Some examples where rheology control is at least partly reported can be found in the literature, like, e.g., capillary oil suspensions containing calcium carbonate particles [134], aggregation in aqueous solutions by working on hydrophobicity and micellization [111], controlled solubility by temperature and photo cleavage [113] and controlled release of nanoparticles [76].

1.6   CHALLENGES FOR ACTIVE CONTROL IN CONCRETE

From a rheological point of view, concrete is a very complex material. It combines features of a colloid and a suspension, while showing at the same time properties of a granular material. In current approaches, reaching desired workability levels in fresh concrete is typically achieved by appropriate mix design, including the use of superplasticizers working on the principle of electrostatic repulsion or steric hindrance [125]. This, however, is a passive way of rheology control, leaving no possibility to adjust workability levels while the concrete is already flowing in a pumping pipe or in a formwork.

Active control is available in some way, as discussed above. Considering the numerous control mechanisms studied in other fields, as summarized in general and related to rheology more specifically, it is clear that new ways could be found to actively control concrete properties in general, and rheological properties of the fresh concrete more specifically. However, many difficulties have to be overcome when applying control mechanisms from other fields to the field of cementitious materials. The authors selected three concrete characteristics that are forming the main obstacles for now:

- High pH
- Low signal conductivity
- Low signal transmission

1.6.1   High pH

The fascinating mechanism displayed by different types of hydrogels is tempting to pursue for fresh concrete. However, the existing stimuli-responsive hydrogels have been switching between swelling and de-swelling at pHs that generally occur in the human body, i.e. pH $\leq$ 7.4. The precarious balance between the fixed charges in the polymer network and the osmotic pressure due to inner mobile ions is sensitive to ion concentration changes in the surrounding fluid. In contrast, a subtle ion concentration change in concrete with a pH of $\pm$12.5 will not be sensed by available weak polybases (pH 7- 11) due to the relative abundance of hydroxyl and other ions. Some applications of hydrogels do exist in concrete industry, like the use of SAP, however, without external control features. They could show some autogenous response, e.g. when SAPs are implemented as a method for self-healing [135]. However, further developments are needed to make them respond to active trigger signals to show on-demand responses.

1.6.2   Low signal conductivity

The application of control signals in cementitious materials also comes with some challenges. Some signals will not be applicable, like light signals, as they will not be able to travel easily through an opaque cementitious material. Electric control of hydrogels or other functional polymers is often based on an isolating dielectric medium suspending the functional polymers [129]. In contrast, fresh cementitious materials contain water with a relatively high ionic conductivity and cannot be considered electrically isolating.

1.6.3   Low signal transmission

Another challenge is posed by the large bulk volume and its inert constituents. As an example, while casting a massive foundation slab or a high wall, will it be possible for the trigger signals to reach the bulk of the concrete? With the intended amplitude and/or frequency? In any location? These questions involve different aspects, including fundamental knowledge regarding the interaction between physical fields and large volume of concrete in a real environment, involving, e.g., pumping pipes and formwork. Furthermore, there are also practical questions concerning the required installations to provoke the intended fields. Safety issues will become important when working with electromagnetic or other signals on a real construction site.

In some cases, however, bulk control will not be required. As an example, the pumping behaviour to a very large extent depends on the properties of the lubrication layer near the pipe surface [136]. In that case, an economical question might emerge, related to the cost of adding (probably relatively expensive) functional material to the bulk of the concrete, while only needing them for control purposes in a limited zone of interest.

Summarizing the challenges, the following aspects will require due attention when developing active control features for fresh and hardened concrete:

- Stability and functioning of the responsive material in a cementitious environment, typically showing higher pH values than in other fields
- Applicability of the control signal in a cementitious material, depending on its physical properties like resistivity, conductivity, opacity, etc.
- Economy, logistics and safety of a control system on a construction site or in precast industry

In spite of the many challenges, a good opportunity might be given by digital fabrication methods, including three-dimensional (3D) printing. Within the relatively limited volume of a printing head, it will be more practical and realistic to provide control signals to modify the rheology of the fresh cementitious material when moving from pumping line to printed layer. As the rheological requirements in a pumping pipe and for a printed layer are significantly different, an active control of the material properties in the printing head would be a very efficient approach. Active control features embedded in 3D printed concrete could open the door to the so-called four-dimensional (4D) printing in construction industry, as already envisioned by Momeni et al. [137].

1.7 SUMMARY

This chapter explains the background and gives a (p)review of the active control of properties of concrete, based on the application of external signals to trigger an intended response in the material, either in fresh or hardened state. Current practices in concrete industry that could be considered as active control have been listed. More advanced active control mechanisms as studied in other fields, e.g., based on hydrogels and other functional polymers, are reviewed and briefly summarized. A specific focus is further given on potential methods for ARC. Based on the concepts developed in other fields, substantial progress could be made in order to achieve active control of fresh and hardened concrete properties. However, several challenges remain, like the stability and functioning of the responsive material in a cementitious environment, the applicability of the control signal in a cementitious material, and the economy, logistics and safety of a control system on a construction site or in precast industry. Finding solutions to these challenges will lead to marvellous opportunities in general, and for 3D and even 4D printing more particularly.

REFERENCES

[1] De Schutter, G., et al., *Self-compacting concrete*. 2008. Dunbeath, Caithness, Scotland: Whittles Publishing, CRC Press, Taylor & Francis Group.

[2] De Schutter, G., Self-compacting concrete after two decades of research and practice. In *9th Int symp on high performance concrete*. 2011. Rotorua, New Zealand.

[3] Van de Voorde, S., *Building in concrete in Belgium (1890–1975). Interplay of knowledge, experiment and innovation*. 2011. Ghent, Belgium: Ghent University.

[4] Kent, D.J., *Edison – The inventor of the modern world*. 2016. New York, USA: Fall River Press.

[5] De Schutter, G. and A. Audenaert, Durability of self-compacting concrete. In *RILEM state-of-the-art reports*. 2007. RILEM. p. 185.

[6] Khayat, K.H. and G. De Schutter, Mechanical properties of self-compacting concrete. In *RILEM state-of-the-art reports*. 2014. RILEM. p. 271.

[7] Roussel, R., *Understanding the rheology of concrete*. 2012. Sawston, UK: Woodhead Publishing.

[8] Shah, S.P. and G.R. Lomboy, Future research needs in self-consolidating concrete. *Journal of Sustainable Cement-Based Materials*, 2015. 4(3–4): p. 154–163.

[9] Wallevik, J.E., Rheological properties of cement paste: Thixotropic behavior and structural breakdown. *Cement and Concrete Research*, 2009. 39(1): p. 14–29.

[10] Roussel, N. and A. Gram, Simulation of fresh concrete flow. In *RILEM state-of-the-art reports*. 2014. RILEM. p. 147.

[11] Lesage, K., *Interactions between cement and combined concrete admixtures. The influence on cement paste rheology*. 2014. Leuven: PhD thesis, KU Leuven; Ghent, Belgium: Ghent University.

[12] Kaplan, D., *Pumping of concretes*. 2001. Paris: Laboratoire Central des Ponts et Chaussées.

[13] Chapdelaine, F., *Fundamental and practical study on the pumping of concrete*. 2007. Québec: University of Laval.

[14] Feys, D., *Interactions between rheological properties and pumping of self-compacting concrete*. 2009. Ghent: Ghent University.

[15] Feys, D., G. De Schutter, and R. Verhoeven, Parameters influencing pressure during pumping of self-compacting concrete. *Materials and Structures*. 2013. 46 : p. 533–555.

[16] De Schutter, G. Thixotropic effects during large-scale concrete pump test on site. In *Advances in construction materials and systems*. 2017. Chennai: RILEM Publications S.A.R.L. p. 1–7

[17] Choi, M.S., et al., Effects of an externally imposed electromagnetic field on the formation of a lubrication layer in concrete pumping. *Construction and Building Materials*, 2014. 31 : p. 18–23.

[18] Goudjil, N., et al., Impact of temperature on the demoulding of concrete elements with a polarization process. *Construction and Building Materials*, 2014. 54 : p. 402–412.

[19] Goudjil, N., et al., Electro-osmosis applied for formwork removal of concrete. *Journal of Advanced Concrete Technology*. 2012. 10: p. 301–312.

[20] Reymann, J., A new production method for precast elements to meet highest demands. *BFT Int*, 2009(03): p. 30–35.

[21] Billberg, P., et al., Field validation of models for predicting lateral form pressure exerted by SCC. *Cement and Concrete Composites*, 2014. 54: p. 70–79.

[22] Tichko, S., et al., Numerical simulation of formwork pressure while pumping self-compacting concrete bottom-up. *Engineering Structures*, 2014. 70 : p. 218–233.

[23] Tregger, N.A., M.E. Pakula, and S.P. Shah, Influence of clays on the rheology of cement pastes. *Cement and Concrete Composites*, 2010. 40: p. 384–391.

[24] LLoret, E., et al., Complex concrete structures – Merging existing casting techniques with digital fabrication. *Computer-Aided Design*, 2015. 60: p. 40–49.

[25] Nadim, W., *Modern methods of construction*. In *Construction innovation and process improvement*, A. Akintoye, J. Goulding, and G. Zawdie, Editors. 2012, Hoboken, New Jersey, USA: Wiley-Blackwell Publishing. p. 209–233.

[26] Akintoye, A., J.S. Goulding, and G. Zawdie, *Construction Innovation and Process Improvement*. 2012. Blackwell Publishing.

[27] Egbu, C., Construction innovation through knowledge management. In *Construction Innovation and Process Improvement*, A. Akintoye, J. Goulding, and G. Zawdie, Editors. 2012, Hoboken, New Jersey, USA: Wiley-Blackwell Publishing. p. 235–250.

[28] Neville, A.M., *Properties of Concrete*. 2011: London, UK: Pearson United Kingdom.

[29] Dils, J., G. De Schutter, and V. Boel, Influence of mixing procedure and mixer type on fresh and hardened properties of concrete: a review. *Materials and Structures*, 2012. 45: p. 1673–1683.

[30] Asghari, A.A., D. Feys, and G. De Schutter, Mix design factors of self-consolidating cement paste affecting the magnitude of variations in rheological properties induced by changes in addition time of PCE-superplasticizer. *Construction and Building Materials*, 2018. 159: p. 269–276.

[31] Feys, D., et al., Changes in rheology of self-consolidating concrete induced by pumping. *Materials and Structures*, 2016. 49: p. 4657–4677.

[32] Han, D. and R.D. Ferron, Effect of mixing speed on rheology of superplasticized Portland cement and limestone powder pastes. *ACI Materials Journal*, 2017. 114(4): p. 559–569.

[33] Van Der Vurst, F., et al., Effect of mix design on the robustness of fresh self-compacting concrete. *Cement and Concrete Composites*, 2017. 82 : p. 190–201.

[34] Asghari, A.A., et al., Which parameters, other than the water content, influence the robustness of cement paste with SCC consistency? *Construction and Building Materials*, 2016. 124 : p. 95–103.

[35] Ferraris, C.F., et al., Role of rheology in achieving successful concrete performance. *Concrete International*, 2017. June: p. 43–51.

[36] Liu, Y.D. and H.J. Choi, Electrorheological fluids: Smart soft matter and characteristics. *Soft Matter*, 2012. 8: p. 11961–11978.

[37] Burns, E.G., et al., The rheology of control flow concrete. *Concrete International*, 2018 (January): p. 43–49.

[38] Grampeix, G., N. Roussel, and J. Dupoirier. Internal vibration and viscous concrete: application and prediction of the radius of action. In *Rheology and processing of construction materials*. 2013. Paris: RILEM.

[39] Klemm, W.A. and R.L. Berger, Accelerated curing of cementitious systems by carbon dioxide. Part I. Portland Cement. *Cement and Concrete Research*, 1972. 2: p. 567–576.

[40] Monkman, S. and M. MacDonald, Carbon dioxide upcycling into industrially produced concrete blocks. *Construction and Building Materials*, 2016. 124: p. 127–132.

[41] Nielsen, P., et al., *Carbonate-bonded construction materials from alkaline residues. RILEM Technical Letters*, 2017. 2: p. 53–58.

[42] De Schutter, G. and L. Taerwe, General hydration model for portland cement and blast furnace slag cement. *Cement and Concrete Research*, 1995. 25(3): p. 593–604.

[43] Nurse, R.W., Steam curing of concrete. *Magazine of Concrete Research*, 1949. 1(2): p. 79–88.

[44] Saul, A.G.A., Principles underlying the steam curing of concrete at atmospheric pressure. *Magazine of Concrete Research*, 1951. 2(6): p. 127–140.

[45] Higginson, E.C., Effect of steam curing on the important properties of concrete. *ACI Journal Proceedings*, 1961. 58(9): p. 281–198.

[46] De Schutter, G. and M. Vuylsteke, Minimisation of early age thermal cracking in a J-shaped non-reinforced massive concrete quay wall. *Engineering Structures*, 2004. 26 : p. 801–808.

[47] Groth, P. and H. Hedlund, Air cooling of concrete by means of embedded cooling pipes – Part II: Application in design. *Materials and Structures*, 1998. 31, 6: p. 387–392.

[48] Hedlund, H. and P. Groth, Air cooling of concrete by means of embedded cooling pipes-Part I: Laboratory tests of heat transfer coefficients. *Materials and Structures*, 1998. 31(5): p. 329–334.

[49] De Schutter, G., Finite element simulation of thermal cracking in massive hardening concrete elements using degree of hydration based material laws. *Computers and Structures*, 2002. 80(27–30): p. 2035–2042.

[50] Rastrup, E., Heat of hydration in concrete. *Magazine of Concrete Research*, 1954. 6(17): p. 79–92.

[51] Kovler, K. and O.M. Jensen, *Internal curing of concrete*. RILEM Report 41, RILEM, Paris, France, 2007.

[52] Feys, D., et al. Influence of air content on the steady state rheological properties of fresh self-compacting concrete, without air entraining agents. In *Design, performance and use of self-consolidating concrete*. 2009. Beijing, China: RILEM.

[53] Dils, J., V. Boel, and G. De Schutter, Vacuum mixing technology to improve the mechanical properties of ultra-high performance concrete. *Materials and Structures*, 2015. 48: p. 3485–3501.

[54] Peters, S., *The influence of powder ultrasound on setting and strength development of cement suspensions*. 2017. Weimar, Germany: Bauhaus University Weimar.

[55] Latham, J., Vacuum treatment of concrete. *Civil Engineering*, 1979.

[56] Yazici, H., et al., Mechanical properties of reactive powder concrete containing mineral admixtures under different curing regimes. *Construction and Building Materials*, 2009. 23(3): p. 1223–1231.

[57] Wang, Y., H. Wei, and Z. Li, Effect of magnetic field on the physical properties of water. *Results in Physics*, 2018. 8: p. 262–267.

[58] Reddy, V.S., A.K. Kumar, and A. Sumanth, Effect of magnetic field treated water on fresh and hardened properties of concrete. *J. Civ. Eng. Env. Techn*, 2017. 4(2): p. 134–138.

[59] Su, N. and C.F. Wu, Effect of magnetic field treated water on mortar and concrete containing fly ash. *Cement and Concrete Composites*, 2003. 25(7): p. 681–688.

[60] Soto-Bernal, J.J., et al., Effects of static magnetic fields on the physical, mechanical, and microstructural properties of cement pastes. *Advances in Materials Science and Engineering*, 2015. article ID 934195: p. 1–9.
[61] Nair, S.D. and R.D. Ferron, Set-on-demand concrete. *Cement and Concrete Research*, 2014. 57: p. 13–27.
[62] Abavisani, I., O. Rezaifar, and A. Kheyroddin, Alternating magnetic field effect on fine-aggregate steel chip-reinforced concrete properties. *J. Mater. Civ. Eng*, 2018. 30(6): p. 04018087.
[63] Rezaifar, O., I. Abavisiani, and A. Kheyroddin, Magneto-electric active control of scaled-down reinforced concrete columns. *ACI Structural Journal*, 2017. 114(5): p. 1351–1362.
[64] Ponsot, I.M.M.M., et al., Magnetic glass ceramics by sintering of borosilicate glass and inorganic waste. *Materials*, 2014. 7: p. 5565–5580.
[65] Joffroy, E., et al., Iron-based particles for the magnetically-triggered crack healing of bituminous materials. *Construction and Building Materials*, 2018. 164: p. 775–782.
[66] Joffroy, E., et al., Iron oxide nanoparticles for magnetically-triggered healing of bituminous materials. *Construction and Building Materials*, 2016. 112: p. 497–505.
[67] Kovtun, M., et al., Direct electric curing of alkali-activated fly ash concretes: a tool for wider utilization of fly ashes. *Journal of Cleaner Production*, 2016. 133: p. 220–227.
[68] Zhang, P., et al., *Alkali-activated cementing material capable of controlling setting time as well as control method and application thereof*. 2016. China. patent application CN105731947A.
[69] Zhang, P., et al., *Concrete construction technique capable of controlling setting time and special equipment therefor*. 2016. patent application WO2016165413A1.
[70] Chess, P.M. and J.P. Broomfield, *Cathodic protection of steel in concrete*. 2003. Boca Raton, Florida, USA: CRC Press.
[71] Pourbaix, M., *Atlas of electrochemical equilibrium in aqueous solutions*. 1976. London: Pergamon.
[72] Xuequan, W., D. Jianbgo, and T. Mingshu, Microwave curing technique in concrete manufacture. *Cement and Concrete Research*, 1987. 17: p. 205–210.
[73] Leung, C.K.Y. and T. Pheeraphan, Microwave curing of Portland cement concrete: experimental results and feasibility for practical applications. *Construction and Building Materials*, 1995. 9(2): p. 67–73.
[74] Huo, M., et al., Redox-responsive polymers for drug delivery: from molecular design to applications. *Polymer Chemistry*, 2014. 5: p. 1519–1528.
[75] Kabiri, M., et al., Smart micro/nanoparticles in stimulus-responsive drug/gene delivery systems. *Chem. Soc. Rev*, 2016. 45: p. 1457–1501.
[76] Kabiri, M., et al., A stimulus-responsive, in situ-forming, nanoparticle-laden hydrogel for ocular drug delivery. *Drug Delivery and Translational Research*, 2018. 8: p. 484–495.
[77] Paris, J.L., et al., *Polymer-grafted mesoporous silica nanoparticles as ultrasound-responsive drug carriers*. acsnano, 2015. 9(11): p. 11023–11033.

[78] Sharma, S., et al., Stimuli-responsive polymeric hydrogel-copper nanocomposite material for biomedical application and its alternative application to catalytic field. *ChemistrySelect*, 2017. 2: p. 11281–11287.

[79] Moghanjoughi, A.A., D. Khoshnevis, and A. Zarrabi, A concise review on smart polymers for controlled drug release. *Drug. Deliv. and Tranl. Res*, 2016. 6(3): p. 333–340.

[80] Sankoh, S., et al., Colloid electrochemistry of conducting polymer: towards potential-induced in-situ drug release. *Electrochimica Acta*, 2017. 228: p. 407–412.

[81] Lin, C.C., et al., Electrically responsive ZnO nanotubes for controlled release of biomolecules. *Ceramics International*, 2017. 43(1): p. S802–S806.

[82] Bijukumar, D., et al., An electro-conductive fluid as a responsive implant for the controlled stimuli-release of diclofenac sodium. *Pharm. Dev. Technol*, 2015. 21(7): p. 875–886.

[83] Zhu, L.Z., et al., Rheological properties of redox-responsive, associative ferrocene-modified branched poly(ethylene imine) and its modulation by beta-cyclodextrin and hydrogen peroxide. *Soft Matter*, 2010. 6(21): p. 5541–5546.

[84] Morales, D., et al., Electro-actuated hydrogel walkers with dual responsive legs. *Soft Matter*, 2014. 10(9): p. 1337–1348.

[85] Xu, B., et al., Electro-responsive polystyrene shape memory polymer nanocomposites. *Nanoscience and Nanotechnology Letters*, 2012. 4: p. 814–820.

[86] Naji, L., M. Safari, and S. Moaven, Fabrication of SGO/Nafion-based IPMC soft actuators with sea anemone-like Pt electrodes and enhanced actuation performance. *Carbon*, 2016. 100: p. 243–257.

[87] Vogel, F., et al., Modeling and simulation of viscous electro-active polymers. *European Journal of Mechanics A/Solids*, 2014. 2(48): p. 112–128.

[88] Yang, C., et al., Reduced graphene oxide-containing smart hydrogels with excellent electro-responsive and mechanical properties for soft actuators. *ACS Appl. Mater. Interfaces*, 2017. 9: p. 15758–15767.

[89] Tu, Y., et al., Self-propelled supramolecular nanomotors with temperature-responsive speed regulation. *Nature Chemistry*, 2017. 9: p. 480–486.

[90] Heinzmann, C., C. Weder, and L. Montero de Espinosa, Supramolecular polymer adhesives: advanced materials inspired by nature. *Chem. Soc. Rev*, 2016. 45 : p. 342–358.

[91] Chen, F., et al., Thermo- and electro-dual responsive poly(ionic liquid) electrolyte based smart windows. *Chem. Commun*, 2017. 53: p. 1595–1598.

[92] Yan, K., et al., Electro-molecular assembly: electrical writing of information into an erasable polysaccharide medium. *ACS Appl. Mater. Interfaces*, 2016. 8: p. 19780–19786.

[93] Koetting, M.C., et al., Stimulus-responsive hydrogels: Theory, modern advances and applications. *Materials Science and Engineering R*, 2015. 93: p. 1–49.

[94] Ullah, F., et al., Classification, processing and applications of hydrogels: A review. *Materials Science and Engineering C*, 2015. 57: p. 414–433.

[95] Chen, Q., et al., Multi-stimuli responsive and multi-functional oligoaniline-modified vitromers. *Chemical Science*, 2017. 8: p. 724–733.

[96]   An, J.S., et al., Swelling-diffusion-interfacial polymerized core-shell types polystyren e/poly(3,4-ethylenedioxythiophene) microspheres and their electro-responsive characteristics. *Polymer*, 2017. 115 : p. 137–145.

[97]   Meng, H. and G. Li, A review of stimuli-responsive shape memory polymer composites. *Polymer*, 2013. 54: p. 2199–2221.

[98]   Petcharoen, K. and A. Sirivat, Magneto-electro-responsive material based on magnetite nanoparticles/polyurethane composites. *Materials Science and Engineering C*, 2016. 61: p. 312–323.

[99]   Wang, Y., J. Ye, and W. Tian, Shape memory polymer composites of poly(styrene-b-butadiene-b-styrene) copolymer/liner low density polyethylene/Fe3O4 nanoparticles for remote activation. *Applied Sciences*, 2016. 6: p. 333.

[100] Wang, D., Y. Jin, and D. Yan, Synthesis and applications of stimuli-responsive hyperbranched polymers. *Progress in Polymer Science*, 2017. 64: p. 114–153.

[101] Shigemitsu, H. and I. Hamachi, Design strategies of stimuli-responsive supramolecular hydrogels relying on structural analysis and cell-mimicking approaches. *Acc. Chem. Res*, 2017. 50: p. 740–750.

[102] Moad, G., RAFT polymerization to form stimuli-responsive polymers. *Polymer Chemistry*, 2017. 8: p. 177–219.

[103] Steinschulte, A.A., et al., Stimulated transitions of directed nonequilibrium self-assemblies. *Adv. Mater*, 2017. 29: p. 1703495.

[104] Jiang, J.Z., et al., Responsive, switchable wormlike micelles for CO_2/N-2 and redox dual stimuli based on selenium-containing surfactants. *Soft Matter*, 2017. 13(37): p. 6458–6464.

[105] Ramtin, A., et al., Cytotoxicity considerations and electrically tunable release of dexame thasone from polypyrrole for the treatment of back-of-the-eye conditions. *Drug Deliv. and Transl. Res*, 2016. 6 : p. 793–799.

[106] Wajs, E., et al., Preparation of stimuli-responsive nano-sized capsules based on cyclodextrin polymers with redox or light switching properties. *Nano Research*, 2016. 9(7): p. 2070–2078.

[107] Sen, S. and G.T.R. Palmore, Stimuli-responsive macromolecular composites: Enhanced stress modulation in polypyrrole with redox-active dopants. *Macromolecules*, 2016. 49: p. 8479–8488.

[108] Durabi, A., P.G. Jessop, and M.F. Cunningham, CO_2-responsive polymeric materials: synthesis, self-assembly, and functional applications. *Chem. Soc. Rev*, 2016. 45 : p. 4391–4436.

[109] Oechsle, A.A., et al., CO_2-switchable cellulose nanocrystal hydrogels. *Chem. Mater*, 2018. 30 : p. 376–385.

[110] Manouras, T. and M. Vamvakaki, Field-responsive materials: photo-, electro-, magnetic- and ultrasound -sensitive polymers. *Polym. Chem*, 2017. 8: p. 74–96.

[111] Du, Z., et al., An end-bifunctionalized hydrophobically modified ethoxylated urethane model polymer: Multiple stimuli-responsive aggregation and rheology in aqueous solution. *Macromolecules*, 2017. 50: p. 1688–1699.

[112] Fang, H., et al., A natural glycyrrhizic acid-tailored light-responsive gelator. *Chem. Asian. J*, 2018. 13 : p. 1192–1198.

[113] He, L., et al., *Influence of cleavage of photosensitive group on thermally induced micellization and gelation of a doubly responsive diblock copolymer in aqueous solutions: a SANS study.* in: Polymer, 2016. (105): p. 25–34

[114] Ma, X. and H. Tian, Stimuli-responsive supramolecular polymers in aqueous solution. *Acc. Chem. Res*, 2014. 47: p. 1971–1981.

[115] Roy, D., J.N. Cambre, and B.S. Sumerlin, Future perspectives and recent advances in stimuli-responsive materials. *Progress in Polymer Science*, 2010. 35: p. 278–301.

[116] Brassinne, J., C.A. Fustin, and J.F. Gohy, Control over the assembly and rheology of supramolecular networks via multi-responsive double hydrophilic copolymers. *Polymer Chemistry*, 2017. 8: p. 1527–1539.

[117] Okuzaki, H. and K. Funasaka, Electro-responsive polypyrrole film based on reversible sorption of water vapor. *Synthetic Metals*, 2000. 108: p. 127–131.

[118] Huang, Y., et al., Micellization and gelatinization in aqueous media of pH- and thermo-responsive amphiphilic ABC (PMMA82-b-PDMAEMA150-b-PNIPAM65) triblock copolymer synthesized by consecutive RAFT polymerization. *RSC Adv*, 2017. 7: p. 28711–28722.

[119] van der Asdonk, P., et al., Spatial and temporal patterning of polymers in electric field responsive LC templates. *J. Mat. Chem*, 2016. 4(35): p. 8263–8269.

[120] Cairns, D.R., et al., Stimulus-responsive fluidic dispersions of rod shaped liquid crystal polymer colloids. *Material Letters*, 2010. 64: p. 1133–1136.

[121] Frechet, J.M., Functional polymers and dendrimers: reactivity, molecular architecture, and interfacial energy. *Science*, 1994. 263(5154): p. 1710–1715.

[122] Jeon, S.J., A.W. Hauser, and R.C. Hayward, Shape-morphing materials from stimuli-responsive hydrogel hybrids. *Acc. Chem. Res*, 2017. 50 : p. 161–169.

[123] Attaran, A., J. Brummund, and T. Wallmersperger, Modeling and simulation of the bending behavior of electrically-stimulated cantilevered hydrogels. *Smart Materials and Structures*, 2015. 24: p. 035021.

[124] Coussot, P., *Rhéophysique: La matière dans tous ses états.* 2012: EDP Sciences.

[125] Aïtcin, P.C. and R.J. Flatt, *Science and technology of concrete admixtures.* 2016. Sawston, UK: Woodhead Publishing.

[126] Barthès-Biesel, D., *Microhydrodynamics and complex fluids.* 2012. Boca Raton, Florida, USA: CRC Press.

[127] Feys, D., et al., Measuring rheological properties of cement pastes: most common techniques, procedures and challenges. *RILEM Technical Letters*, 2017. 2: p. 129–135.

[128] Vega, J.F. and A. Santamaria, *Small*-Amplitude Oscillatory Shear Flow Measurements as a Tool To Detect Very Low Amounts of Long Chain Branching in Polyethylenes. *Macromolecules*, 1998. 31: p. 3639–3647.

[129] Zheng, C., et al., Enhanced stimuli-responsive electrorheological property of poly(ionic liquid)s-capsulated polyaniline particles. *Polymers*, 2017. 9: p. 385.

[130] Gao, C.Y., et al., Hollow submicron-sized spherical conducting polyaniline particles and their suspension rheology under applied electric fields. *Polymer*, 2018. 140: p. 80–88.

[131] Filipcsei, G., et al., Magnetic field-responsive smart polymer composites. *Adv. Polym. Sci*, 2007. 206: p. 137–189.

[132] Zhang, W.L. and H.J. Choi, Stimuli-responsive polymers and colloids under electric and magnetic fields. *Polymers*, 2014'. 6: p. 2803–2818.

[133] Winslow, W.M., Induced fibration of suspensions. *J. Appl. Phys*, 1949. 20: p. 1137–1140.

[134] Das, A.A.K., et al., Thermally responsive capillary suspensions. *ACS Appl. Mater. Interfaces*, 2017. 9: p. 44152–44160.

[135] Snoeck, D., *Self-healing and microstructure of cementitious materials with microfibres and superabsorbent polymers*. 2015. Boca Raton, Florida, USA: PhD thesis, Ghent University.

[136] De Schutter, G. and D. Feys, Pumping of fresh concrete: insights and challenges. *RILEM Technical Letters*, 2016. 1: p. 76–81.

[137] Momemi, F., et al., A review of 4D printing. *Materials and Design*, 2017. 122: p. 42–79.

Chapter 2

Rheology and processing of cementitious materials

Karel Lesage, Khadija El-Cheikh, and
Geert De Schutter

CONTENTS

2.1	Introduction	38
2.2	Rheology	38
	2.2.1 Shear stress	39
	2.2.2 Shear rate	39
	2.2.3 Flow curve	40
	2.2.4 Viscosity	41
	2.2.5 Yield stress	41
	2.2.6 Structural build-up and thixotropy	42
	2.2.7 Visco-elasticity	42
2.3	Rheometry	43
	2.3.1 Parallel plate geometry	44
	2.3.2 Protocols	45
	2.3.3 Steady flow properties	46
	2.3.4 Yield descriptors	47
	2.3.5 Thixotropy descriptors	50
2.4	Superplasticizers	50
2.5	Rheology of self-compacting concrete	53
	2.5.1 Mix design in relation to the key requirements of SCC	54
	2.5.2 SCC placing challenges	56
2.6	Processing techniques	57
	2.6.1 Pumping	57
	2.6.2 Formwork casting (top-down and bottom-up)	62
	2.6.3 Extrusion of cement-based materials	63
	2.6.4 3D Printing	66
2.7	Benefits of active rheology control	69
2.8	Summary	70
References		70

DOI: 10.1201/9781003289463-2

2.1 INTRODUCTION

The flow behaviour (or rheology) of concrete is a key factor in the ease of transporting, pumping, casting, consolidating and finishing of fresh concrete. Besides its economic effects during processing, the rheology of concrete can also affect its performance at hardened state. Therefore, it becomes increasingly common to quantify the concrete rheology in practice, sometimes by means of a series of empirical tests (e.g. slump, slump flow, V-funnel and Vebe). However, it is still possible that two mixtures with the same slump value will show a different flow behaviour [1]. Furthermore, the only way to obtain the desired rheological properties of a mixture is either by modifying the proportion of its raw materials or by adding admixtures. This strategy showed unguaranteed results and requires high expertise in concrete rheology. It is this kind of uncertainty that is feeding the need to develop more performant processing techniques, such as Active Rheology Control (ARC).

In this book, most of the chapters deal with the rheological behaviour of cement paste instead of concrete, for good reason. Obviously, the granular skeleton (i.e. sand and aggregates) will have a significant impact on the concrete rheology. However, in the context of ARC, the current activating mechanisms are taking place at the paste level. It is tempting to imagine a mechanism at the level of the aggregates but due to their size and mass the intended mechanism is likely to require extensive control equipment or rare kinds of aggregates. By intervening in the concrete rheology at the level of cement paste, the current ARC techniques enable control without the massive replacement of traditional concrete components. Simultaneously, ARC remains available for applications at both small scale (e.g. grout injection) and large scale (e.g. concrete casting).

2.2 RHEOLOGY

In practice, the term 'workability' is often used to indicate the handling properties of a fresh concrete mixture. It literary refers to how good a fresh mixture is 'to work with' in a certain situation. Similarly, the concrete's 'rheology' will describe this flow behaviour in a more fundamental way. Rheology itself is defined as the science of deformation and flow of matter under the influence of stresses. Therefore, knowing the concrete's rheology allows to describe its flow behaviour but now for a multitude of process steps.

Determining the rheology of a fluid can be done by means of rheometrical tests. From these experiments a relation is found between shear stress and shear rate. This relation expresses how a fluid would flow under a certain shear stress of interest (e.g. pouring vs. pumping). The relation is often mathematically expressed by an equation and some of the coefficients will be referred to as 'rheological parameters'. In the following sections, the main rheological parameters of interest and the ways we determine them are briefly explained.

2.2.1   Shear stress

We can imagine a material contained between two plates separated by a gap 'y', as shown in Figure 2.1. The material could be imagined to exist out of a number of small fictitious layers. When a horizontal force is applied to the upper plate while the bottom plate remains fixed, the material deformation occurs by relative deformation (i.e. slipping) of the fictitious layers. Therefore, the total horizontal translation of each layer will be a cumulated sum of all relative deformations of the layers below. Similarly, as this total deformation occurs in a constant time interval, the velocity will be high for the top layer and zero for the bottom layer.

This configuration corresponds with the theoretical case of simple shear and laminar flow. The force on the upper plate enforces a constant speed and causes the different material layers to transfer a part of the induced frictional force. We say the material experiences a 'shear stress' equal to F/S, with F being the magnitude of the force and S being the magnitude of the plate's surface area. The shear stress is usually expressed with equation 2.1, reporting its value in the Pascal unit.

$$\tau = \frac{F}{S} \tag{2.1}$$

2.2.2   Shear rate

The deformation enforced by the top plate during a time interval is represented by 'x' in Figure 2.1. This means that an intact material volume would experience a shear strain γ equal to x/y, for an infinitesimally small x. While this parameter is useful for more solid-like materials, it does not

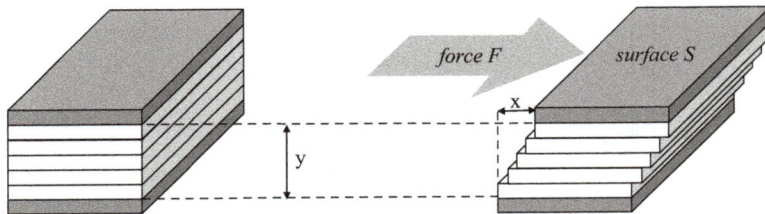

force F   surface S

Figure 2.1 **Shear applied to a layered volume.**

describe well the stress response of liquid-like materials as they tend to keep flowing beyond the time interval of shear application. Therefore, a more interesting parameter is the change in shear strain with time, considering a few consecutive time intervals. To that end, the 'shear strain rate' – or shortly 'shear rate' – $\dot{\gamma}$ is defined as the time derivative of the shear strain, as seen in equation 2.2. Doing so, the shear rate can also be written as a function of the horizontal velocity 'v', which is the derivative of the translation of x to time. The unit of shear rate then becomes the inverse of time (i.e. s^{-1}).

$$\dot{\gamma} = \frac{d\gamma}{dt} = \frac{dx}{dt} \cdot \frac{1}{y} = \frac{v}{y} \tag{2.2}$$

2.2.3   Flow curve

The macroscopic measurement of both shear stress and shear rate allows to collect data points which, together, constitute the flow curve of a material. For that purpose, a series of shear rates (or shear stresses) will be imposed on the material by a rheometer and the corresponding stresses (or rates) will be measured. Assuming proper rheometrical care is taken, the collection of data points is usually fitted with one out of several models which characterizes the material's flow behaviour the best. The most frequent models in the field of cementitious materials are summarized in Figure 2.2.

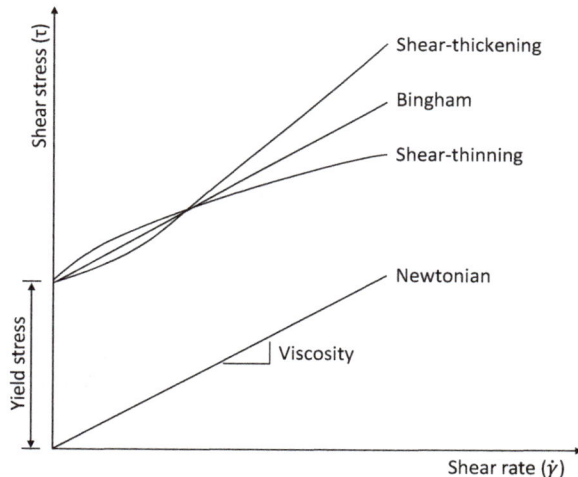

Figure 2.2 **Shear curves different materials.**

2.2.4   Viscosity

The most simple flow behaviour model is given by the Newtonian model that prescribes a linear relationship between shear stress and shear rate. It is observed in simple fluids like water which can be fully characterized by its viscosity. The Newtonian model is given in equation 2.3 and shows the viscosity η as the coefficient of the shear rate. The viscosity unit is Pa.s.

$$\tau = \eta \, \dot{\gamma} \tag{2.3}$$

It is noteworthy that viscosity does not have to remain constant over the entire shear rate range. For example when a fluid like ketchup is being stirred, the viscosity tends to decrease for higher stirring rates. We call this a shear-thinning fluid as it seems to get 'thinner' under shearing. Similarly, fluids like a corn starch solution tend to increase in viscosity when being stirred more vigorously and therefore we call them shear-thickening fluids.

2.2.5   Yield stress

When comparing water with ketchup, for example, it is clear that a portion of ketchup is able to keep its shape once deposited on your plate. This means that, for a zero shear rate, ketchup is able to withstand gravitational loads while water is not. This inherent material resistance shown in the rheological models as the intercept with the vertical axis is called yield stress (τ_0). We find a zero yield stress for Newtonian fluids, while a non-zero yield stress (with constant viscosity) is used to describe Bingham fluids. In this case, we call the viscosity *plastic viscosity, μ_p*, which is sometimes also indicated with simply μ.

$$\tau = \tau_0 + \mu_p \dot{\gamma} \tag{2.4}$$

The Bingham model (Eq. 2.4) is commonly used to describe the rheological behaviour of fresh cementitious materials [1–3]. The yield stress is said to originate from the spatial network of cement particles and the paste behaves elastically under applied stresses below the yield stress [4]. Depending on the shear rates and/or admixtures, a non-constant viscosity might be observed for higher shear rates and another rheological model will be used [5]. The Herschel–Bulkley model (Eq. 2.5) has an exponent for the shear rate 'n' called flow index and a consistency factor 'K' as a substitute viscosity parameter.

$$\tau = \tau_0 + K \dot{\gamma}^n \tag{2.5}$$

2.2.6   Structural build-up and thixotropy

In the above, the influence of time was limited to the definition of the shear rate. However, for complex fluids like cement paste there are also other microstructural effects that depend on time. These effects influence the value of the yield stress or the viscosity depending on how long the material has been at rest or flowing, respectively. This time-dependent phenomenon is often called *thixotropy*. For example, for a constant shear rate the viscosity of cement paste will likely decrease over time until a steady-state viscosity is found. Subsequently interrupting that shearing for one minute will lead to an again higher viscosity upon re-shearing. In fact, during that minute the cement particles will have started re-flocculating due to colloidal attractive forces, hydrogen/ionic bonding and nucleation of hydration products [4, 6].

Obviously, that flocculation mechanism will also work for a cement paste at rest, explaining the increasing yield stress over time. However, a yield stress might never be fully recovered like a viscosity because of the large share of irreversible particle connections that constitute the yield stress. This difference between reversible and irreversible breakdown was first introduced by Tattersall and Banfill [1].

In this book the above concepts are generally referred to as structural build-up. In the interpretation of results, the permanent changes in rheological properties with time are defined as workability loss [7] and fully reversible changes are considered as pure thixotropy. So, when measured structural build-up is partly irreversible, a combination of thixotropy and workability loss is at hand.

2.2.7   Visco-elasticity

In order to probe the effect of micromechanical mechanisms at cement particle level, like time-dependent flocculation or network structuring, the measurement of yield stress and viscosity will result in the flow of the sample. During the onset of this flow a large part of the microstructure will have been inevitably erased. Moreover, a new sample has to be loaded into the rheometer for every time step of interest to describe a time evolution, which is rather impractical. Therefore, the interest in non-destructive and more informative tests (cf. irreversibility) has risen over the past few decades.

As most yield stress fluids usually behave as visco-elastic materials in their solid regimes [8], cement paste can be probed in its elastic regime without permanently destroying it. At sufficiently low strain, the elastic modulus G is defined as $\frac{\tau}{\gamma}$. However, when strained, these materials also exhibit a viscous behaviour and their elastic response is not entirely instantaneous. To distinguish viscous from elastic effects, a strain oscillation (Eq. 2.6) with frequency ω and amplitude γ_0 can be applied while measuring the stress response $\tau(t)$.

$$\gamma(t) = \gamma_0 \sin \omega t \qquad\qquad (2.6)$$

When the strain amplitude is within the linear regime, the stress response $\tau(t)$ is of the form (Eq. 2.7):

$$\tau(t) = G'(\gamma_0 \sin \omega t) + G''\left(\gamma_0 \cos \omega t\right) \qquad\qquad (2.7)$$

The elastic modulus G' (or storage modulus) is an indication for all microstructural effects that are in phase with the strain (i.e. elastic effects). The viscous modulus G'' (or loss modulus) represents all effects that occur in phase with the strain rate (i.e. viscous effects). A summarizing parameter is found in the phase angle δ (tan δ = G''/G'). The measurement of both moduli occurs simultaneously as they are each other's complement. As cement paste shows time-dependent rigidification or solidification, the evolution of G' will be of primary interest to quantify the structural build-up under all kinds of conditions. Similarly, the contribution of supplementary powders or admixtures to this behaviour can be easily quantified in this non-destructive test.

2.3   RHEOMETRY

Measuring rheological properties of cement pastes is a more fundamental way of assessing fresh properties compared to standard workability tests [7]. However, measuring the rheological properties of these materials is not a straightforward task and substantial care must be taken prior, during and after the measurements [9]. There are a number of difficulties arising when measuring on cement pastes:

i. The pastes contain a large number of particles, with sizes around the boundary between colloidal and non-colloidal particles [1, 10, 11]. The colloidal particles are subjected to inter-particle interactions, such as Van der Waals dispersion and electrostatic repulsion forces [10];
ii. The pastes undergo a chemical reaction with time [1, 11, 12], changing its rheological properties;
iii. A majority of the developed chemical admixtures are designed to either change the properties of the suspending medium, the interaction forces between the particles, or to change the rate at which the chemical reaction occurs [13].

Besides these material challenges, each rheometrical geometry also has its own peculiarities that can trouble swift measurement execution. In this book all rheometrical tests were performed with parallel plate geometries. Therefore, the next section will discuss the concerns related to this setup only [7].

2.3.1 Parallel plate geometry

The parallel plate geometry, as in Figure 2.3, typically consists of two parallel circular plates with radius R separated by a gap width h in which the sample will be contained by surface tension [14–17]. Usually the upper plate is lowered upon the deposited sample, squeezing it until the pre-set gap width is reached. Paste excesses are subsequently trimmed manually. Literature values for the gap width are between 700 and 1000 µm, although it is still not clear what would make a value appropriate [18].

During the measurement, the upper plate rotates to create a shear velocity profile between the plates, as shown in Figure 2.3. The shear rate is not constant over the radius. It depends on the considered height in the squeezed sample and is calculated using equation 2.8 [14]. The shear stress can be calculated using equation 2.9, also depending on the shear rate [14–17]. In these equations, the radius R is the maximum of the radial distance r, the rotational velocity equals Ω and the torque equals T.

$$\dot{\gamma}(r) = \frac{\Omega r}{h} \tag{2.8}$$

$$\tau(r) = \frac{T}{2\pi.r^3}\left[3 + \frac{d\ln T}{d\ln \dot{\gamma}_R}\right] \tag{2.9}$$

The advantages of the parallel plate geometry are as follows: (1) a steady laminar flow and (2) shear rate and strain can be altered also by changing the gap width (h). The disadvantages are the following: (1) a non-homogeneous flow field; (2) settling can be a problem with the upper plate rotating on the suspension's medium only; (3) surface drying at the outer sample rim; and (4) the loading procedure of the sample is considerably more sensitive to system settings [19].

As an alternative to the parallel plates, a cone and plate geometry can be suggested. This has the particular advantage that the shear rate is

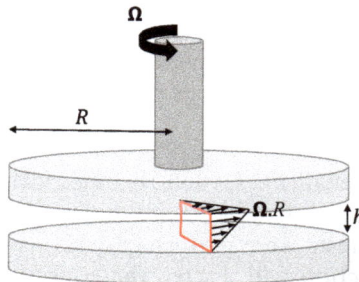

Figure 2.3 Parallel plate geometry schematically. Source: after [7].

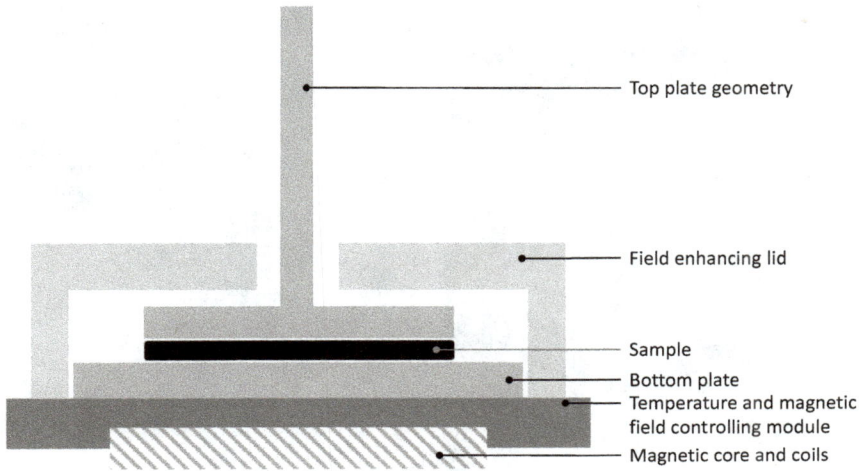

Figure 2.4 Schematic cross-section of the magneto-rheological device (Anton Paar®).

homogeneously distributed in contrast to the parallel plates [14]. However, below the central downwards facing tip of the cone, the gap width is too small for cementitious particles to flow smoothly, causing blockages and stress peaks.

An additional advantage of the parallel plate configuration is the possibility to add a magnetic field to the sample, as shown in Figure 2.4. The vertical magnetic field lines are distributed over the entire sample with acceptable homogeneity and its strength can be controlled within the rheometer software. In this way, it is possible to alter the field strength together with the steps in the rheometrical protocol.

Similarly, any other type of flat sensor device can be fixated at the bottom plate and serve to apply other kinds of fields to the sample during measurement. For example, multiple screen-printed electrodes have been embedded in a bottom plate in Figure 2.5. By three-dminesional (3D) printing the sensor holder, a smooth and continuously flat surface has been created that does not interfere with the sample flow. Through their individual wire connections the sensors remain controllable with dedicated equipment externally to the rheometer. The sensors allow at least partial exposure of the sample to an electrochemical field.

2.3.2 Protocols

The rheometrical protocol is chosen based on the investigated material properties, i.e. steady flow, thixotropic or visco-elastic behaviour in rotational or oscillatory shear [7]. Initially, the material response towards

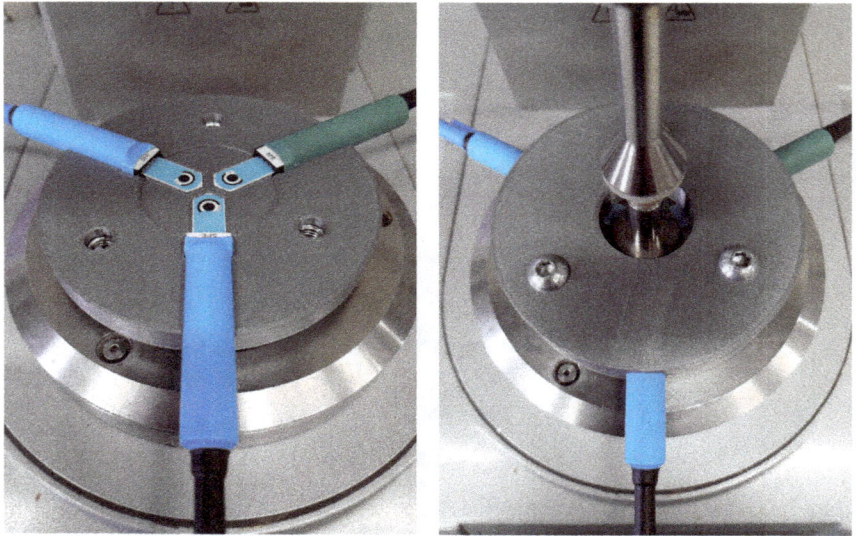

Figure 2.5 Electrochemical sensors embedded in the bottom plate (left) and top plate positioning (right).

deformation is analysed in a series of preliminary tests. Subsequently, the specific testing regime, e.g. the magnitude of the applied shear strain, shear stress, shear rate and frequency, is defined. Also part of the preliminary test is assuring the proper shearing of the sample for the chosen gap width and composition.

In order to set up the material in a controlled state independent of its previous history, the usual technique consists of imposing a very rapid flow (pre-shear) and then leaving the material at rest for some time (time of rest) [20]. The time of rest following pre-shear can play several roles: (1) possible inertia effects resulting from the previous rapid rotation of the rheometrical tools and the material can be avoided; (2) a viscoelastic material has time to return to its solid structure; and (3) in the case of thixotropy, different times of rest provide in principle different homogeneous states of restructuration of the material.

2.3.3   Steady flow properties

The steady flow properties of cement pastes refer to viscosity and yield stress obtained by means of the so-called flow curves, as shown in in Figure 2.6 [7]. The test is either performed in steps with a couple of seconds for each shear

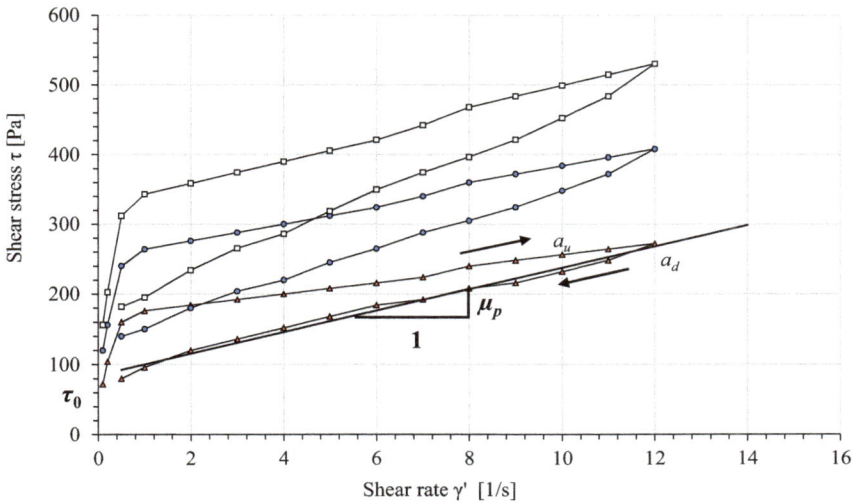

Figure 2.6 Shear stress versus shear rate hysteresis cycles at different material age showing the ascending a_u and descending a_d branches and a linear approximation of the descending branch to determine the Bingham parameters yield stress τ_0 and plastic viscosity μ_p. Source: after [7].

stress or shear rate increment, or in a linear fashion. Generally, either the up and down parts of the loop curve are completed, or solely the decreasing part is imposed. The downward branch is used to calculate the rheological properties depending on the applied model [21]. In this book the used models are the Bingham (Eq. 2.4) or Herschel–Bulkley (Eq. 2.5) model.

2.3.4 Yield descriptors

The yield stress obtained as a fitting parameter of the downward curve in the previous section actually denotes an approximation of the *dynamic* yield stress. It is the self-supporting stress when a dynamic system comes to rest. It should be distinguished from the *static* yield stress for which a system at rest is put into motion. The latter is typically obtained from slow sequential strain increments, which eventually result in stresses deviating from the linear behaviour, as shown in Figure 2.7. This allows to record a peak stress often referred to as the static yield stress [22]. Alternatively, one could gently increase the stress and measure the deformation response, which is expected to deviate from the linear behaviour as well. This is illustrated in Figure 2.8.

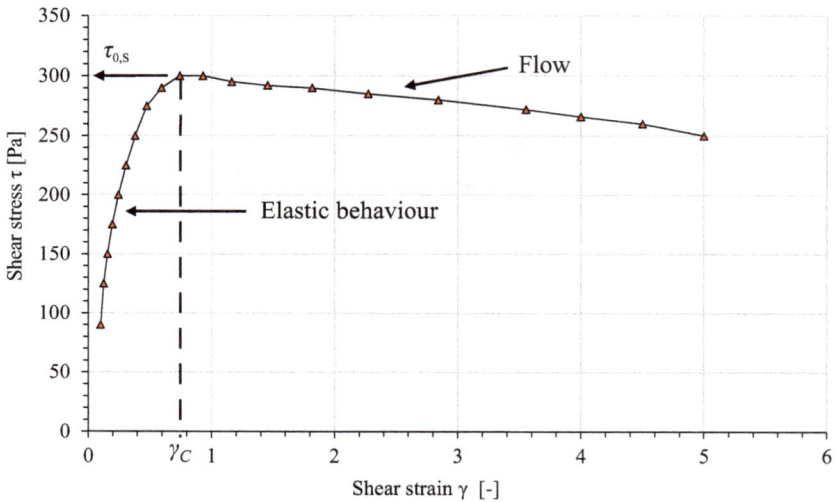

Figure 2.7 Stress–strain diagram with peak value. Source: after [7].

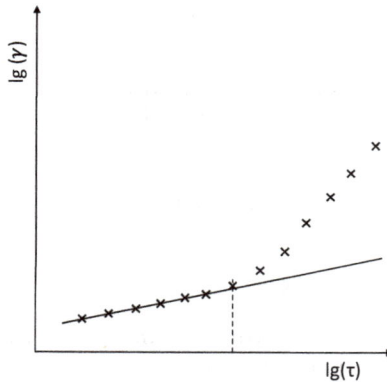

Figure 2.8 Strain–stress diagram with inflection point.

The main downside of these latter two static yield descriptors is that their value heavily relies on the speed of application, regardless of whether it is deformation or stress controlled. Perhaps more robust is applying a series of constant stresses to the sample while plotting the deformation as a function of time (i.e. creep test), as shown in Figure 2.9. The start of non-saturated

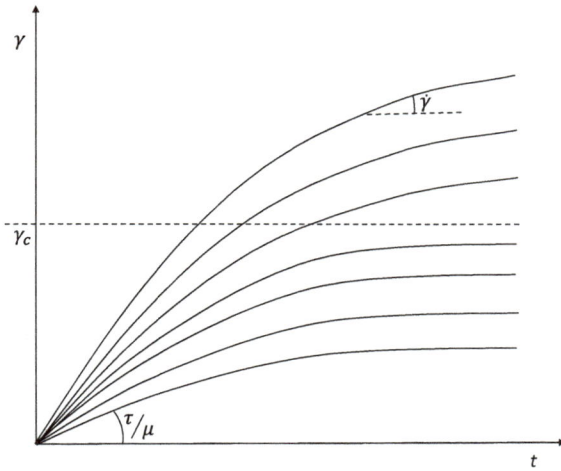

Figure 2.9 Creep curves for a series of constant applied stresses.

deformation behaviour then indicates that the yield stress of the materials has been exceeded and therefore indicates the yield stress. Of course, this test is more labour intensive and requires multiple samples and repetitions only to get an idea of the yield stress.

A second group of yield descriptors can be found within the small amplitude oscillatory shear (SAOS) measurements. Here, rotational shear is not a prerequisite and the presence of *flow* is ambiguous. As can be seen in Figure 2.10, the end of the linear visco-elastic region is indicated by τ_y, also referred to as the linearity limit [23]. Beyond this limit the storage modulus is no longer constant. Further down the curve, the cross-over point between G' and G'' is designated with τ_f, sometimes called the flow point. Beyond this point, the viscous behaviour becomes dominant over the elastic solid-like behaviour of the paste. Note that you still might not see any macroscopic flow at the outside of the sample. Both oscillatory values depend on the measuring conditions, e.g. the pre-set angular frequency. This makes it hard to correlate them unilaterally with the yield stress obtained from rotational shear experiments.

Due to all above practical concerns and the wide availability of (simple) rotational devices, preference is often given to the yield stress fit parameter following rotational shear tests mentioned in the previous section. Unless mentioned otherwise, this will also be the case for the majority of the test results reported in this book.

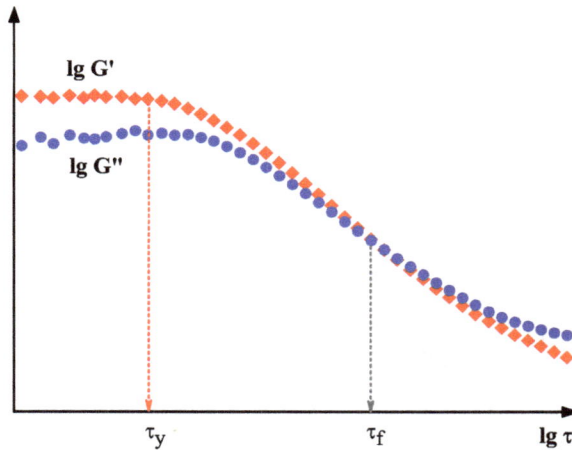

Figure 2.10 Evolution of dynamic moduli as a function of shear stress.

2.3.5   Thixotropy descriptors

Numerous techniques and strategies can be employed to determine the thixotropy of which the hysteresis loops and static yield stress measurements are deemed most common in the literature.

The hysteresis loops shown in Figure 2.6 are often applied as preliminary attempts to 'quantify' the thixotropy of the material [7, 20, 24, 25]. The entire area in between the up and down curves, and their evolution with time, is one way to assess thixotropy. Other approaches only consider a part of the area in between the up and down curves for the thixotropic index [26]. However, one must address such results rather as subjective due to a high risk of inaccurate interpretation, since the area and the shape of the hysteresis loops depend very much on the testing parameters [1, 27, 28].

An alternative technique is to determine the static yield stress as above by increasing shear strain at low or very low shear rates in the range between 10^{-1} and 10^{-3} s^{-1}. The result is the static yield stress τ_{0s}, which corresponds to a critical shear strain γ_C (dashed line in Figure 2.7), partly related to the thixotropic build-up of the material [22]. When plotting the static yield stress values as a function of resting time, the slope (A_{thix}) is a strong indicator for structural build-up [24, 29].

2.4   SUPERPLASTICIZERS

Superplasticizers are chemical additives in concrete that are used to (1) reduce the water/cement ratio and (2) prevent particle agglomeration of cement particles. The first function is particularly useful for reducing the pore water volume and increasing the strength and durability of the hardened concrete. The second function enhances particle movement during the handling of the fresh concrete which results in an improved concrete workability [30, 31].

From a chemical point of view, superplasticizers are organic water-soluble polymers belonging to the family of polymeric dispersants. They act by adsorption at the solid–liquid interface between the cement particles and the pore solution. After adsorption, repulsive forces between the superplasticizer polymers lead to the dispersion of the cement grains. Another important effect is that the superplasticizer may delay the normal hydration reactions by its presence on the particle surface [32, 33].

Different kinds of superplasticizer families have been applied in the history of plasticized concrete. The following main groups of superplasticizers are often mentioned [30, 34]:

- Polycondensates of sulfonated naphtalene formaldehyde (SNF);
- Polycondensates of sulfonated melamine formaldehyde (SMF);
- Modified lignosulfates (LS);
- Polycarboxylate ethers (PCE).

All of these groups have in common that they adsorb on the charged cement grain surface through the charges embedded in their own molecular structure. The mechanism of particle repulsion, however, is expected to result from different origins. The first three families repel particles by electrostatic forces, while the polycarboxylate polymers repel mainly by steric hindrance. These mechanisms are illustrated, respectively, in Figure 2.11 and Figure 2.12. The steric hindrance is induced by the polyethylene oxide chains swirling into the liquid phase. The length and the amount of chains on the surface are responsible for the effectiveness of the steric stabilization of a PCE in cement suspensions [8, 35, 36].

Regardless the working mechanism, the PCE superplasticizers have become the most promising family because of their high adaptability, leading to nearly infinite product formulations that can be designed for

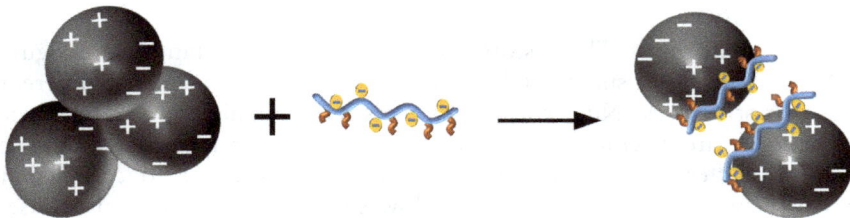

Figure 2.11 Dispersion through electrostatic repulsion (for reasons of clarity, cement particles and superplasticizer molecules are drawn at different scales).

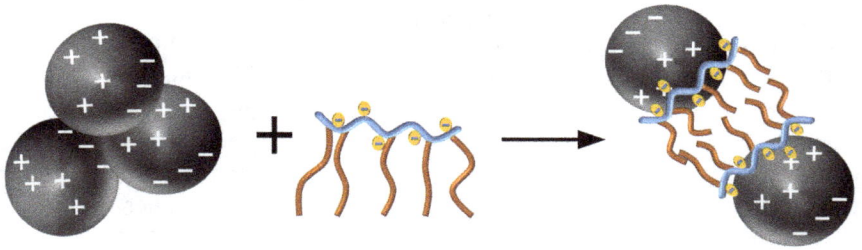

Figure 2.12 Dispersion through steric hindrance (for reasons of clarity, cement particles and superplasticizer molecules are drawn at different scales).

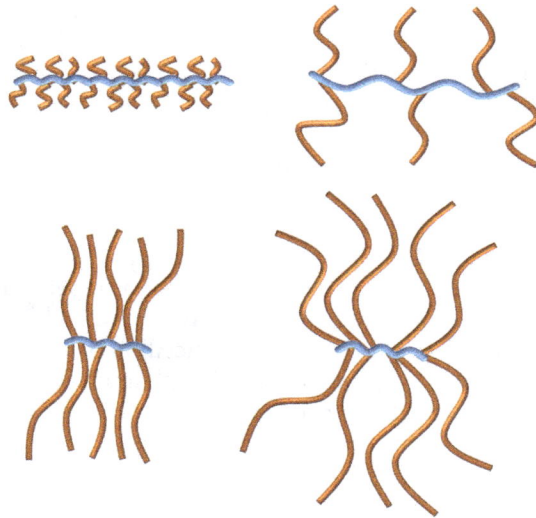

Figure 2.13 Schematic representation of different PCE architectures. Source: after [46].

specific applications. The customizable asset of the PCE family (see Figure 2.13) has led to a substantial amount of research focusing on different polymer structures. Numerous researchers have highlighted the influence of polymer architectures and their interaction with cementitious systems [37–45]. Different relations were found but overall it was conceived that polymers with short side chains and low side chain density – and consequently a high polymer charge – create strong adsorption especially on positively charged particle surfaces.

The adsorption of PCE superplasticizer has also gained a lot of attention in the literature because it is closely related with the polymer effectivity. Only adsorbed polymer that is not intercalated by hydration products can function optimally as a dispersive agent in-between particles. If the polymer adsorbs too fast, it will be quickly covered with hydration products and too soon the concrete will lose its workability. If the polymer adsorbs too slowly, it is possible that it will be overdosed to obtain sufficient workability during the induction period, resulting in uneconomic practice.

In this book, the applied superplasticizers are of the polycarboxylic ether type. That means most superplasticizers are composed of an anionic backbone made of mainly carboxylic groups and side chains of polyethers covalently attached to the backbone. Alternative groups can be added as backbone or side chains groups but this will be specifically mentioned.

In practice, the main monomer building blocks for the anionic component are acrylic acid, methacrylic acid and maleic acid [8]. As side chains, polyethylene oxide (or polyethylene glycol) is used in most of the cases. Polyethylene oxide makes up almost 90% of the mass of a polycarboxylic ether. These side chains are connected to the backbone through an ester, amide or ether bond. The amide or ester group results from the esterification or the amidation of acrylic, methacrylic or maleic units by mono-functional polyethylene oxides. The ether bond results from the copolymerization of mono-allyl or vinyl or iso-prenyl polyethylene oxide macro-monomers.

The main parameters that influence the properties of a PCE polymer are [8]:

- Chemistry of the backbone (acryl, methacryl, vinyl, allyl, malein, etc.);
- Length of the backbone (degree of polymerization);
- Amount of anionic groups (mostly carboxylic acids or salts) with respect to the side chains and with respect to the backbone;
- Number of side chains with respect to ionic groups and with respect to the backbone;
- Length of the side chains;
- Type and stability of the bond between side chain and backbone;
- Overall charge density of the polymer.

2.5   RHEOLOGY OF SELF-COMPACTING CONCRETE

The earlier development and implementation of SCC is an interesting precedent for any future implementation of admixture-driven innovation in concrete – especially considering the challenging rheological aspects of SCC that have been coped well in pre-cast industry but not so much in ready-mix industry. The material-related aspects and challenges will be recapitulated here and the process-related challenges will be covered in following sections.

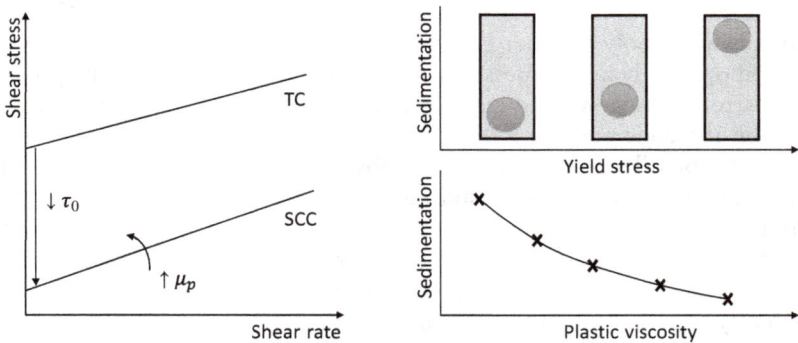

Figure 2.14 Rheological requirements for fresh SCC and the inverse relationship between yield stress and plastic viscosity and the rate of aggregate sedimentation. Source: after [49].

SCC has been defined as a concrete that is able to flow and consolidate under its own weight, without the need of any external compaction effort, completely filling the formwork even in the presence of congested reinforcements, while remaining homogeneous [47].

There are three key properties of fresh SCC [48]:

- Filling ability: the ability to flow under its own weight and completely fill all the spaces in the formwork;
- Passing ability: the ease to flow through confined and constricted spaces, narrow openings and between reinforcement;
- Stability or resistance to segregation: the ability of concrete to maintain its original distribution of constituent materials during transport, placing and (self-)compaction.

These key properties have consequences for the rheological requirements for SCC. Using the Bingham model introduced earlier, we can state that the flowability of SCC, i.e. both the filling and passing ability, is related to the yield stress τ_0, while the resistance to segregation is controlled mostly by the plastic viscosity μ of the SCC mixture. These rheological requirements for fresh SCC and the inverse relationship between yield stress or plastic viscosity and the rate of aggregate sedimentation are illustrated in Figure 2.14 [49, 50].

2.5.1   Mix design in relation to the key requirements of SCC

The way of controlling the workability requirements of SCC is by altering the concrete mix design to ensure a good balance between flowability and resistance to segregation. Hereafter, the effect of both the mix design and the main mix constituents on each key property will be explained in more detail [50].

Considering the filling ability or flowability of the concrete, the most relevant component is the flowability of the cement paste. Its fluid constituent, water, has the effect of reducing both yield stress and viscosity of the cement paste. The decrease in viscosity can seem beneficial but also puts limits on the cohesiveness and segregation resistance. Unlike water addition, the addition of high-range water-reducing admixture (HRWRA) mainly reduces yield stress [51, 52]. In this way, a highly flowable concrete can be obtained without significant reduction in cohesiveness.

Another important mix parameter in SCC is the water-to-powder ratio (W/P). If the W/P ratio is too low, the cement paste might appear too stiff or sticky, while a too high W/P ratio will again challenge the cohesiveness of both paste and mortar. This balance is also influenced by the interparticle friction between the solids (coarse and fine aggregates and powder materials) since solid-to-solid friction increases the internal resistance to flow [51].

When concrete flows through smaller openings, the interparticle friction increases resulting in higher shear stresses possibly blocking the flow. If the solid-to-solid friction (per unit mass) can be low, the risk of reduced flowability is prevented. In SCC this is mainly achieved by reducing the coarse aggregate content to increase the average interparticle distance. Other factors contributing to maintaining good flowability are round-shaped aggregate and continuous grading.

Given a good flowability, the passing ability of the concrete prevents blockage resulting from flow through narrow spaces. The passing ability is then mainly provided by an increased cohesion of the concrete mixture. This can be achieved by reducing the W/P ratio or incorporating a viscosity modifying admixture (VMA) or both. Given a proper cohesiveness, the passing ability can further be increased by reducing the maximal coarse aggregate diameter with respect to the most congested reinforcement zone in the formwork.

Together with providing flowability and passing ability, the mixture must also be prevented to segregate. This balance is usually achieved by making the matrix outside the aggregates more viscous and by making the viscosity of the entire concrete as low as possible. Increasing the matrix viscosity is usually done by reducing the W/P ratio or by the addition of VMA. The viscosity of the entire concrete is usually decreased by restricting the aggregate volume and its maximum aggregate diameter.

It is striking how many interactions are created among all the components of the fresh SCC. Not all of them are equally sensitive to change, but it is clear how small perturbations in one piece of the puzzle will have large consequences for the entire puzzle. Moreover, the stability of the fresh concrete is also required to guarantee homogeneous properties of the hardened concrete, including strength, stiffness, bond to the embedded reinforcement and durability. So, with the precarious balance in component interactions established after production, the fresh SCC will still have to be transported and placed into the formwork without disturbing the balance. These final steps are therefore worth giving the proper attention and will pose some additional challenges that we will cover in the next section.

2.5.2 SCC placing challenges

Although SCC is apparently more flowable than traditional concrete, recent studies have shown that the pumping behaviour of SCC is more complex, involving more significant shear deformation in the material, along with the effect of the lubrication layer near the pipe surface. Considering potential shear-thickening behaviour of powder-type SCC, high pumping rates can lead to much higher pumping pressures than in the case of traditional concrete. Furthermore, the additional role of thixotropy during pumping is not entirely clear yet [53].

At this moment, a good understanding of pumping of cementitious materials is available. Figure 2.15 shows the velocity profiles in a concrete pumping pipe for the case of conventional vibrated concrete (CVC) and SCC. Due to the lower yield stress of SCC, the central plug flow zone is smaller compared to CVC, showing that more material is sheared. Due to the higher viscosity of SCC, and due to the shear-thickening behaviour which might occur, pumping of SCC requires higher pressures. Despite this fundamental knowledge, questions remain concerning the pumping

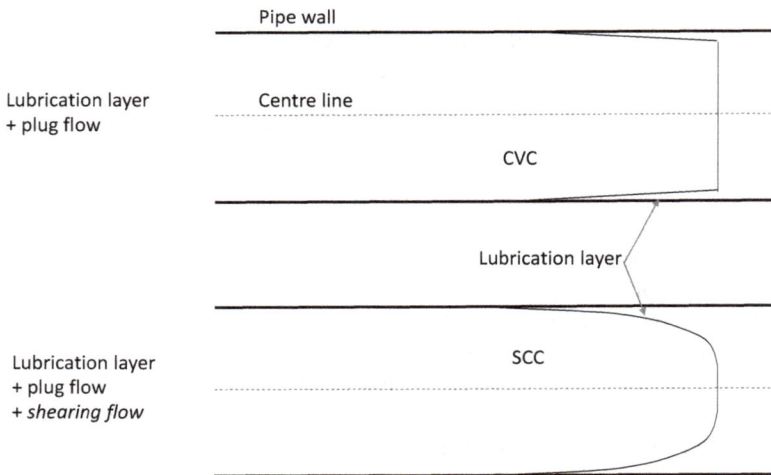

Figure 2.15 Velocity profiles of pumped conventionally concrete (CVC) and self-compacting concrete (SCC).

behaviour of specific types of concrete in which the formation of a lubrication layer is not induced, or in which internal friction and Coulomb effects become prevalent (e.g. in case of ultra-high performance concrete). During current pumping operations, the only option is to control the pumping pressure or the discharge rate while operating the pump. Short interruptions of the pumping process can lead to major difficulties in resuming pumping operations due to the sometimes tremendous effect of internal structural build-up or thixotropy.

Similarly to pumping, formwork jobs completed with SCC show still some challenges. In general, contractors refer to formwork pressure and leakage problems as one of the main reasons not to use SCC. In practice, formwork tightness requirements are defined in normative documents, e.g. limiting the extent of honeycombs. However, no fundamental study is available on the relationship between formwork design and detailing on the one hand, and concrete rheology and pressure on the other hand.

Some relevant state-of-the-art knowledge is available concerning formwork pressures during top-down casting [54], but the situation in case of bottom-up pumping is considered in a too simplistic way in the literature, typically assuming formwork pressures equal to hydrostatic pressures. In reality, the bottom-up pumping process can induce very high pumping pressures exceeding hydrostatic pressures in case of high viscosity, dense reinforcement and high filling rates [55]. Furthermore, in current practice, due to the lack of robustness of SCC, small changes in concrete composition can lead to increased flowability and reduced segregation resistance, increasing the risk of formwork leakage.

The balance between material-related benefits and process-related challenges holds for many more concretes than just SCC. For all special concrete types this balance needs to be figured out and guarded, sometimes even for every new batch. In the next section, we shortly recapitulate the most relevant concrete processing techniques to demonstrate the importance of rheology during processing of concrete.

2.6   PROCESSING TECHNIQUES

The processing techniques considered here are pumping, formwork casting, extrusion and 3D printing. All of them have their own particular interference with the rheology of the fresh concrete, e.g. through their geometry, mechanics of shear application and time dependency. The following subsections will treat each processing technique one by one.

2.6.1   Pumping

Concrete pumping is an efficient and economical mode of transporting and placing a large volume of fresh concrete into a desired location on a site. This technique has been used since the 1960s. A fresh concrete batch is fed into a pump through a hopper and transported under specific flow rate through a pipeline to a specific location. This technique requires fewer labourers, less time and less effort compared to manual concrete pouring. For hard access areas, not reachable by a concrete lorry or ready-mix truck, pumping becomes the only adapted way to deliver concrete. Moreover, concrete pouring by pumping is suitable for large volumes where an uninterrupted concrete supply is needed.

Regarding the pumped concrete, the properties of the placed concrete may differ from the initially designed mixture after being agitated and pushed through the pump. Thus, the concrete mixture must be properly designed with specific properties to be suitable for this particular application. On the one hand, a very fluid concrete may require less pressure to be pumped but it is more exposed to segregation, especially after being submitted to pressure from the pump. At this stage the pumping process has significantly high risk of blockage. On the other hand, a stiff concrete exerts a high friction at the concrete-wall pipe interface which may lead as well to blockage.

In reality, concrete does not flow homogenously in a pipe. Depending on the rheological properties, concrete can be totally or partially sheared inside the pumping circuit [56, 57]. Indeed, once concrete is pumped in the pipeline, a lubricating layer composed of water and fine particles (cement and sand) is formed at the interface between the concrete and the pipe surface [58-62]. The formation of the lubrication layer has been explained by two phenomena. First, the driving force created from the shear gradient over the pipe radius promotes the coarse particles, initially uniformly suspended in the mixture, to migrate towards the centre of the pipe (low shear region) [60, 63]. The paste composed of water and fine particles present in the suspension fills the places left by the coarse aggregates near the pipe surface. As a consequence, a layer depleted of coarse aggregates forms near the pipe surface. The second phenomenon is the wall effect of a random loose packing defined by a zone of thickness $d/2$ at the pipe surface with high void fraction, where d is the maximal particle size. The void fraction near the surface is filled with paste.

As a consequence of combined phenomena, the flow profile inside a pipe can be divided into two or three zones of different properties [56, 57]. The composition of the concrete in the central part of the pipe differs from sheared and lubrication layers as shown in Figure 2.16 [63]. In the plug zone (central part), the concrete is not sheared and it flows at constant velocity. This zone is formed mainly due to a yield stress higher than the applied shear stress. In the sheared zone, the shear stress is equal to or exceeds the yield stress of the cementitious material near the pipe surface. At the interface, the friction resistance is mainly governed by the rheological properties of the lubrication layer. This latter plays a key role in the pumping process.

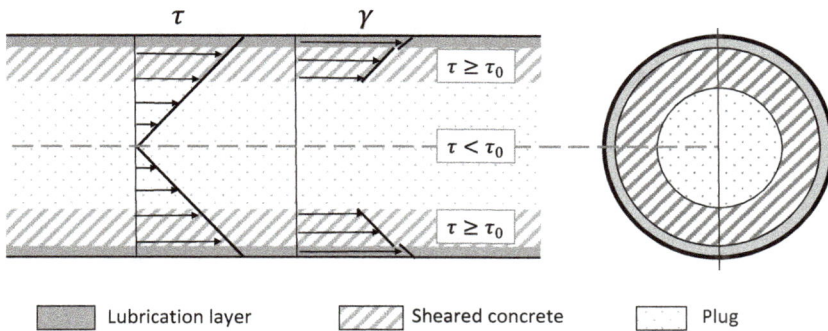

τ γ

$\tau \geq \tau_0$

$\tau < \tau_0$

$\tau \geq \tau_0$

Lubrication layer   Sheared concrete   Plug

Figure 2.16 Flow profile in a pipe. Source: after [63].

It lubricates the pipe surface, reduces the friction at the interface and eases the pumping of concrete. In practice, a cement mortar is pumped in the pipes circuit before concrete pumping in order to facilitate the formation of the lubrication layer. However, questions remain about how to extract and characterize the lubrication layer. The flow behaviours of CVC and SCC have been widely validated experimentally in a full-scale pumping circuit [53, 61, 64, 65].

The fundamental problem encountered during concrete pumping is blockage of concrete in the pipeline mainly due to excessive segregation, high interface friction or underestimation of the required pressure to pump the concrete. Concrete blocking may also occur when resuming the pumping operation after a short break mainly due to a delay in concrete supply delivery. Indeed, cement hydration and flocculation form an internal structural build-up in fresh concrete which can be partially broken by shear. This gives to concrete the thixotropic behaviour, and as a aresult the concrete becomes stiffer after a resting time and consequently higher pumping pressure is required [4].

The blockage phenomenon can be clearly detected by a sharp increase in the pressure at the pump inlet shown on the pressure gauge. For this reason it is highly important to maintain the stability of the concrete during pumping and predict the needed pressure to pump the concrete to the desired location. The determination of the required pumping pressure depends on both the pumping setup configuration and the rheological properties of the pumped concrete. The main setup factors affecting the pumping pressure are pumping rate, hose length and diameter, horizontal and vertical distances and the use of reducers. It is indeed recommended to pump along a low height, straight and short distance. In certain complex situations angled bends and high-level pumping become necessary.

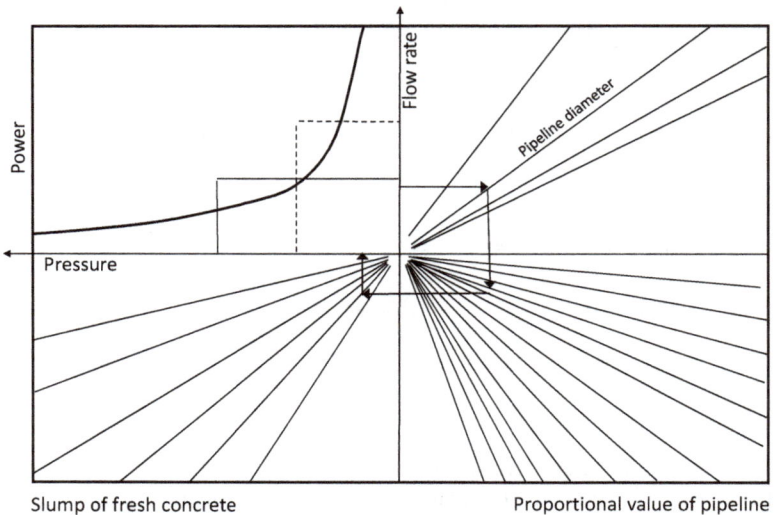

Figure 2.17 Concept example predicting pumping pressure for conventional concrete.

Design charts obtained from experimental results have been developed to predict the minimum required pressure to pump a CVC [66]. The charts, with an example presented in Figure 2.17, are based on the pumping configuration, including diameter, length of the pipe and desired flow rate and the properties of the fresh concrete basically determined by the slump or the spread test. Based on these charts, the ability of the concrete to be pumped is determined according to its slump value. However, these charts do not allow the prediction of pumping pressure of very flowable concrete, such as SCC. It is important to highlight that very flowable concretes may require higher pressure to be pumped due to their high viscosity which is not taken into account in the design charts [53, 56].

Many fundamental models have been developed by applying principles of energy equilibrium, momentum conservation and rheology in order to predict the required pressure to pump concrete [56, 63, 67, 68]. In these models concrete is considered as saturated (suspension of solids in water) which has been proved more advantageous for pumping than unsaturated (solid-like material) [59]. In fact, for stiff concrete where the volume of coarse aggregates is considerably high, the inter-particle friction is dominating and the concrete behaves as a solid-like material [69]. Thus, this type of concrete is more susceptible to blocking. On the contrary, concrete can be considered as saturated when the paste can fill the spaces between the coarse aggregates reducing the inter-particle friction.

The difficulty to find the most adapted model remains in taking into account and measuring the thickness and rheological properties of the lubrication layer, the rheological properties of the bulk concrete and the pumping circuit geometry. For example, if a CVC is assumed to behave as a Newtonian fluid by neglecting the yield stress, the flow regime of the concrete inside a typical pipeline can be considered as a fully laminar flow (Reynolds number ≤ 2100). For laminar flow of a Newtonian fluid, the shear stress is simply proportional to the velocity gradient. This means that the velocity decreases from the pipe centreline to the pipe wall as the shear stress profile inside a circular pipe is linear with zero value at the centre and maximum at the boundary of the pipe. In this case, the Poiseuille equation can be used to obtain the relationship between the pressure drop and the flow rate [70].

$$Q = \frac{\pi \Delta p_{tot} D^4}{128 \mu L} \tag{2.10}$$

where Q is flow rate, ΔP is the pressure loss, μ the is fluid viscosity, and D and L are the diameter and the length of the pipe, respectively.

From the Poiseuille equation applied for Newtonian fluids, it can be concluded that any increase in viscosity of concrete or length of the pipe leads to an increase in the pressure loss for a constant flow rate. The Poiseuille equation also shows that the discharge rate increases with the fourth power of the diameter or, inversely, the pressure loss decreases with the fourth power of the diameter.

In reality, concrete behaves more likely as Bingham material where the yield stress is not neglected. Especially for CVC, the ratio between the yield stress and the plastic viscosity is relatively high. For Bingham materials, the Buckingham–Reiner equation enables predicting the pressure loss for a given flow rate and rheological properties of concrete [71]. In the Q-ΔP relationship, the effect of both yield stress and viscosity on pressure loss is taken into account. For very flowable concrete (like SCC) the pressure loss is more dependent on the viscosity as the value of the yield stress is considerably low. For this reason a higher pressure may be required to pump an SCC and the effect of viscosity on the pressure loss must be taken into account in the charts used to predict the required pressure [53, 56].

$$Q = \pi \frac{3\Delta p_{tot}{}^4 R^4 + 16\,\tau_0{}^4 L^4 - 8\tau_0 L R^3 \Delta p_{tot}{}^3}{24\,\Delta p_{tot}{}^3 \mu L} \tag{2.11}$$

Regarding the velocity profile, due to the yield stress, a Bingham material flows as a plug with uniform velocity at the central part of a pipe, while the velocity profile sharply increases from the pipe surface to the interface with the plug zone. The magnitude of each zone (plug and shear) depends mainly on the rheological properties of the material and the pipeline configuration [56, 57].

Both cases (i.e. Newtonian and Bingham material) lead to overestimation of the required pressure to pump concrete because the particle migration inside the pipe, the thickness and rheological properties of the lubrication layer are not considered. The first attempt to obtain flow rate–pressure loss relationship by a full-scale concrete pumping circuit has been made by Kaplan et al. in 2005 [57]. The model proposed by Kaplan showed that the rheological properties of the lubrication layer must be taken into account to accurately predict the required pumping pressure. After Kaplan, many attempts have been made to develop test methods in order to extract and characterize the lubrication layer [1, 58, 61, 67, 68, 72]. Other test methods have been developed to directly predict the pumpability of concrete [59, 62, 73, 74].

Despite the fundamental research performed on predicting flow behaviour and pumpability of concrete, significant challenges remain, particularly in extracting and characterizing the lubrication layer and pumping modern concrete mixtures containing new types of additives and admixtures. At this point, research is still ongoing on predicting and ensuring a pumpable concrete without being blocked or clogged during the pumping operation. A more detailed study of the lubrication layer and its importance for accurate prediction of concrete pumping pressures is beyond the scope of this book.

2.6.2   Formwork casting (top-down and bottom-up)

Cast-in-place concrete, also known as poured-in concrete, is a technique widely used on site for concrete element construction, such as columns, beams and retaining walls. After transport by a ready mix concrete truck, fresh concrete is poured into pre-designed removable formworks. Once the concrete is sufficiently cured and hardened, the formworks can be removed.

The quality of the cast concrete depends on the rheological properties of the fresh concrete and the placing method. A proper casting method consists of pouring the concrete continuously into the formwork with uniform filling rate and minimum formwork pressure. The cast concrete must be dense and homogeneous with neither sign of segregation. In the traditional placement method, formwork is filled from the top by pumping concrete through pipes. The concrete is poured in horizontal layers of uniform thickness. The layers are vibrated to promote its consolidation before pouring the next layer. This step is necessary when pouring a CVC to compact the concrete and eliminate honeycombed areas and entrapped air. It is highly important to control the rate of layers pouring in order to avoid placing a fresh layer on top of a layer that is already set. Delays in concrete layer pouring can create weak cold joints which result in undesirable consequences, such as poor mechanical properties. Practical considerations must be taken, such as placing with controlled drop height or limiting the horizontal movement of placed concrete to avoid segregation.

SCC has many benefits in terms of placement compared to CVC. Its high flowability feature allows the concrete to fill the form of the mould under its own weight without additional mechanical compaction and to pass easily through dense reinforcement. Moreover, due to its fluid-like behaviour, SCC offers faster placement compared with CVC. Furthermore, the high flowability of SCC enables a new casting method by filling the formwork bottom-up in a single pumping operation in contrast to the traditional top-filling process. The bottom-up casting technique consists of pumping SCC mixtures into the formwork through an inlet inserted at the bottom part of the formwork, connected to a pump. In this way, the formwork is filled by upward movement of the concrete. The filling rate can be controlled by the discharge rate of the used pump and the resulting formwork pressure can be monitored to ensure a good casting.

Each filling process (top-down or bottom-up filling) has its own benefits and disadvantages. When filling from the top, the surface finish quality is not affected by the local inlet duct used for pumping bottom-up, while the quality of bottom-up cast is better compared with top-filling. Indeed, pumping bottom-up limits the entrapped air and the risk of segregation. The continuous bottom-up cast avoids weak interfaces and reduces the filling time. However, the SCC must remain flowable during the filling process and the methodology results in high total formwork pressures especially associated with high filling rates.

The research conducted by Tichko et al. showed that when filling bottom-up, the formwork pressures can be higher than the hydrostatic formwork pressure, due to hydraulic pressure losses [75]. Therefore, the capacity of the formwork to retain the total pressure must be considered. Additionally, pumping bottom-up is considered costly as it requires a more rigid and complex formwork compared to the typical formwork panels used for top filling. In order to ensure a good filling process, guidelines of good practice for top and bottom filling formwork have been established for SCC [76, 77]. For example, it was highlighted that when pumping bottom-up, the pumping point should be as close as possible to the middle of the cast element to reduce the flow distance of SCC and consequently limit the risk of segregation.

2.6.3   Extrusion of cement-based materials

Extrusion of mineral materials to produce building elements such as bricks finds its application widely in the construction sector. Although the extrusion of cementitious materials is adopted in the precast construction industry, the use of cement-based material extrusion in the construction sector has been limited to specific applications, such as hollow core slabs and roadside kerbs [78]. However, the extrusion process of cement-based materials has grown in importance steadily over recent years due to concrete 3D printing [79, 80].

An extrusion system is mainly composed of an extruder barrel and a die which has a designed shape depending on the desired extrudate form. After being prepared and placed in the barrel, a formable cementitious material is pushed forward and then extruded gradually or abruptly through a narrower extrusion die. Two techniques can be used to extrude cementitious materials: ram or screw extrusion [81]. In the ram extrusion, the material is pushed by a ram towards the die system, while in screw extrusion the material is driven by a screw. Because of its simplicity ram extrusion is more convenient for laboratory and numerical studies such as flow properties and parameters affecting the extrudability of materials, while screw extrusion is more used in industry as it is more efficient allowing a continuous extrusion process. Feeding the extruder by mean of a concrete pumping line can be compared to the situation of ram extrusion.

During extrusion, the cementitious material undergoes different forces. Additional to a frictional force acting along the nozzle surface, a forming force exhibits in the shaping zone, evolving gradually or abruptly towards the die system, as shown in Figure 2.18. Thus, for a successful extrusion, it is vital to estimate the required total force. This, in turn, requires understanding the extrusion flow behaviour.

Cementitious materials are known to behave as visco-plastic materials described likely by Bingham or Herschel–Bulkley models depending on their rheological properties. In fact, when the material is pushed by a ram, the

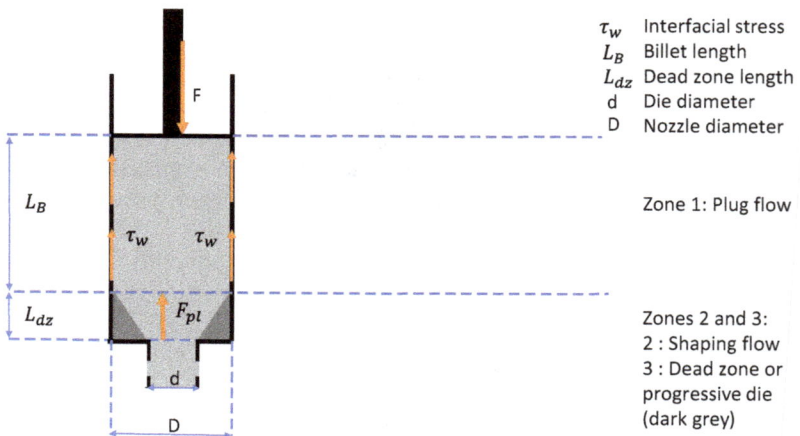

Figure 2.18 **Forces and flow zones in ram extrusion. Source: after [81].**

material behaves as it is pumped through a pipe. This means, as discussed in the pumping section, that the material inside the nozzle can be partially or totally sheared depending on its rheological properties (yield stress and viscosity), formation of lubrication layer (thickness and composition) and extrusion rate.

The extrudable material has conflicting requirements; it must be flowable to flow easily through the die system, while it must be dense, cohesive and stiff to retain its shape after extrusion. High yield stress materials, concentrated suspensions with high solid volume content, are frequently used for an extrusion process due to their ability to retain their shapes.

The study conducted by Perrot et al. [82] showed that ram extrusion flow of high yield stress material can be divided into three zones. In the plug zone, the material moves at a constant velocity and the frictional surface stress depends mainly on the rheological properties of the lubrication layer. Once arrived at the shaping zone, the material is plastically deformed due to reduction in geometry (from D to d). Many studies have been based on the forming force model proposed by Benbow and Bridgwater [83] to describe the flow extrusion in the forming zone [84, 85]. However, the intrinsic material parameters are not involved in the proposed model. In their study on the behaviour of high yield stress fluids at low strain rate, Perrot et al. [15] have successfully used the improved approach proposed by Basterfield et al. on the interpretation of orifice extrusion data for viscoplastic materials [85].

Three test methods are mainly used to evaluate the extrusion ability of cementitious materials: penetration resistance, ram extrusion and squeeze tool test. The penetration resistance test consists in measuring the force and the penetration depth of a needle in fresh cementitious materials after determined intervals. The extrudability window is defined in terms of penetration resistance (upper and lower limits) [86]. The ram extrusion test is a direct test which consists of pushing the material into a cylindrical tube by a piston. The force acting along the surface is measured and the different mechanisms can be studied [87]. In the squeezing test [14], a cylindrical fresh material is pressed between two paralleled plates. The curve of reduction of specimen height versus vertical compression load can be plotted and rheological properties can be obtained in order to evaluate the extrusion ability. This test allows to predict if the material will likely behave as pure plastic or drained frictional plastic material. An overview of the existing rheological test methods for the characterization of extrudable cement-based materials is given in [84].

A lower interface friction is required not only to reduce the extrusion force but also to ensure a good external extrudate aspect. One way to reduce the interface frictional force is promoting the formation of lubrication layer by applying difference in electric potential between screw and nozzle surface [88]. Vibration at the die system has been proved as well to reduce the required force to extrude the material by reducing the frictional stress in this area [89]. In order to limit drainage during extrusion, optimized cementitious mixtures with low permeability by adding admixtures and fine particles may be designed [90]. Although the axial force acting in screw extruder has been proved to be equal to the sum of the barrel frictional and forming forces [91], the studies conducted on cementitious materials using screw extrusion are limited due to the complexity of the system geometry [92].

2.6.4   3D Printing

3D printing, also known as additive manufacturing, refers to processes used to create from a digital design an object by superimposing successive layers of material with the help of a 3D printer. Alternative terms are also in use, such as large-scale additive manufacturing (LSAM) or freeform construction (FC). Once limited to only particular materials in the manufacturing industry – such as plastic – 3D printing covers today a wide range of materials like metal, wax, etc. and more recently cementitious materials in the construction sector. The term '3D concrete printing' nowadays is often used to refer to the mortar printing. Indeed, studies conducted on 3D concrete printing with a maximum aggregate size larger than 5 mm are very limited [93].

The main 3D printing methods used in the construction sector are extrusion, powder bonding and additive welding. Currently, 3D printing by extrusion is the most popular process used in additive construction. This method has gained more interest because of its compatibility with a wide range of building materials, such as mortar, clay, etc. This technology receives enormous recognition due to many advantages, such as reduction of labour cost and construction materials waste. The most important benefits of the use of 3D concrete printing in the construction sector result from generating a concrete element without the use of a traditional formwork and placing only functionally required material.

3D concrete printing by extrusion consists of pumping a prepared mixture through a hose pipe towards a printhead. The material is then extruded through a nozzle attached to the end of the printhead. After being extruded, successive layers are deposited on top of each other on a printing platform. The desired element is therefore generated by superimposing layers of concrete. Thus, 3D printing by extrusion does not require a single process but rather a combination of different processes (e.g. mixing, pumping, extrusion and deposition). The quality, mechanical and durability performances of the printed element depend on both the rheological properties of the concrete (workability, consistency, etc.) and the pre-programmed machine parameters (printing speed, flow rate, layer thickness and nozzle shape).

For each stage of the printing process, the material must have appropriate rheological properties allowing, on the one hand, the pumping of the material without any blockage or segregation, and on the other hand the stability of the layers after deposition without any collapse or excessive deformation. 'Pumpability', 'extrudability', 'buildability' and 'open time' are terms used as properties to describe the ability of a concrete mixture to be printed, or in short the 'printability'. Pumpability describes the ability of the concrete to move from the pump to the printhead without any sign of blockage or segregation, whereas extrudability is the ability of the concrete to be uniformly extruded from the nozzle. Besides the rheological properties of the material, extrudability depends on the nozzle shape and the ratio between the maximum grain diameter and the nozzle opening. Regarding buildability, this term is defined as the ability of a freshly printed layer to retain its shape under the extrusion pressure, its own weight and the weight of the deposited layers without an excessive deformation. Open time indicates the time frame in which concrete conserves its printability. Many researches have been carried out to develop test methods for predicting the extrudability or buildability of 3D concrete printing [87, 94, 95].

For a successful concrete printing, it is important to understand and control the required rheological properties of the concrete mixture passing through the different printing stages from mixing to deposition of successive layers. Pumpability and extrudability can be guaranteed by a flowable concrete (low yield stress and low viscosity), while buildability requires a more stiff material (high structural build-up). In order to determine the required rheological properties at each stage, advanced studies have been conducted on understanding and characterizing the rheological properties in terms of yield stress, viscosity, elastic modulus, critical strain and structuration rate of printable concrete mixtures [80, 96]. During pumping, concrete displays a viscous behaviour as already discussed in the pumping section described by a lubrication layer at the interface concrete/pipe with a sheared and plugged zone towards the centre of the pipe. Once extruded from the printhead, as the material is mainly sheared in the pipe, it spreads to take its final shape. However, after deposition, the material is at rest and at this stage its elasto-plastic behaviour dominates. During printing, more attention must be given to the deposition process and more precisely to the ability of the material to build up an internal structure at rest (flocculation and hydration). Once the layer is deposited, its rheological properties evolve at rest with time and the material becomes more stiff with increasing yield stress. A high yield stress with limited spread flow must be reached in a short time after deposition of the layer.

The challenge here remains in the transition in a short time from a low yield stress (pumping and extrusion) to a high yield stress to enhance the buildability (increasing the structural build-up of the material) of a flowable concrete. Technically, this can be done by use of nanoparticles in the mixture or by injection of accelerator in the printhead. Other physical and chemical options are presented in [97]. However, a too fast structural build-up is not recommended. Indeed, during deposition, special attention must be given to the bonding between layers when the material exhibits a fast structural build-up – the latter leads to 'cold joints' where weak zones are identified. Thus, the structural build-up of the deposited layer must be fast enough to prevent the collapse of the printed element and slow enough to enhance the bonding between layers.

As mentioned previously, when plotting the static yield stress values as a function of resting time, the slope (A_{thix}) is a strong indicator of structural build-up [24, 29]. Models for structuration rate are discussed in [6, 98]. It is important to highlight that measuring the structural build-up of cementitious materials using a rheometer is ambiguous and complex. Indeed, an undisturbed sample is required to measure a static yield stress at a given time. Thus, measuring the evolution of the yield stress requires many samples left at rest. Unique sample techniques can also be used to assess the evolution of the structuration but these techniques are either limited to some applications, such as oscillation rheometry for only fine particles, or not fully investigated, such as ultrasound measurement. A guideline for understanding the rheological requirements for printable concretes is given by Roussel [80].

Furthermore, a single layer is able to bear its gravity and retain its shape when its yield stress is larger than the gravity stress $\rho g h_0$ (where h_0 is the thickness of the layer) induced by its deposition. By depositing more layers, each layer is able to retain its shape if its yield stress is higher than its own weight and the gravity-induced stresses of the deposited layers. As a consequence, gravity-induced stresses in the bottom layer are at the origin of yielding when $\rho g H / \sqrt{3}$ (where H is the final object height) reaches the yield stress value. Based on this principle, Roussel identified target rheological properties within a time window as shown in Figure 2.19. It is also shown that immediately after deposition flocculation leads to a supporting structure. Over longer timeframes the growth of hydration products at the grain contact points dominates the structural build up [80, 96].

Nowadays, numerous concrete structures have been constructed using 3D printing by concrete extrusion. Despite recent important advances, the adequate rheological properties are still not fully identified due to the conflicting rheological requirements and complex parameters. Indeed, practical guidelines for preparing adequate mixtures are needed. Moreover, standard methods and prediction models to evaluate pumpability, extrudability, buildability and open time must be established in order to assess the ability of the tested material to be printed. Finally, an active control of the rheology of the printed material can be considered to enable co-existence of conflicting rheological properties required at any step of the printing process.

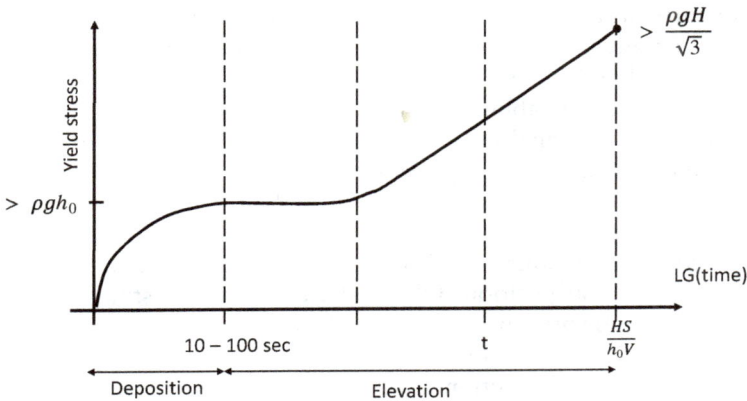

Figure 2.19 General behaviour of fresh cement-based systems at rest described through yield stress evolution. Source: after [80].

2.7   BENEFITS OF ACTIVE RHEOLOGY CONTROL

Considering the importance of rheology and its interference with the processing technique, as illustrated in previous sections, it is now clear what would be the benefit from being able to control the rheology as a function of the process step.

During current pumping operations, the only traditional option is to control the pumping pressure or the discharge rate while operating the pump. Short interruptions of the pumping process can lead to major difficulties in resuming pumping operations due to the sometimes tremendous effect of internal structural build-up or thixotropy. Once the concrete enters the pump, the operator can only passively consider the evolution of the rheological properties of the cementitious material, and has no means to adjust the material properties. So, current pumping operations do not allow to actively control concrete rheology during pumping.

ARC in this case could be achieved by introducing responsive additives in the concrete, enabled to interact with an externally applied signal such as an electromagnetic field. This signal could be applied in the cross-section or the wall surface of the pumping pipe in order to locally activate the admixture's interaction with the lubrication layer based on real-time in-line flow measurements. As we saw earlier, the lubrication layer plays a major role in the pumpability and pumping rate of the mix and might be one way to control the pumping process in a more active way.

Similarly, once fresh concrete is cast in its final position, the flowability of the material becomes obsolete and even unwanted as it contributes to the high formwork pressure. A fast stiffening process can reduce the formwork risks. In current practice, the setting process due to hydration of the cement or binder is gradually converting the cementitious material from a liquid to a solid material and there is no way to accelerate this process in situ. With responsive additives enabling active triggering of setting or stiffening once the material is in its final position, the immediate reduction of formwork pressure and leakage becomes possible.

In the 3D printing process, which is a kind of extrusion process, a stiffening strategy applies similar to the formwork, with the difference that the concrete remains only briefly in the external signal before being deposited in its final position. Obviously, an increased stiffening speed is desirable during 3D printing but increased control of the stiffening is equally interesting when considering local cold joints or aesthetically desirable layer shapes. Indeed, after deposition a well-activated layer segment could maintain its initial shape while a subsequent un-activated layer segment could be allowed to change shape, giving the final component a surface texture that cannot be achieved otherwise. We refer to this possibility to purposely alternate the time-dependent behaviour in real time as '4D concrete printing'.

2.8 SUMMARY

Mastering the interaction between fresh concrete properties and the processing step applied to this concrete proves to be particularly challenging. Advanced rheometrical setups and duly considered measuring protocols help explore ways forward to innovative practices. This will allow to give back control to the concrete user, even with the highly desired but sometimes confusing superplasticizer incorporation. This chapter describes the many facets of processing cementitious materials and stipulates the remaining challenges in traditional and future practices.

REFERENCES

[1] Tattersall, G.H. and P.F.G. Banfill, *The rheology of fresh concrete*. 1983. Montclair, New Jersey, USA: Pitman Advanced Pub. Program.

[2] Tattersall, G.H., *Workability and quality control of concrete*. 1991. Boca Raton, Florida, USA: CRC Press.

[3] Wallevik, O.H. Rheology – a scientific approach to develop self-compacting concrete. In *Proceedings of the 3rd international RILEM symposium on self-compacting concrete*. Reykjavik, Iceland. 2003.

[4] Roussel, N., et al., The origins of thixotropy of fresh cement pastes. *Cement and Concrete Research*, 2012. 42(1): p. 148–157.

[5] Banfill, P., *The rheology of fresh cement and concrete – a review*. 2003.

[6] Jiao, D., et al., Thixotropic structural build-up of cement-based materials: A state-of-the-art review. *Cement and Concrete Composites*, 2021. 122: p. 104152.

[7] Feys, D., et al., *Measuring rheological properties of cement pastes: most common techniques, procedures and challenges*. RILEM technical letters, 2018. 2: p. 129–135.

[8] Roussel, N., *Understanding the rheology of concrete*. 2011. Amsterdam, the Netherlands: Elsevier.

[9] Shaughnessy Iii, R. and P.E. Clark, The rheological behavior of fresh cement pastes. *Cement and Concrete Research*, 1988. 18(3): p. 327–341.

[10] Roussel, N., et al., Steady state flow of cement suspensions: A micromechanical state of the art. *Cement and Concrete Research*, 2010. 40(1): p. 77–84.

[11] Wallevik, J.E., Rheological properties of cement paste: Thixotropic behavior and structural breakdown. *Cement and Concrete Research*, 2009. 39(1): p. 14–29.

[12] Struble, L.J. and W.-G. Lei, Rheological changes associated with setting of cement paste. *Advanced Cement Based Materials*, 1995. 2(6): p. 224–230.

[13] Flatt, R. and I. Schober, 7 - Superplasticizers and the rheology of concrete. In *Understanding the Rheology of Concrete*, N. Roussel, Editor. 2012, Woodhead Publishing. p. 144–208.

[14] Toutou, Z., N. Roussel, and C. Lanos, The squeezing test: a tool to identify firm cement-based material's rheological behaviour and evaluate their extrusion ability. *Cement and Concrete Research*, 2005. 35(10): p. 1891–1899.

[15] Perrot, A., et al., Use of ram extruder as a combined rheo-tribometer to study the behaviour of high yield stress fluids at low strain rate. *Rheologica Acta*, 2012. 51(8): p. 743–754.

[16] Perrot, A., D. Rangeard, and A. Pierre, Structural built-up of cement-based materials used for 3D-printing extrusion techniques. *Materials and Structures*, 2016. 49(4): p. 1213–1220.

[17] Macosko, C.W., *Rheology principles, measurements, and applications.* 1994.

[18] Mezger, T.G., *The rheology handbook*, second ed. 2006. Hannover, Germany: Vincentz Network GmbH & Co. KG.

[19] Hoang, Q.G., et al., A new methodology for characterizing segregation of cement grouts during rheological tests. *Construction and Building Materials*, 2015. 96: p. 119–126.

[20] Coussot, P., Experimental procedures and problems in paste viscometry. In *Rheometry of pastes, suspensions, and granular materials.* 2005. Hoboken, New Jersey, USA: John Wiley & Sons, p. 81–152.

[21] Barnes, H.A. and Q.D. Nguyen, Rotating vane rheometry – a review. *Journal of Non-Newtonian Fluid Mechanics*, 2001. 98(1): p. 1–14.

[22] Yuan, Q., et al., On the measurement of evolution of structural build-up of cement paste with time by static yield stress test vs. small amplitude oscillatory shear test. *Cement and Concrete Research*, 2017. 99: p. 183–189.

[23] Mezger, T.G., *Applied rheology: With Joe Flow on rheology road.* 2015. Vienna, Austria: Anton Paar.

[24] Roussel, N., A thixotropy model for fresh fluid concretes: Theory, validation and applications. *Cement and Concrete Research*, 2006. 36(10): p. 1797–1806.

[25] Petkova, V. and V. Samichkov, Some influences on the thixotropy of composite slag Portland cement suspensions with secondary industrial waste. *Construction and Building Materials*, 2007. 21(7): p. 1520–1527.

[26] Ferron, R., et al., Rheological method to evaluate structural buildup in self-consolidating concrete cement pastes. *ACI Materials Journal*, 2007. 104: p. 242–250.

[27] Mewis, J. and N.J. Wagner, Thixotropy. *Advances in Colloid and Interface Science*, 2009. 147–148: p. 214–227.

[28] Barnes, H.A., Thixotropy – a review. *Journal of Non-Newtonian Fluid Mechanics*, 1997. 70(1–2): p. 1–33.

[29] Billberg, P., Form pressure generated by self-compacting concrete: influence of thixotropy and structural behaviour at rest. In *Trita-BKN. Bulletin*. 2006, Stockholm: Byggvetenskap. p. xiv, 91.

[30] Spiratos, N., et al., *Superplasticizers for concrete: Fundamentals, technology and practice*. 2006. Quebec: Marquis.

[31] Lesage, K., *Interactions between cement and combined concrete admixtures*. 2014. UGent.

[32] Aïtcin, P.-C., et al., Retardation effect of superplasticizer on different cement fractions. *Cement and Concrete Research*, 1987. 17(6): p. 995–999.

[33] Jansen, D., et al., Change in reaction kinetics of a Portland cement caused by a superplasticizer – Calculation of heat flow curves from XRD data. *Cement and Concrete Research*, 2012. 42(2): p. 327–332.

[34] Chen, W.F., *Handbook of structural engineering*. 2010: Taylor & Francis.

[35] Flatt, R.J., et al., Conformation of adsorbed comb copolymer dispersants. *Langmuir*, 2009. 25(2): p. 845–855.

[36] T. Nawa, H.I. and M. Kinoshita, Influence of Temperature on Fluidity of Cement Paste Containing Superplasticizer with Polyethylene Oxide Graft Chains. *ACI Symposium Publication*, 2000. 195: p. 181–194.

[37] Plank, J., et al., Synthesis and performance of methacrylic ester based polycarboxylate superplasticizers possessing hydroxy terminated poly(ethylene glycol) side chains. *Cement and Concrete Research*, 2008. 38(10): p. 1210–1216.

[38] Winnefeld, F., et al., Effects of the molecular architecture of comb-shaped superplasticizers on their performance in cementitious systems. *Cement and Concrete Composites*, 2007. 29(4): p. 251–262.

[39] Plank, J. and C. Winter, Competitive adsorption between superplasticizer and retarder molecules on mineral binder surface. *Cement and Concrete Research*, 2008. 38(5): p. 599–605.

[40] Ferrari, L., et al., Multi-method approach to study influence of superplasticizers on cement suspensions. *Cement and Concrete Research*. 41(10): p. 1058–1066.

[41] Houst, Y.F., et al., Design and function of novel superplasticizers for more durable high performance concrete (superplast project). *Cement and Concrete Research*, 2008. 38(10): p. 1197–1209.

[42] Zingg, A., et al., Interaction of polycarboxylate-based superplasticizers with cements containing different C3A amounts. *Cement and Concrete Composites*, 2009. 31(3): p. 153–162.

[43] Schröfl, C., M. Gruber, and J. Plank, Preferential adsorption of polycarboxylate superplasticizers on cement and silica fume in ultra-high performance concrete (UHPC). *Cement and Concrete Research*, 2012. 42(11): p. 1401–1408.

[44] Kjeldsen, A.M., R.J. Flatt, and L. Bergström, Relating the molecular structure of comb-type superplasticizers to the compression rheology of MgO suspensions. *Cement and Concrete Research*, 2006. 36(7): p. 1231–1239.

[45] Janowska-Renkas, E., The effect of superplasticizers' chemical structure on their efficiency in cement pastes. *Construction and Building Materials*, 2013. 38(0): p. 1204–1210.

[46] Ferrari, L., et al., Multi-method approach to study influence of superplasticizers on cement suspensions. *Cement and Concrete Research*, 2011. 41(10): p. 1058–1066.

[47] *The European guidelines for self-compacting concrete.* BIBM, CEMBUREAU, ERMCO, EFCA and EFNARC , 2005.: p. 563

[48] Schutter, G.D., et al., *Self-compacting concrete.* 2008. Gaithness: Whittles Publishing.

[49] Bonen, D. and S.P. Shah, Fresh and hardened properties of self-consolidating concrete. *Progress in Structural Engineering and Materials*, 2005. 7(1): p. 14–26.

[50] Heirman, G., *Modelling and quantification of the effect of mineral additions on the rheology of fresh powder type self-compacting concrete.* PhD thesis, 2011. Leuven, Belgium: KU Leuven.

[51] Khayat, K.H., Workability, testing, and performance of self-consolidating concrete. *ACI Materials Journal.* 96(3): p. 346–353.

[52] Wallevik, O., *Rheology of cement based particle suspensions such as cement paste, mortar and concrete.* 2004. Reykjavik, Iceland: IBRI-Rheocenter.

[53] Feys, D. *Interactions between rheological properties and pumping of self-compacting concrete.* 2009. Ghent, Belgium: PhD thesis, Ghent University.

[54] Billberg, P.H., et al., Field validation of models for predicting lateral form pressure exerted by SCC. *Cement and Concrete Composites*, 2014. 54: p. 70–79.

[55] Tichko, S., *Hydrodynamic modelling of the flow of self-compacting concrete in formworks.* 2016. Ghent, Belgium: PhD thesis, Ghent University.

[56] Feys, D., G. De Schutter, and R. Verhoeven, Parameters influencing pressure during pumping of self-compacting concrete. *Materials and Structures*, 2013. 46(4): p. 533–555.

[57] Denis Kaplan, F.d.L. and S. Thierry, Design of concrete pumping circuit. *ACI Materials Journal.* 102(2): 110–117.

[58] Choi, M., et al., Lubrication layer properties during concrete pumping. *Cement and Concrete Research*, 2013. 45: p. 69–78.

[59] Bamforth, R.D.B.a.P.B., Tests to establish concrete pumpability. *ACI Journal*, 1977. 74: p. 193–203.

[60] Jacobsen, S., et al., Flow conditions of fresh mortar and concrete in different pipes. *Cement and Concrete Research*, 2009. 39(11): p. 997–1006.

[61] Le, H.D., et al., Effect of lubrication layer on velocity profile of concrete in a pumping pipe. *Materials and Structures*, 2015. 48(12): p. 3991–4003.

[62] Choi, M.S., Y.J. Kim, and S.H. Kwon, Prediction on pipe flow of pumped concrete based on shear-induced particle migration. *Cement and Concrete Research*, 2013. 52 : p. 216–224.

[63] Khatib, R., *Analysis and prediction of pumping characteristics of high-strength self-consolidating concrete*. 2013: Sherbrooke, Canada: PhD thesis, Université de Sherbrooke.

[64] Choi, M.S., et al., Effects of an externally imposed electromagnetic field on the formation of a lubrication layer in concrete pumping. *Construction and Building Materials*, 2014. 61: p. 18–23.

[65] Secrieru, E., et al., Formation of lubricating layer and flow type during pumping of cement-based materials. *Construction and Building Materials*, 2018. 178: p. 507–517.

[66] ACI Committee 304, *Placing Concrete by Pumping Methods*, ACI 304.2R-96. 2008. Michigan, USA: American Concrete Institute, Farmington Hills.

[67] Jolin, M., et al. *Pumping concrete: A fundamental and practical approach*. 2006.

[68] Feys, D., et al., Prediction of pumping pressure by means of new tribometer for highly-workable concrete. *Cement and Concrete Composites*, 2015. 57: p. 102–115.

[69] Yammine, J., et al., From ordinary rhelogy concrete to self compacting concrete: A transition between frictional and hydrodynamic interactions. *Cement and Concrete Research*, 2008. 38: p. 890–896.

[70] Malvern, L.E., *Introduction to the mechanics of continuous medium*. 1969. Englewood Cliffs, New Jersey: Prentice-Hall, Inc.

[71] Buckingham, E., On plastic flow through capillary tubes. *Proc. Am. Soc. Testing Materials*, 1921. p. 1154–1156.

[72] Ngo, T.T., et al., Use of tribometer to estimate interface friction and concrete boundary layer composition during the fluid concrete pumping. *Construction and Building Materials*, 2010. 24(7): p. 1253–1261.

[73] Ngo, T.-T., et al., Practical tribometer to estimate pumpability of fresh concrete. *Journal of Asian Architecture and Building Engineering*, 2010. 9(1): p. 229–236.

[74] Mechtcherine, V., V.N. Nerella, and K. Kasten, Testing pumpability of concrete using sliding pipe rheometer. *Construction and Building Materials*, 2014. 53: p. 312–323.

[75] Tichko, S., et al., Influence of the viscosity of self-compacting concrete and the presence of rebars on the formwork pressure while filling bottom-up. *Engineering Structures*, 2015. 101: p. 698–714.

[76] *The European guidelines for self compacting concrete specification, production and use*. BIBM, CEMBUREAU, ERMCO, EFCA and EFNARC, 2005.

[77] De Schutter, G., et al., *Self-compacting concrete*. 2008. Dunbeath, Scotland, UK: Whittles Publishing. 296.

[78] McKinnon, P.M., *Curb forming and extruding apparatus*. 1994. USA. patent application US5354189A.

[79] Buswell, R.A., et al., 3D printing using concrete extrusion: A roadmap for research. *Cement and Concrete Research*, 2018. 112: p. 37–49.

[80] Roussel, N., Rheological requirements for printable concretes. *Cement and Concrete Research*, 2018. 112: p. 76–85.

[81]   Perrot, A., et al., Extrusion of cement-based materials - an overview. *RILEM Technical Letters*, 2019. 3(0): p. 91–97.

[82]   Perrot, A., et al., Extrusion criterion for firm cement-based materials. *Applied Rheology*, 2009. 19(5): p. 53042-1–53042-7.

[83]   Benbow, J. and J. Bridgwater, *Paste flow and extrusion*. 1993. Oxford, UK: Oxford University Press. P. 153.

[84]   Alfani, R. and G.L. Guerrini, Rheological test methods for the characterization of extrudable cement-based materials – A review. *Materials and Structures*, 2005. 38(2): p. 239–247.

[85]   Basterfield, R.A., C.J. Lawrence, and M.J. Adams, On the interpretation of orifice extrusion data for viscoplastic materials. *Chemical Engineering Science*, 2005. 60(10): p. 2599–2607.

[86]   Chen, Y., L.J. Struble, and G. Paulino, Extrudability of cement-based materials. *American Ceramic Society Bulletin*, 2006. 85(6) .

[87]   Nerella, V.N., et al., Inline quantification of extrudability of cementitious materials for digital construction. *Cement and Concrete Composites*, 2019. 95: p. 260–270.

[88]   Mortreuil, F., et al., Utilisation des propriétés électriques des pâtes céramiques pour leur extrusion. *Industrie Céramique et Verrière*, 2000. p. 88–95.

[89]   Perrot, A., et al., Vibro-extrusion: a new forming process for cement-based materials. *Advances in Cement Research*, 2009. 21(3): p. 125–133.

[90]   Kuder, K.G. and S.P. Shah, Rheology of extruded cement-based materials. *ACI Materials Journal*, 2007. 104(3): p. 8.

[91]   Perrot, A., Conditions d'extrudabilité des matériaux à base cimentaire. PhD thesis. 2006, INSA, Rennes, France.

[92]   Mu, B., et al., Cementitious composite manufactured by extrusion technique. *Cement and Concrete Research*, 1999. 29: p. 4.

[93]   Le, T.T., et al., Hardened properties of high-performance printing concrete. *Cement and Concrete Research*, 2012. 42(3): p. 558–566.

[94]   Rahul, A.V., et al., 3D printable concrete: Mixture design and test methods. *Cement and Concrete Composites*, 2019. Amsterdam, the Netherlands: Elsevier. 97: p. 13–23.

[95]   Tay, Y.W.D., Y. Qian, and M.J. Tan, Printability region for 3D concrete printing using slump and slump flow test. *Composites Part B: Engineering*, 2019. 174: p. 106968.

[96]   Reiter, L., et al., The role of early age structural build-up in dee fabrication with concrete. *Cement and Concrete Research*, 2018. 112: p. 86–95.

[97]   Marchon, D., et al., Hydration and rheology control of concrete for digital fabrication: Potential admixtures and cement chemistry. *Cement and Concrete Research*, 2018. 112: p. 96–110.

[98]   Lecompte, T. and A. Perrot, *Non-linear modeling of yield stress increase due to SCC structural build-up at rest. Cement and Concrete Research*, 2017. 92: p. 92–97.

Chapter 3

Active rheology control of cementitious materials with responsive mineral particles

Dengwu Jiao, Mert Yucel Yardimci, Karel Lesage, and Geert De Schutter

CONTENTS

3.1 General introduction   78
3.2 Theoretical background and rheology control mechanism   82
   3.2.1   Clustering   82
   3.2.2   Micro-vibration and particle translation   86
3.3 Magneto-rheological behaviour of cement paste with nano-Fe_3O_4 particles   93
   3.3.1   Physical properties of nano-Fe_3O_4 particles   93
   3.3.2   Typical magneto-responsive structural build-up   95
   3.3.3   Influence of cement paste medium   98
   3.3.4   Influence of nano-Fe_3O_4 concentration   102
   3.3.5   Influence of particle size of nano-Fe_3O_4 104
   3.3.6   Influence of magnetic field types   108
   3.3.7   Magneto-responsive flowing behaviour   111
3.4 Magneto-rheological behaviour of fly ash-cement paste   112
   3.4.1   Magnetic properties of fly ash   113
   3.4.2   Influence of fly ash volume fraction   115
   3.4.3   Influence of fly ash type   117
   3.4.4   Relationship between magneto-rheological effect and magnetic properties of FA   118
3.5 Other potential materials   119
   3.5.1   Steel fibres, metal scraps and chips   120
   3.5.2   Specific types of slags   122
3.6 Short summary of state of the art and outlook   126
References   129

DOI: 10.1201/9781003289463-3

3.1 GENERAL INTRODUCTION

Active rheology control is an essential way to advance construction technology and make pumping, casting and three-dimesional (3D) concrete printing processes more reliable. One approach is the use of newly developed responsive superplasticizer controlled by an electromagnetic field [1], which will be described in Chapter 4. Another potential approach to achieve the active rheology control is by adding magnetic (nano- or micro-)particles in combination with exploiting an external magnetic field [2].

In the literature, some applications of magnetic fields in cement-based materials are given. Available experimental results seem to suggest that cementitious materials (cement paste, mortar and concrete) prepared by using magnetized water exhibit higher flowability [3], stronger compressive strength [4] and better shrinkage cracking resistance [5] compared to those prepared with normal tap water. However, why water could maintain its magnetization after the removal of the magnetic field is still not clear [6], and the reported improvement is heavily debatable. For fresh steel fibre-reinforced concrete, applying a magnetic field could vibrate the steel fibres or chips in fresh concrete and facilitate the fibres orientation, and thus the efficiency of reinforcement can be significantly increased [7–9]. During pumping process, the properties of the lubrication layer seem to be somewhat increased by applying an electromagnetic field, although the mechanism is not fully clear, and the pumpability of fresh concrete is improved [10]. Besides, fresh cement pastes exposed to an external magnetic field seem to have larger amount of C-S-H gel, denser morphology and less porosity [11].

Diamagnetism, paramagnetism and collective magnetism are a rough classification of magnetism based on the magnetic susceptibility of the materials. Magnetic susceptibility, which is the term to describe how susceptible the material is to magnetization, differs from material to material. Materials exhibiting a weak negative magnetic susceptibility are called diamagnetic materials. In diamagnetic materials, an external magnetic field induces a magnetic moment that is oriented antiparallel to the external field; therefore, a diamagnetic material is repelled by a permanent magnet. This class of materials includes nearly all organic substances, mercury, gold, silver, lead, silicone and water. All materials possess this property; however, it is obscured by larger types of magnetic effects in other types of materials. The relative permeability, which is the ratio of the permeability of a specific medium to the permeability of free space, of diamagnetic materials is smaller than 1 but very close to 1. Except for the superconducting materials having a magnetic permeability equal to zero, the diamagnetic materials present very weak magnetic properties that could even be assumed to be nonmagnetic [12]. Materials whose relative permeability is slightly higher than 1 are called paramagnetic materials. The existence of permanent magnetic dipoles in the material, which are oriented in the magnetic field, is a fundamental precondition of paramagnetism [13]. In these materials, the orbital and spin moments do not cancel each other and the atoms have a net magnetic moment even in the absence of an external magnetic field. However, since these moments are oriented randomly, the net external field is almost zero. Some common examples of paramagnetic materials are aluminium, palladium, tungsten and air [12].

Collective magnetism is classified into three groups: ferromagnetism, ferrimagnetism and antiferrimagnetism. Ferromagnetic materials have a relative magnetic permeability much higher than 1, and can be thousands or even higher. Ferromagnetic materials have some unpaired electrons causing their atoms to have a net magnetic moment. These magnetic moments of a relatively large number of atoms are aligned parallel to each other and create strongly magnetized areas, which are called magnetic domains (first mentioned by P. Weiss in 1907 [13]). In un-magnetized conditions, these magnetic domains are randomly oriented and the net magnetic field strength in the material is zero. For this reason, two ferromagnetic materials do not attract each other when they are brought close together. However, when a ferromagnetic material is exposed to a magnetic field, some or all of these magnetic domains align, and the material becomes a magnet. The magnetic field becomes stronger in the material as more domains align, and the ferromagnetic material becomes magnetically saturated when all of the domains are aligned. After the ferromagnetic material is magnetically saturated, a further increase in external magnetization will not cause any increase in the magnetization of the material. Some ferromagnetic materials retain the magnetization after the removal of the external magnetic field. The domain structure of ferromagnetic materials is schematically shown in Figure 3.1.

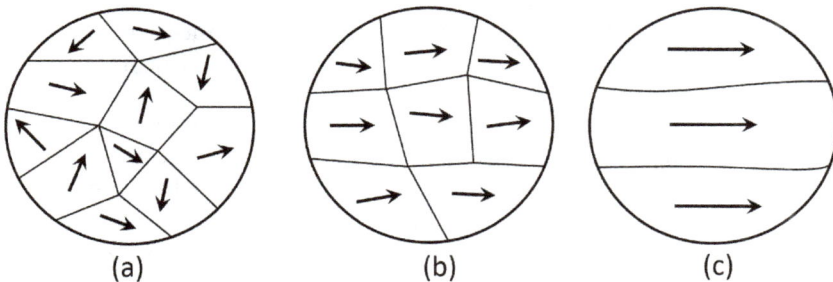

Figure 3.1 Domain model of magnetization in ferromagnetic materials. (a) Random orientation of domains. (b) The external field causes aligned domains to grow. (c) Aligned domains occupy all or most of the volume. Source: after [12].

It should be noted here that collective magnetism is a temperature-dependent phenomenon. The collective magnetism materials present paramagnetism above some critical temperatures (Curie temperature for ferromagnetic materials) with the corresponding characteristic behaviour of inverse susceptibility [13].

Ferrimagnetic materials have an appreciable net magnetization, but their atomic magnetic moments are antiparallel in contrast to ferromagnetic materials. They have a remanent magnetization. Typical examples of ferrimagnetic materials are magnetite (Fe_3O_4), maghemite (γ-Fe_2O_3) and several minerals with spinel structures [14]. Above the Curie temperature (about 570 °C), ferromagnetic materials present paramagnetism. Undoubtedly, ferrimagnetic and ferromagnetic materials are the most important materials with respect to the subject of this book as they present strong magnetic properties that are far better than the other types of materials presenting the other magnetic properties.

Ferromagnetic particles that could be used in ARC generally refer to particles containing iron or iron oxides such as magnetite (Fe_3O_4) or maghemite (γ-Fe_2O_3) which can have a strong response to an external magnetic field. When applying a magnetic field to a magnetic material, the magnetic dipoles will align along the direction of the field to create a response magnetization. The degree of the magnetic properties of a magnetic material can be described by magnetization hysteresis curve. A typical full hysteresis curve for a ferromagnetic material is shown in Figure 3.2. Saturation magnetization (Ms) is the magnetization value in which increasing the applied external magnetic field strength cannot further increase the magnetization of the material. Remanent magnetization (Mr) is the magnetization left behind in a material when the external magnetic field is removed. Coercive field (Hc) is the ability of a material to withstand a magnetic field without becoming demagnetized. The saturation magnetization Ms is a very important characteristic of magnetic materials. Typical values of Ms of different magnetic materials are summarized in Table 3.1.

The most popular magnetic particles used in cement-based materials include carbonyl iron powder (CIP) and nano-Fe_3O_4. CIP, invented by BASF in 1925 [16], is a highly pure iron possessing both electric and magnetic properties. The most widespread application of CIP in cement-based materials is acting as electromagnetic or microwave absorption material for microwave interference shielding [17]. Besides, the addition of CIP has negligible influences on the chemical hydration and mechanical properties of cement-based materials [18]. The applications of nano-Fe_3O_4 particles in cementitious materials also attract extensive interest due to its supermagnetic properties and ultra-high specific surface area [19]. The nano-Fe_3O_4 particles, on the one hand, fill the voids between cement particles, leading to a denser suspension system. On the other hand, the nanoparticles possibly provide nuclei for cement hydration products [20]. Consequently,

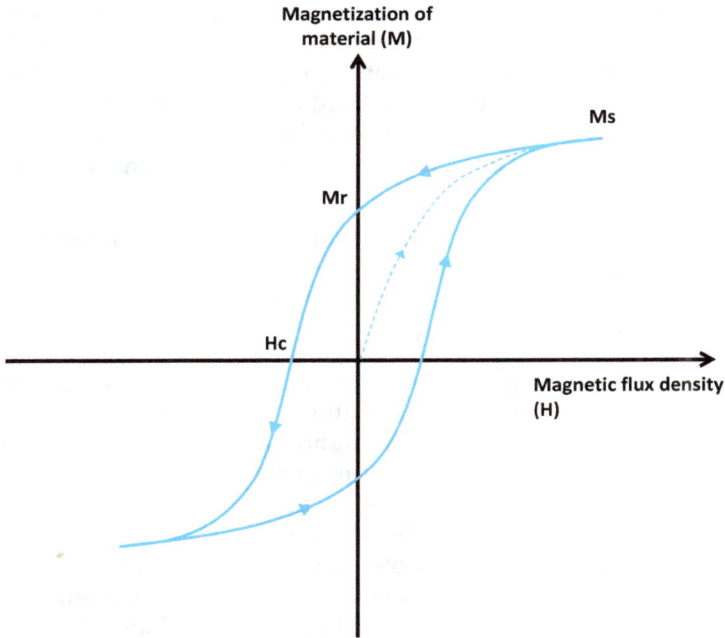

Figure 3.2 Typical magnetization hysteresis curve of a ferromagnetic material. Ms is the maximum magnetization (Am^2/g or emu/g), Mr is the magnetization left behind in a material after an external magnetic field is removed (Am^2/g or emu/g), and Hc is the ability of a material to withstand a magnetic field without becoming demagnetized (Oe).

Table 3.1 Typical values of Ms of different magnetic materials

Material	Ms (emu/g)
Magnetite (Fe_3O_4)	90–92
Maghemite (γ-Fe_2O_3)	84–88
$CoFe_2O_4$	~75
Iron (α-Fe)	217.9
Cobalt	162.7
Nickel	57.5

Source: [15].

the hydration reaction of cementitious materials is somewhat accelerated with the addition of nano-Fe_3O_4 particles, and thus the early age mechanical properties are enhanced [21, 22]. Moreover, a hydration product with fairly good hydraulic property, designated as Fe-ettringite [23], can be formed by the interaction of nano-Fe_3O_4 particles with free $Ca(OH)_2$. The by-product accumulates in the pores and voids between cement particles, resulting in an enhanced microstructure and higher electrochemical stability [24]. Mansouri et al. [25] found that the replacement of 2% cement with nano-Fe_3O_4 particles dramatically reduced the percentage of water absorption and improved the resistance against chloride ion permeation, providing a possible improvement approach of service life of submerged structures.

3.2   THEORETICAL BACKGROUND AND RHEOLOGY CONTROL MECHANISM

3.2.1   Clustering

The controllable rheology of cementitious materials by applying a magnetic field is based on the theoretical foundations of magneto-rheological (MR) fluids. MR fluid, first introduced by Rabinow in 1948 [26], is a kind of smart material consisting basically of magnetic particles and carrier fluid. Without magnetic field, the magnetic particles distribute randomly in the carrier fluid and MR fluid exhibits Newtonian behaviour. After applying an external magnetic field, the dipoles inside the magnetic particles become magnetized, and the magnetic particles are promoted to align along the direction of the magnetic field and form chain- or column-like structures [27–29]. As a consequence, MR fluid transforms from liquid to semi-solid state within milliseconds, where the viscosity could increase about 10^5–10^6 times and the field-induced yield stress increases up to 100 kPa [30]. After removing the external magnetic field, the magnetic particles are demagnetized and the MR fluid reversibly changes from semi-solid to liquid, as shown in the schematic diagram in Figure 3.3. Therefore, the controllable MR fluid by applying external magnetic field is widely used in many engineering applications, such as dampers, brakes, clutches, polishing, valves, medical devices, etc. [31, 32]. However, due to the big difference of density between the magnetic particles and the carrier fluid, sedimentation is the greatest challenge for widespread use of conventional MR fluids in industrial applications. Many stabilization methods, such as coating particles, adding chemical additives and using yield-stress fluid as carrier, have been proposed to improve the stability of MR fluids. Fresh cementitious paste is usually regarded as a concentrated suspension with yield stress [33–35]. A stable MR fluid could be obtained when using fresh cementitious paste as the carrier fluid. Conversely, the rheology and stiffness of cementitious paste containing magnetic particles would become controllable.

In the case of cement-based MR fluid upon application of an external magnetic field, the magnetic particles can move to form clusters or agglomerates if the magnetic force between neighbouring magnetic particles overcomes the resistance of the suspension. For the case of nano-Fe_3O_4 particles in

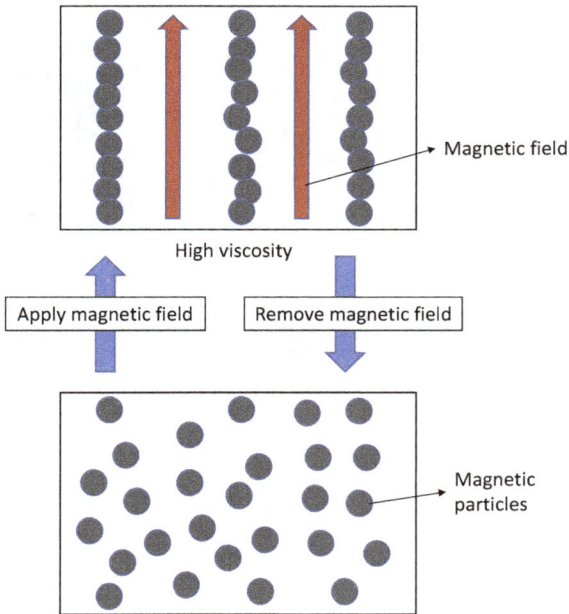

Figure 3.3 Schematic diagram illustrating the reversibility of MR fluids.

cement paste, as an example, the magnetic force between neighbouring nanoparticles and their movement velocity can be estimated [36]. Assuming that all the nanoparticles have the same constant dipole moment under magnetic field [37], the magneto-dynamic force between two neighbouring nanoparticles when their particle centres align along the direction of the magnetic field can be expressed as:

$$F_{magnetic} = \frac{\pi d^6 (\rho M)^2 \mu_0}{24 r_{ij}^4} \tag{3.1}$$

where $F_{magnetic}$ is the magnetic force (N), ρ and M are the density (kg/m^3) and the magnetization per unit mass (Am^2/kg) of the magnetic nanoparticles, respectively, d is the average particle size of the nanoparticles (m), μ_0 is the magnetic permeability of the medium (N/A^2), r_{ij} is the centre distance between two magnetic nanoparticles (m). In case of cement paste as the medium, μ_0 can be considered to be equal to the value of vacuum (4$\pi\times$10^{-7} N/A^2). Assuming that all the nanoparticles are randomly arranged in the voids between cement particles in simple cubic order with the same inter-particle distance (h (m)), as presented in Figure 3.4, the centre distance between two magnetic nanoparticles is expressed by:

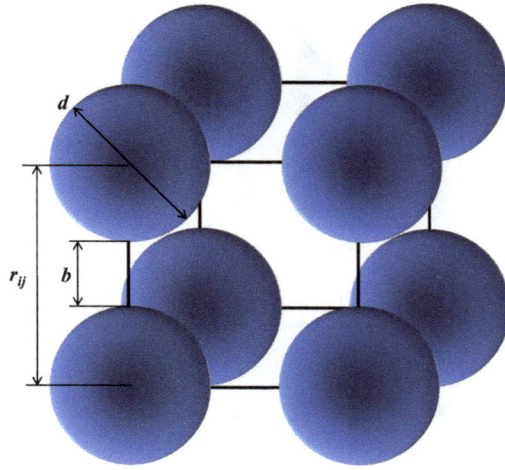

Figure 3.4 Schematic diagram of magnetic nanoparticles arranged as simple cubic order. Source: after [36].

$$r_{ij} = d + h = d \cdot \left(\frac{\mu}{6\varphi_{MNPs}} \right)^{\frac{1}{3}} \tag{3.2}$$

With the given expression, an estimation value is obtained that can be used as a quantitative parameter in further evaluation of the magneto-rheological behaviour of cement paste including magnetic nanoparticles.

Comparing the magneto-dynamic force to the resistance force induced by the viscoelastic stress of the suspension, a dimensionless parameter defined as magnetic yield parameter [38] can be obtained:

$$Y_M = \frac{F_{magnetic}}{F_{viscous}} = \frac{F_{magnetic}}{\pi d^2 \cdot \tau_{ys}} = \frac{\mu_0 \left(\rho \varphi_{MNPs} \right)^{\frac{4}{3}}}{\pi} \tag{3.3}$$

where τ_{ys} is the viscoelastic stress of the suspension (Pa), and φ_{MNPs} is the volume fraction of the nanoparticles in the voids between cement particles, which can be calculated by the following equation:

$$\varphi_{MNPs} = \frac{V_{MNPs}}{\left(1 - \varphi_C \right) \cdot V_{Total}} \tag{3.4}$$

where φ_C is the volume fraction of cement particles and V_{MNPs} and V_{total} are the volume of nanoparticles (m^3) and total paste (m^3), respectively.

Theoretically, when $Y_M > 1$, the magneto-dynamic force is higher than the resistance force, magnetic chains or clusters can be formed and the suspension will show significant magneto-rheological response, while $Y_M < 1$ means that the suspension prevents the formation of chains or cluster structure. It can be clearly seen that the magnitude of the magnetic yield parameter depends on the concentration and physical properties of the nanoparticles and on the viscoelastic stress of the suspension.

For a spherical particle settling in a Bingham plastic fluid under gravitational force, the viscous drag equation can be expressed as [39]:

$$F_D = 3\pi d^2 \left[\frac{\mu_{pl} v}{d} + \tau_{pl} \right] \tag{3.5}$$

The equilibrium velocity of the spherical particle can be calculated by equating the gravitational and buoyant force to the viscous drag force. In the case of nano-Fe_3O_4 incorporated cement paste under an external magnetic field, the magnetic force plays a comparable role to the gravitational force. When the magnetic force is higher than the resistance induced by the viscoelastic stress of the suspension ($Y_M > 1$), similarly, the magnetic nanoparticles in equilibrium can be considered to move at constant velocity in the interstitial medium. Assuming that the flow behaviour of cement paste induced by moving nanoparticles follows Bingham law, the equilibrium velocity can be estimated by equalizing the magnitude of the magneto-dynamic force and the drag force:

$$\frac{\pi d^2 \mu_0 (\rho M)^2}{24} \cdot \left(\frac{6\phi_{MNPs}}{\pi} \right)^{\frac{4}{3}} = 3\pi d^2 \left(\frac{\mu_{pl} v}{d} + \tau_{pl} \right) \tag{3.6}$$

Or in an explicit way as:

$$v = \frac{d}{\mu_{pl}} \left[\frac{\mu_0 (\rho M)^2}{72} \cdot \left(\frac{6\phi_{MNPs}}{\pi} \right)^{\frac{4}{3}} - \tau_{pl} \right] \tag{3.7}$$

where v is the movement velocity of nanoparticles (m/s), and τ_{pl} and μ_{pl} are the Bingham yield stress (Pa) and the plastic viscosity (Pa.s) of the suspension, respectively. It can be seen that the estimated movement velocity of a nanoparticle in cement-based suspensions is dependent on the properties of the nanoparticles (particle size, density, concentration and magnetic properties) and the Bingham yield stress and plastic viscosity of the cementitious paste. It should be mentioned that the magnetic yield parameter and the movement velocity are considered as conceptual indicators representing the movement of nanoparticles, not the actual evolving force or velocity while moving in the paste [36, 40].

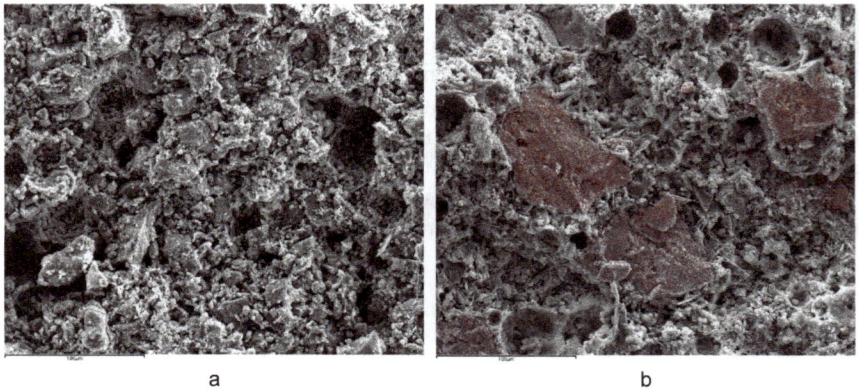

a   b

Figure 3.5 SEM-EDX images with Fe element mapping of cement paste (w/c=0.4, MNPs=3 wt.%). (a) Zero magnetic field and (b) 0.5 T magnetic field. Red dot indicates Fe-element distribution. (Courtesy Dengwu Jiao).

The formation of magnetic chains or clusters is often monitored in conventional MR fluids or fluids with yield stress carrier [27, 38, 41, 42]. The formation of magnetic clusters can also be observed in cementitious paste after applying a sufficiently high magnetic field. Based on the Fe-element distribution map obtained from scanning electron microscopy (SEM) employing energy dispersive X-ray (EDX) spectroscopy, the clustering of nano-Fe_3O_4 particles in cementitious suspension can be monitored and quantified [43]. Representative SEM images with Fe element distribution for cement paste (w/c=0.4, MNPs=3 wt.%) treated under 0 T and 0.5 T are shown in Figure 3.5. At the same magnification and scanning times, the Fe element distribution was approximately uniform for the cementitious paste under 0 T, and only some small Fe element clusters were observed. By contrast, obvious large clusters of Fe element were recognized for the cementitious paste exposed to a magnetic field with 0.5 T. The main source of iron element in the cementitious paste is provided by the nano-Fe_3O_4 particles. The Fe element aggregates indicate that the nano-Fe_3O_4 particles agglomerate to form magnetic clusters in the fresh cement-based suspensions under an external magnetic field. Further in-depth analysis in view of studying the chain or cluster formation in a more objective and quantitative way can be found in [43].

3.2.2   Micro-vibration and particle translation

As summarized before, fibrous structures are formed due to the head-to-tail arrangement of the dipolar particles in the direction of the magnetic field. Therefore, the magneto-rheological properties are obtained due to the proportionality of the yield stress and the force required to break these columnar structures [38]. Those magneto-rheological properties, which are based on the formation of the columnar structures under a uniform magnetic field, usually allow obtaining an on-demand structuration of magneto-rheological suspensions. Furthermore, other types of magnetic responses of the magnetic particles, such as particle vibration and particle translation, could provide new opportunities for active rheology control of cementitious materials. Magneto-rheology of magnetically responsive cementitious mixtures under magnetic field types causing a micro-vibration and particle translation has not been well established and documented in the literature yet. However, several studies dealing with the particle vibration and some theorized ideas developed during the SmartCast project will be summarized below.

The main mechanism of magnetic particle vibration is related to the multi-domain structure of the magnetic particles. The magnetic force leads to a parallel arrangement of the domains of ferromagnetic material, which results in the formation of a temporary magnetic dipole in the ferromagnetic particle [44]. Alternating the magnetic field causes the flipping of the ferromagnetic particle due to the alternation of its dipoles. This could involve a magneto-mechanical effect that could potentially cause some changes in the surrounding matrix rheology. To obtain an effective magneto-mechanical effect induced by the vibration or oscillation of the particles, the applied magnetic field must be in a rotating or alternating mode. This effect is mainly originated from the magnetic anisotropy, which is the term used for the dependence of the magnetic properties on the direction of the applied field with respect to the crystalline lattice of the magnetic particles. There are directions in space in which a magnetic material is easier to magnetize than other ones [13]. The crystal direction requiring a small amount of magnetic field to reach saturation is called the easy axis of the material, while the one reaching saturation only with a stronger magnetic field is called as hard axis of the material. As the particles tend to rotate until their easy axis aligns with the direction of the applied magnetic field, the alternation of the applied magnetic field results in the continuous rotation of the particle inducing a vibration [45].

The vibration of magnetic particles in response to an alternating magnetic field is observed even in nano-scale magnetic particles. However, a stronger magnetic field is required as the nanoparticle size decreases [46]. Alternating magnetic field-induced vibration is more pronounced when the shape of the polycrystalline material is not spherical, as more specific directions represent easy magnetization which is solely caused by the shape (shape anisotropy) [13]. To have the highest vibration effect on the magnetic particles under an alternating magnetic field, the applied magnetic field should preferably be close to the saturation magnetic field level of the magnetic particle, and the magnetic particle should be highly anisotropic. With a strong vibration effect of magnetic particles, the mechanical energy is transferred to the surrounding environment to make some local mechanical stresses and heat generation. Normally low-frequency mechanical vibrations generate local stresses without significant heat generation, while high-frequency alternating magnetic field applications to magnetic nanoparticles generate heat and are used for obtaining magnetic hyperthermia. It should be noted here that the amplitude of mechanical vibration decreases with an increase in alternating magnetic field frequency [45].

The micro-vibration of the magnetic phases of a magnetically responsive cement paste has been observed by a special setup (Figure 3.6) developed during the SmartCast project. In this setup, a responsive cement paste with high amount of magnetite phase inside was diluted by distilled water with 1:50 weight ratio and a drop of dilution was put on an electromagnet. The electromagnet was positioned underneath an optical microscope and operated by a function generator providing sinusoidal pulsating electric current.

This setup allows for the application of an alternating magnetic field to the paste sample (a droplet) over the electromagnet by a function generator and to monitor the movement of the magnetic particles under the

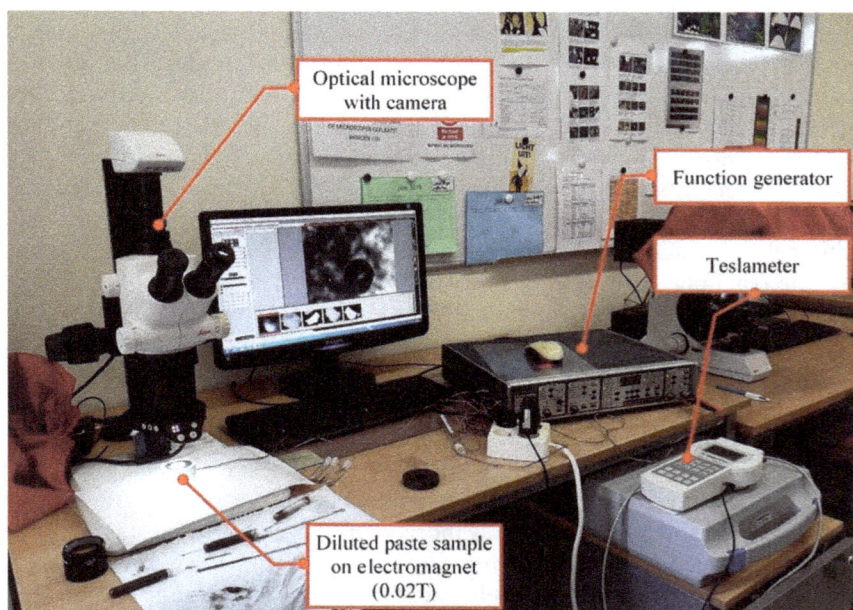

Figure 3.6 General appearance of the setup for magnetic particle vibration observation in diluted responsive cement paste. (Courtesy Mert Yucel Yardimci).

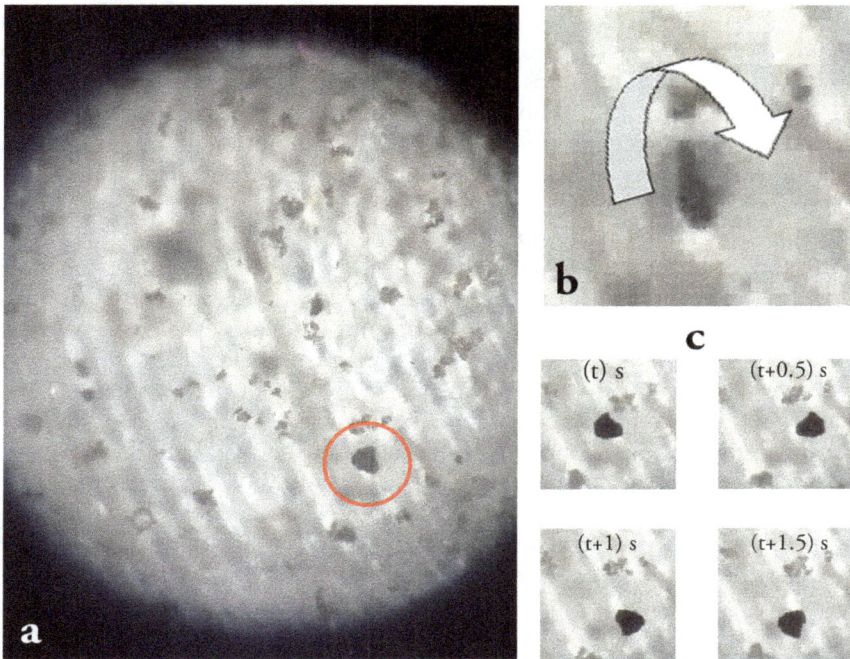

Figure 3.7 Optical microscope view of diluted responsive cement paste under alternating magnetic field (a), the captured flip-over moment of the red-circled magnetic particle (b) and flip-over movement of the red-circled magnetic particle under an alternating magnetic field at 2 Hz (c). (Courtesy Mert Yucel Yardimci).

applied magnetic field by a camera-aided optical microscope continuously. Figure 3.7a is a frame of a video record showing the vibration of magnetic particles inside a responsive cement paste over the electromagnet energized by an alternating current at 2 Hz with the setup shown in Figure 3.6. The vibration of magnetic magnetite particles in the responsive cement paste originated from the turning – rotating (rollover or flip-flap) type of movement of magnetic particles under the alternating magnetic field at 2 Hz. In Figure 3.7b, the captured flip-over movement of the red-circle-marked magnetic particle in Figure 3.7a is shown. Figure 3.7b clearly shows the reversible flip-over movement of the particle over time.

This observation of the continuous-reversible rotation of the magnetic particles in a cement paste with high magnetite content might potentially induce an internal vibration in the cement paste. On the other hand, it should be kept in mind that the laboratory observation given in Figure 3.7 has been obtained by a diluted cement paste of which the yield stress and viscosity are far lower than the actual paste, therefore probably the magnetic torque of a small particle with a very low mass would not be enough to overcome the drag forces to initiate the flip-over movement. Even if one can think that the pore solution rheology is somewhat similar to that of the diluted sample case, the flip-over type of movement of magnetic particles without a translation may not be solely enough to disaggregate the surrounding cement particles for a meaningful internal vibration effect. In Figure 3.7 it was shown that the alternating magnetic field (from several Hz to higher frequencies) causes the vibration of the magnetic particles presented in cement paste due to fast flip-over of the particle. In the test setup shown in Figure 3.6, the magnetic field provided by an electromagnet was alternated by the function generator, but the regions of the applied magnetic field over the exposed area did not change as the electromagnet position was not changed. Under such a magnetic field, the magnetic particle reversibly flips over around its axis (vibrates) but does not translate. This is because of the fact that the magnetic force is caused by an inhomogeneous magnetic field and the materials tend to move along the direction of the highest field gradient when exposed to an inhomogeneous magnetic field. The magnetic force acting on the magnetic particle itself is directly proportional to the magnetic flux density (B), the gradient of the magnetic flux density ($gradB$) and the magnetic susceptibility of the material (χ), while being inversely proportional to the magnetic permeability of the material in a vacuum [47] meaning that the higher the magnetic gradient, the higher the magnetic force acting on the magnetic material. If there is no magnetic gradient in the field, meaning a lack of differentiation of magnetic field in magnitude or direction between two points in space, in the applied magnetic field (homogeneous magnetic field) the magnetic particle will only tend to rotate in the direction of an alternating magnetic field due to the anisotropy but will not experience any translation. Therefore, such vibration may not effectively provide disaggregation of surrounding cement particles. A special type of magnetic field application providing a particle vibration and translation together may be required for an internal movement of magnetic particles to disaggregate cement particles. Therefore, alternating the magnetic field will cause the flipping over of the magnetic particles and hence the vibration, but this vibration probably will not provide any beneficial magneto-mechanical effect for the rheology control mechanism in cement paste medium due to the lack of translation of magnetic particles. This was observed in the test setup given in Figure 3.8.

In this test setup, an electromagnet, which was positioned in the rheometer, was operated by a DC power supply connected to a relay to provide a pulsating magnetic field. Magnetite containing cement paste with a w/c ratio of 0.32 was placed on the electromagnet and the generated shear stress during a continuous shearing at 20 s⁻¹ shear rate was recorded over time under the pulsating magnetic fields in on/off states in different frequencies. The recorded shear stress evolution over time under different pulsating

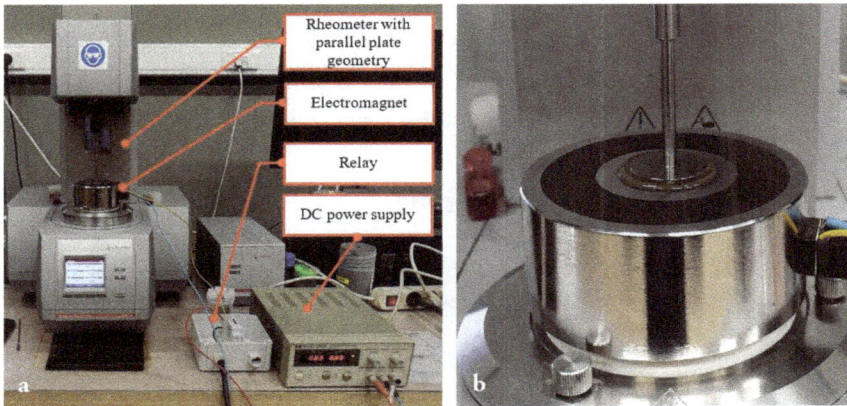

Figure 3.8 Test setup for rheology measurement with parallel plate geometry (gap is 1 mm) under pulsating magnetic field (a), enlarged view of a sheared sample under magnetic field (b). (Courtesy Mert Yucel Yardimci).

magnetic fields, which were provided by the setup in Figure 3.8, is shown in Figure 3.9.

As can be seen from Figure 3.9, without a magnetic field, the shear stress of the high-amount magnetite containing cement paste with a w/c ratio of 0.32 shows a progressive decrease until a steady-state value over time is reached due to reaching a more de-structured state after experiencing a continuous shearing for a while as a typical response of a material reaching a new state of the structure with shear rate [48]. When a pulsating magnetic field in different frequencies was applied in on/off mode, shear stress spikes were recorded for the instant times when the electric current was on, reflecting the contribution of the columnar structures of magnetic particles during the on mode to the shear resistance of the material. This shear contribution diminished in the instant times when the current is off, showing no residual columnar structures left during the instant time without magnetic field, probably due to the applied magnetic field, which is lower than the remanent magnetization of the responsive cement because of the limitations in the electronics of the setup shown in Figure 3.8. On the other hand, the reversible formation of chain or columnar magnetic particle structures was evidenced by the test setup shown in Figure 3.8 for a real cement paste case (not diluted). The magnitude of the shear stress spikes decreases with an increase in the frequency because of the decreased magnetic field strength in high pulsating mode due to the limitations of the electronics of the setup in Figure 3.8. This test shows two important outcomes: first, the pulsating magnetic field application in the perpendicular direction to the shearing plane

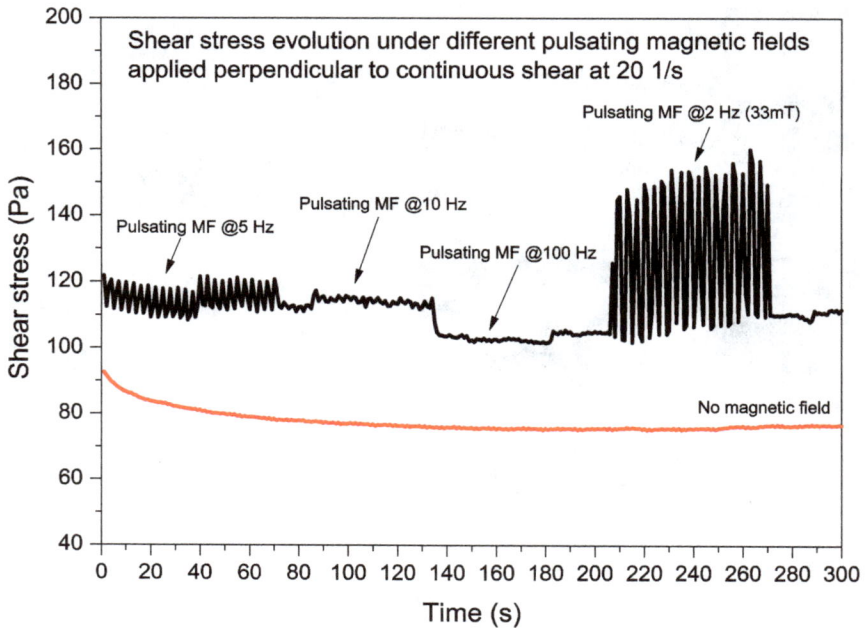

Figure 3.9 Shear stress versus time curves of a cement paste with high magnetite content under pulsating magnetic field in different frequencies during a continuous shearing at 20 1/s shear rate. (Courtesy Mert Yucel Yardimci).

causes an instantaneous and reversible formation of columnar structures, as evidenced by a sudden increase and decrease in shear stress. Second, even the rapid formation and de-formation of magnetic columnar structures and associated particle vibration due to flipping over around a centre on themselves may not be enough to make a significant change in the rheology of magneto-rheological cement paste, as no shear stress decay was observed in the shear stress during or after pulsating magnetic field.

On the other hand, the magneto-mechanical effect of particles is a growing field of research on cancer treatment based on triggering cancer cell death by low-frequency mechanical vibrations of magnetic nano-particles without heat generation [45]. In addition, many novel biomedical science approaches, such as cancer treatment with magnetic hyperthermia, drug release, controls of gene expression and neurons, or activation of chemical reactions, have been introduced based on high-frequency alternating magnetic field applications [49]. There are also several examples of applying an alternating magnetic field to construction materials. It has been reported that applying an alternating magnetic field at a high frequency (typically >100 kHz to several MHz) induces the heating of magnetic particles that could be used for crack healing of bituminous materials [50, 51]. It has been reported that the application of an alternating magnetic field (with AC current at 50 Hz) perpendicular to the placing direction of concrete cylindrical samples provides somewhat internal small vibrations, possibly due to the movement of charged particles, such as silica [52]. Although the authors did not report any improvement in fresh state properties of the concrete mixture due to the mentioned internal vibration, they claim some improvements in mechanical properties of concrete exposed to a magnetic field during sample preparation because of the micro-vibrations influencing the microstructure by making it less porous. The beneficial effect of the alternating magnetic field is more apparent in metallic fibre-incorporated cementitious mixtures. Magnetic field application on pulsating mode provides a better consolidation and more preferential alignment of metallic fibres as described in detail in section 3.5.1.

In magneto-rheological fluids, as shown by Jorge et al. [53], if the magnetic field is rotated, the chains which were already oriented in the direction of the magnetic field also rotate to remain oriented in the field. On the other hand, rotating magnetic fields lead to additional hydrodynamic forces that oppose head-to-tail columnar structures (aggregates). The size of the aggregates remains almost constant for a low-frequency rotating magnetic field, while the aggregates are divided into fragments due to the hydrodynamic drag under the high-frequency rotating magnetic field [54]. Such rotational movements, coupled with the vibration of the magnetic particles of the chains in magneto-rheological fluids, may present some new opportunities for cement paste medium with magneto-rheological properties. Such type of formation changes on columnar structures due to the rotating magnetic field may facilitate another type of internal vibration based on a somewhat whipping effect of magnetic clusters in magnetically responsive cement paste. On the other hand, more research is needed to prove the hypothesized potential of columnar structures of magnetically responsive cement pastes to trigger an on-demand rheology change in magnetically responsive concretes.

3.3   MAGNETO-RHEOLOGICAL BEHAVIOUR OF CEMENT PASTE WITH NANO-FE$_3$O$_4$ PARTICLES

3.3.1   Physical properties of nano-Fe$_3$O$_4$ particles

Nano-Fe$_3$O$_4$, i.e., nano-magnetite, is a common magnetic iron ore with Fe^{2+} (ferrous) and Fe^{3+} (ferric). The crystal structure of magnetite is an inverse spinel structure with alternating octahedral and tetrahedral-octahedra layers [55]. The effective surface area of nano-Fe$_3$O$_4$ particles with particle size around 50 nm is determined as approx. 100 m^2/g [56]. The density of magnetite nanoparticles does not significantly change with the particle size, and its morphology is assumed to be nonporous in nature [57]. The melting and boiling points of magnetite are 1590 °C and 2623 °C, respectively. Generally, nano-Fe_3O_4 particles exhibit high electronic conductivity because of the inverse spinel structure [58]. Due to the high specific surface area and good reactivity, magnetite nanoparticles have been applied in biological, chemical and industrial areas [59].

From the viewpoint of magnetic properties, magnetite shows high Curie temperature (~850 K). This means that nano-Fe_3O_4 is a kind of ferromagnetic material at room temperature, where the magnetic moments can be aligned along the direction of external magnetic field. At ambient temperature higher than the Curie temperature, the ferromagnetic behaviour is diminished and magnetite will show superparamagnetic behaviour. The magnetic properties vary with the particle size due to differences of the crystalline structure. In the case of cube-like nano-Fe_3O_4 particles, crystallite size is similar to particle size with a diameter ranging from 10 nm to 76 nm, indicating a single-domain structure. With the further increase of particle size, however, the crystallite size remains unchanged, which means a multi-domain structure [60]. The crystallite size of spherical-like nano-Fe_3O_4 particles is smaller than the cube-like nano-Fe_3O_4 particles. The saturation magnetization (Ms) increases with increasing particle size, regardless of the crystallite size and particle shape of the nano-Fe_3O_4 particles [60–62]. By contrast, magnetic coercivity is significantly influenced by the crystalline structure, where Li et al. [60] found that the magnetic coercivity increased with increasing particle size from 10 nm to 76 nm, while further increasing particle size reduced the magnetic coercivity value. The magnetization curves and X-Ray Diffraction (XRD) patterns of three typical nano-Fe_3O_4

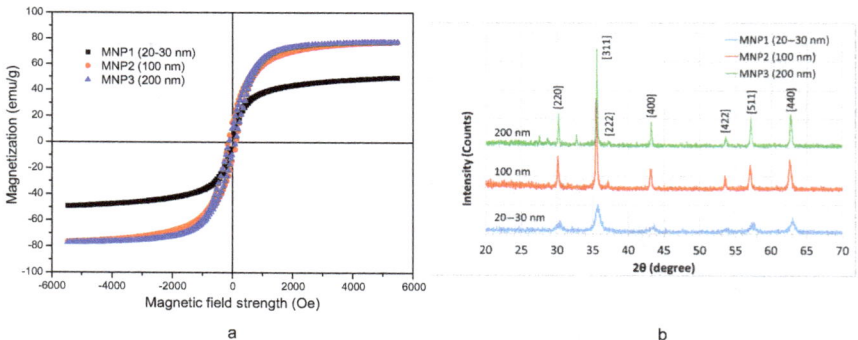

Figure 3.10 (a) Magnetization curves and (b) XRD patterns of three different nano-Fe_3O_4 particles. Source: after [63].

particles are shown in Figure 3.10. Strong and sharp peaks indicate highly crystalline nature of magnetite, and thus highly magnetic particles.

3.3.2 Typical magneto-responsive structural build-up

The structural build-up of cement paste can be described by the evolution of storage modulus, loss modulus and phase angle obtained from small-amplitude oscillatory time sweep test [64]. The effect of magnetic field on the evolution of storage modulus and loss modulus of cementitious paste with w/c of 0.4 and nano-Fe_3O_4 (20–30 nm) content of 3% (by mass of cement paste) is presented in Figure 3.11. In the absence of magnetic field, the storage modulus is always higher than the loss modulus, indicating that the solid-like property dominates the liquid-like behaviour. Both the storage modulus and the loss modulus increase over time, but the storage modulus shows relatively higher increase rate. This points to a gradual enhancement in the solid-like properties due to multiple effects of flocculation, thixotropy and slight chemical hydration reactions [64–66].

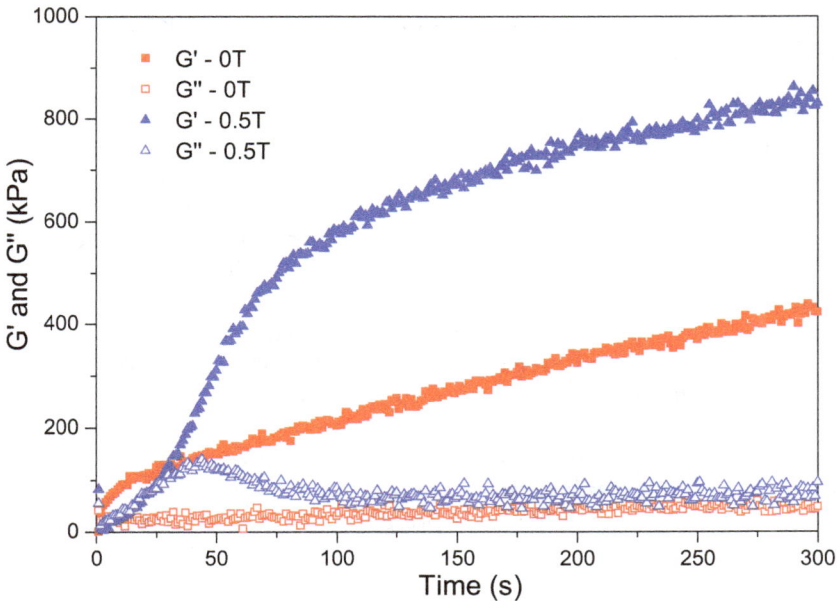

Figure 3.11 Evolution of storage modulus and loss modulus for cementitious paste with w/c of 0.4 and nano-Fe_3O_4 (20–30 nm) content of 3%, in the absence and presence of magnetic field of 0.5 T. Source: after [36].

In the presence of an external magnetic field with 0.5 T, the cementitious paste shows obvious magneto-responses, as applying the magnetic field evidently affects the evolution of the storage modulus and loss modulus. Specifically, the storage modulus shows a slight increase immediately after applying the magnetic field and then dramatically increases with magnetization time. This can most probably be explained by the formation of chains or clusters of magnetic nano-Fe_3O_4 particles [43]. The increase rate of the storage modulus slows down after about 150 s, nevertheless reaching G' values that are significantly higher than those obtained without magnetic field. The linear increase of the storage modulus at steady state represents the enhancement of flocculated structure due to chemical hydration of cement particles [67, 68]. The same increase rate of the storage modulus at steady state indicates that applying an external magnetic field has negligible influence on the intensity of C-S-H links between cement particles. From the viewpoint of loss modulus, without the magnetic field, it slightly increases with increasing magnetization time. In the presence of the external magnetic field, however, the loss modulus increases first and then gradually decreases. Moreover, the loss modulus of the cement paste is even higher than the storage modulus at very early age. This can be attributed to the micro-movement of the nanoparticles under external magnetic forces [36]. To illustrate the magneto-responses structural build-up more clearly, the relationship between storage modulus, loss modulus and phase angle for the cementitious paste is established, as shown in Figure 3.12.

At the initiation of the external magnetic field, i.e. t_0, the loss modulus is higher than the storage modulus and the phase angle is close to 80°. This indicates that the cement paste shows almost totally liquid-like behaviour immediately after initiation of the magnetic field, which is consistent with the findings of Nair and Ferron [69]. This can be attributed to the displacement and movement of magnetic nanoparticles in cement paste medium under external magnetic field. Without the external magnetic field, the magnetic nanoparticles can be regarded as randomly distributed in cement paste medium. When the magnetic field is applied, the nano-Fe_3O_4 particles will be magnetized and have a potential to form chains or clusters in a very short time. During their displacement to form clusters, the nanoparticles create a sort of mechanical micro-agitation, destroying the bridges of early hydration products (mainly C-S-H bridges [65]) between cement particles, as schematized in Figure 3.13 (a). Moreover, some possibly entrained water in the agglomerated cement clusters is released, which increases the content of free water. As a result, a more dispersed state is obtained after initiation of the external magnetic field. Note that most probably not all the nano-Fe_3O_4 particles will contribute to the formation of chains or clusters due to resisting viscoelastic properties of the cement-based medium.

With increasing magnetization time, the cement paste gradually transitions from viscous state to elastic state, where the capacity of energy

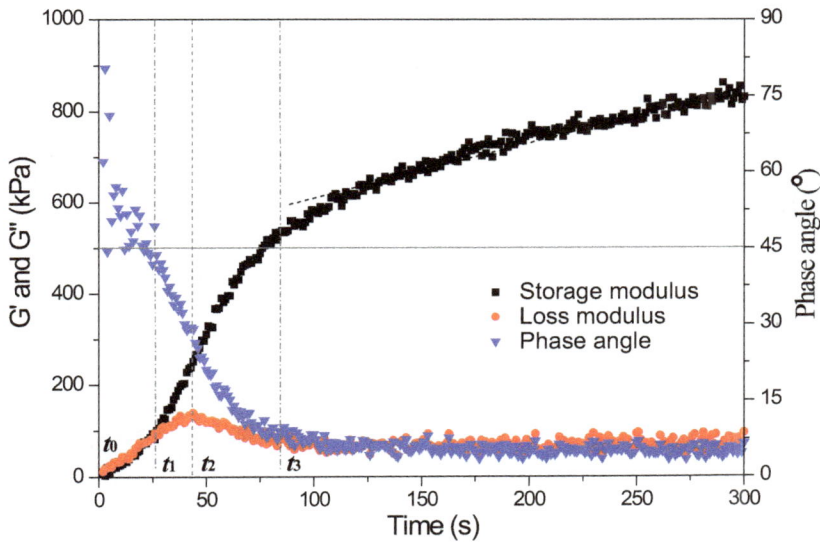

Figure 3.12 Evolution of storage modulus, loss modulus and phase angle for cementitious paste with w/c of 0.4 and nano-Fe_3O_4 (20–30 nm) content of 3%, in the presence of a magnetic field of 0.5 T. Source: after [36].

stored gradually increases and simultaneously part of the mechanical energy dissipates. This is reflected by the rapid increase in the storage modulus, the decrease in the phase angle and the slight increase in the loss modulus. On the one hand, part of the remaining nanoparticles in the interstitial solution try to move to join the clusters under the magnetic force, and meanwhile, the cement particles disturbed by the moving nanoparticles will try to reach new equilibrium positions, as depicted in Figure 3.13 (b), resulting in a slight increase in the loss modulus. On the other hand, the interactions between the magnetic chains or clusters are improved, and new bridges and links between cement particles are gradually re-formed due to the colloidal interactions and cement hydration, which enhances the elastic property. At t_1, the storage modulus is equal to the loss modulus and the phase angle is 45°, indicating that the elastic-solid property starts to dominate the viscous-liquid property.

When the elastic-solid property reaches a threshold (t_2), the number of nanoparticles contributing to the chains or clusters reaches a maximum, and the cement particles arrive at their final equilibrium positions, as represented in Figure 3.13 (c). Therefore, the energy dissipation becomes limited, and the loss modulus reaches a maximum. With continuous magnetization, the

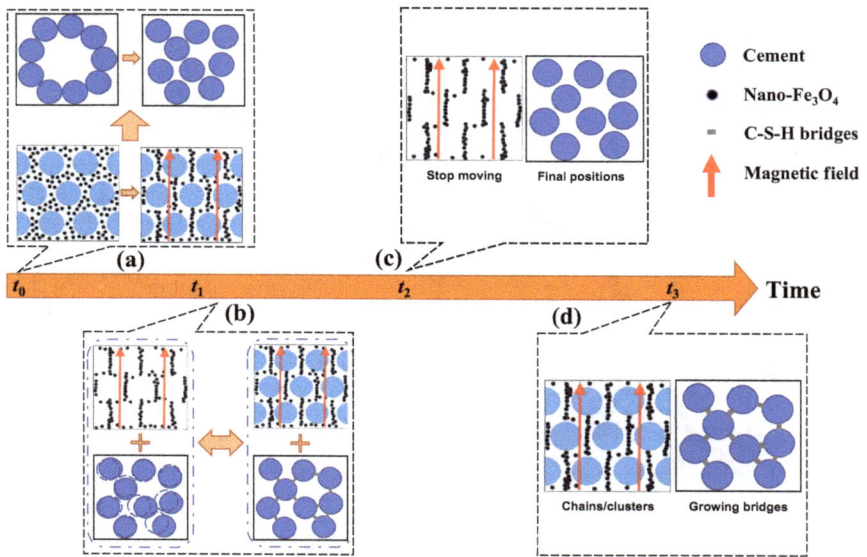

Figure 3.13 Schematic diagram of distribution state of magnetic nanoparticles in cement paste with magnetization time. Source: after [36].

solid-like behaviour evolves and the loss modulus starts to decline. After a sufficiently long period of magnetization (*e.g.*, t_3), the cement-based suspensions possess enough elastic property and the deformation energy is hardly dissipated. This is reflected by the slight decrease in phase angle and loss modulus. Since the links and bridges of hydration products gradually evolve over time, as shown in Figure 3.13 (d), the storage modulus gradually increases with magnetization time. Overall, for the cementitious paste containing nano-Fe_3O_4 particles, the application of an external magnetic field improves the liquid-like properties at very early age, and enhances the solid-like properties after a sufficiently long period of magnetization.

3.3.3   Influence of cement paste medium

The viscoelastic properties and stiffness of the cement paste medium have a large influence on the movement and distribution of magnetic nanoparticles when applying an external magnetic field, and thus on the magneto-rheological response. The relationship between magneto-rheological response, paste stiffness without magnetic field and magnetic yield parameter for cementitious pastes with fixed nano-Fe_3O_4 (100 nm) concentration

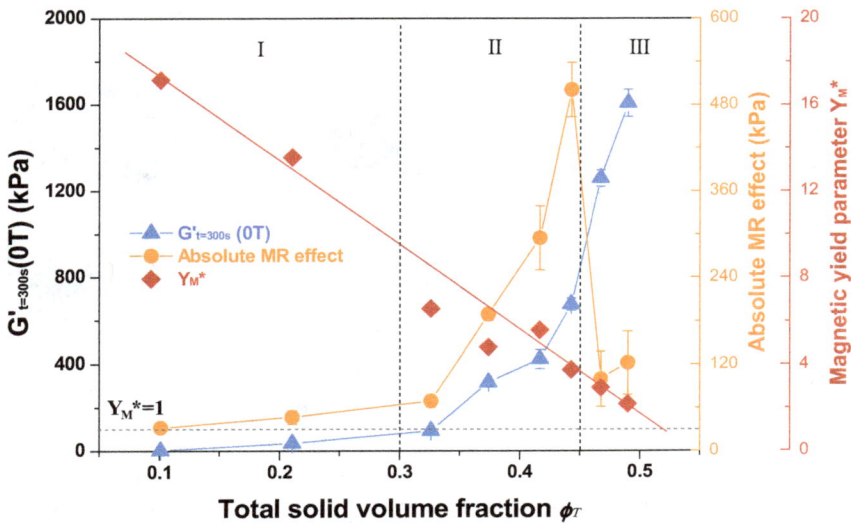

Figure 3.14 Correlations between G't=300s (0 T), difference of G't=300s, and magnetic yield parameter for cementitious pastes with fixed nano-Fe3O4 (100 nm) concentration of 3% by mass of cement + water. φT is the total solid volume fraction. Error bar indicates the standard deviation. Source: after [70].

of 3% and various total solid volume fractions ϕ_T is presented in Figure 3.14, where the magneto-rheological effect is described by the difference of storage modulus at 300 s ($G'_{t=300s}$) between 0 T and 0.5, and the paste stiffness without magnetic field is characterized by the $G'_{t=300s}$ obtained without magnetic field, i.e. $G'_{t=300s}$ (0 T), respectively.

At relatively low solid volume fractions (e.g. $\phi_T < 0.3$), the cementitious paste without magnetic field acts as a dilute suspension, as shown in Figure 3.15 (a), having very high fluidity and liquid-like properties. After applying an external magnetic field of 0.5 T, the magnetic force between adjoining nanoparticles is much larger than the resistance induced by the viscoelastic stress of the suspension, and the nano-Fe_3O_4 particles can contact with each other to form clusters in the suspension. However, the formed magnetic clusters probably cannot connect the neighbouring cement particles due to the large interparticle distance. In this case, the cementitious paste after applying an external magnetic field is still regarded as a dilute suspension, and the stiffness cannot increase obviously. Therefore, insignificant increase in the difference of $G'_{t=300s}$ is observed at relatively low solid volume fractions.

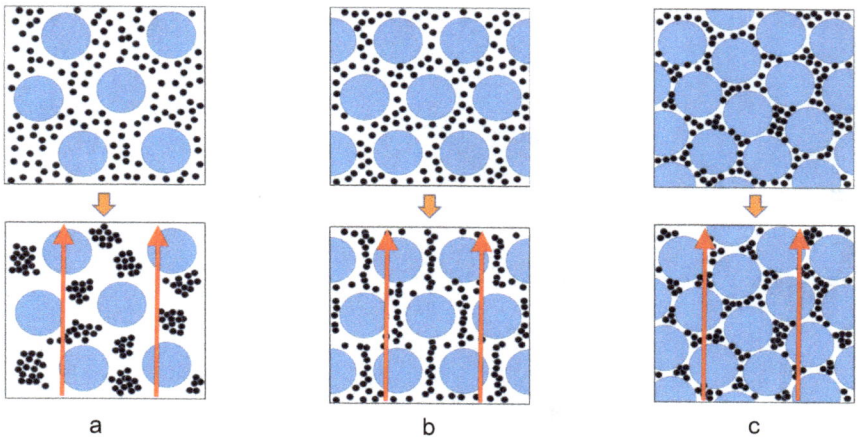

Figure 3.15 Schematic diagram of particle distribution at various solid volume fractions. (a) Low ϕ_T; (b) moderate ϕ_T; and (c) high ϕ_T. Blue dots indicate cement particles, dark dots are nano-Fe_3O_4 particles and red arrow is magnetic field. Source: after [70].

At moderate solid volume fractions (e.g. $0.3 < \phi_T < 0.45$), the stiffness of the cementitious paste without magnetic field gradually increases, behaving as a concentrated suspension. After applying a magnetic field of 0.5 T, the magnetic force between nanoparticles is higher than the resistance of the suspension. The formed magnetic clusters can fill the voids between cement particles, as presented in Figure 3.15 (b), connecting the solid particles and thus increasing the stiffness. The difference of $G'_{t=300s}$ between 0 T and 0.5 T gradually increases with the increase of the total solid volume fraction, until an optimal fraction with maximum difference value of $G'_{t=300s}$. At relatively high solid volume fractions (e.g. $\phi_T > 0.45$), the cementitious paste shows very high solid-like behaviour. After applying a magnetic field of 0.5 T, the magnetic force between nanoparticles might overcome the resistance of the paste, and the nanoparticles could move in the suspension. Nevertheless the dense particle packing hinders the formation of magnetic clusters, as illustrated in Figure 3.15 (c), and thus the increase of the storage modulus is limited. Thereby, the difference of $G'_{t=300s}$ between 0 T and 0.5 T gradually decreases at relatively high solid volume fractions. It can be predicted that very dense particle packing can totally prevent the movement of nanoparticles to form clusters, and in this case, the cementitious paste will not show any rheological response to an external magnetic field.

Polycarboxylate ether (PCE) superplasticizer is widely applied in cement-based materials due to their high water reduction rate and good slump

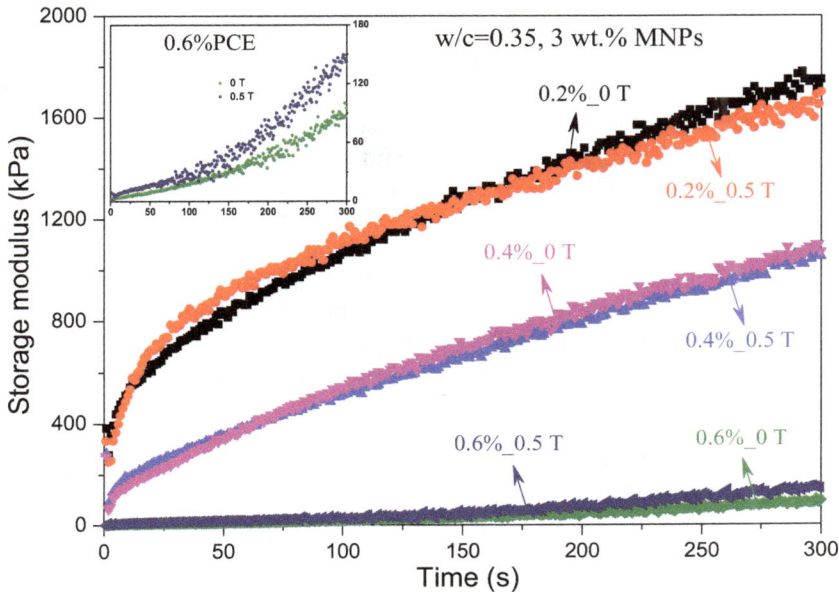

Figure 3.16 Effect of PCE dosage on the evolution of storage modulus in the absence and presence of magnetic field (w/c=0.35, 3 wt.% nano-Fe$_3$O$_4$ (20–30 nm)). Source: after [36].

retention at low dosage [71]. The incorporation of PCE has a large influence on the viscoelastic properties of cementitious suspension, and thus on the response to an external magnetic field. Figure 3.16 shows the influence of PCE dosage on the evolution of storage modulus of cementitious paste with w/c of 0.35 and nano-Fe$_3$O$_4$ (20–30 nm) content of 3 wt.%. Viewed as a whole, all the cementitious pastes with PCE seem to exhibit unapparent responses to the external magnetic field with 0.5 T. This can be explained by the high elastic limit yield stress and viscosity of the cementitious pastes [36]. As a result, no clusters or limited magnetic clusters can be formed and the difference in magneto-rheological response is not significant. In the case of cement paste with high PCE additions, e.g. 0.6%, as can be seen from the inserted enlarged view in Figure 3.16, the storage modulus obtained under 0.5 T is higher than in case of zero magnetic field, especially after a longer period of magnetization. This means that the cementitious paste with 0.6% PCE shows slight magneto-rheological response and magnetic clusters could be formed under magnetic field. The slight increase of the storage modulus between 0 T and 0.5 T is probably due to the very high liquid-like properties of the suspension.

3.3.4 Influence of nano-Fe$_3$O$_4$ concentration

From equations (3.3) and (3.7), it can be clearly seen that the magnitude of the magnetic yield parameter and movement velocity depend on the concentration and physical nature of the nanoparticles and the viscoelastic properties of the suspension. The early structural build-up illustrated by the evolutions of storage modulus (G') and phase angle (δ) for cementitious pastes with nano-Fe$_3$O$_4$ (20–30 nm) concentrations of 0%, 0.25%, 1% and 3% is presented in Figure 3.17. To assess the structural build-up rate quantitatively, the increase rate of the storage modulus (ΔG') at steady period was calculated. The application of an external magnetic field with 0.5 T shows negligible influences on the viscoelastic properties of the reference cement paste, which is consistent with [18]. In the absence of external magnetic field, the magnitude and increase rate of the storage modulus increase with the addition of nano-Fe$_3$O$_4$ particles, especially at higher concentration, indicating higher and faster structural build-up of the cementitious paste. This can be attributed to the reduced particle distance, increased inter-

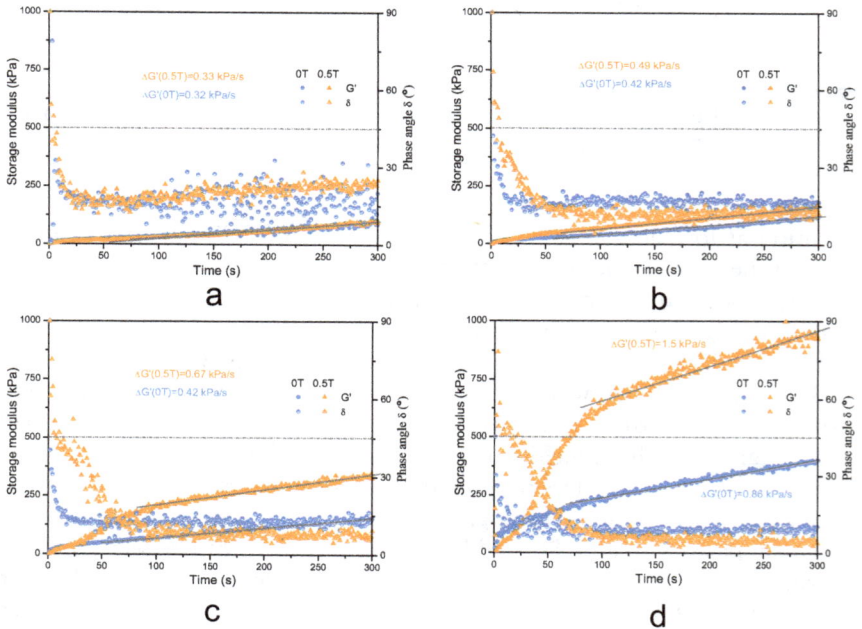

Figure 3.17 Influence of particle concentration on the early structural evolution of cementitious paste (w/c=0.4 and nano-Fe$_3$O$_4$ of 20–30 nm). (a) 0% nano-Fe$_3$O$_4$; (b) 1% nano-Fe$_3$O$_4$; (c) 2% nano-Fe$_3$O$_4$ and (d) 3% nano-Fe$_3$O$_4$. ΔG' is the increase rate of storage modulus at steady state. Source: after [63].

particle contacts [63], and possible agglomeration of nano-Fe_3O_4 particles due to their high magnetic properties [72, 73]. Despite the high stiffness of the cementitious pastes with nanoparticles, it seems that higher nanoparticles addition results in slightly shorter percolation time. This implies that the cementitious pastes with nano-Fe_3O_4 particles have faster formation rate of flocculation structures compared to the reference cement paste due to the improved colloidal interactions.

After applying a magnetic field with 0.5 T, all the cementitious pastes containing nano-Fe_3O_4 particles exhibit obvious rheological response. The intensity of the magneto-rheological response magnifies with the concentration of the nano-Fe_3O_4 particles. Specifically, the differences of the storage modulus under 0 T and 0.5 T are enlarged with increasing concentration of the nanoparticles, regardless of the magnetization time. In the presence of the external magnetic field, with the increase of nano-Fe_3O_4 particles content, the phase angle at steady state gradually decreases and the increase rate of the storage modulus increases. Moreover, a higher nano-Fe_3O_4 concentration results in a longer time for the phase angle to reach 45°, a higher increase rate of the loss modulus at early age as well as a larger peak value of the loss modulus, as presented in Figure 3.18. These results

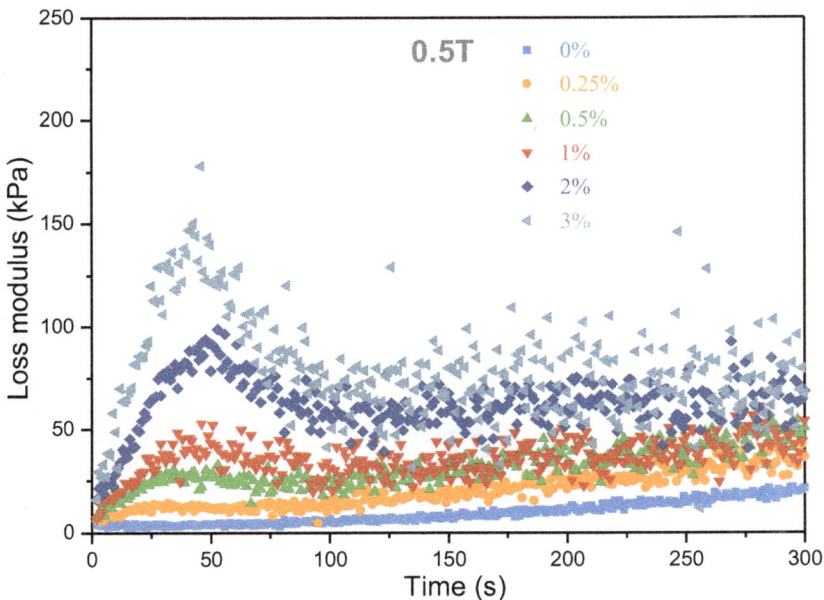

Figure 3.18 Influence of particle concentration on the evolution of loss modulus of cementitious paste (w/c=0.4 and various concentrations of nano-Fe_3O_4 with 20–30 nm) under an external magnetic field of 0.5 T. Source: after [63].

indicate that increasing nano-Fe_3O_4 concentration increases the intensity of liquid-like properties immediately after introduction of the magnetic field and enhances the stiffness of the cementitious paste after undergoing longer magnetization. More detailed explanations can be found in [63]. It should be mentioned that the time for the loss modulus to reach the peak does not seem to show a linear correlation with the concentration of the nano-Fe_3O_4 particles. Instead, under the magnetic field of 0.5 T, increasing nano-Fe_3O_4 particles content from 0.25% to 1% leads to a slight increase in peak time, while it shows negligible change when the nano-Fe_3O_4 particles content further increases to 3%.

3.3.5 Influence of particle size of nano-Fe_3O_4

The particle size of the nanoparticles plays a significant role in the movement velocity of nanoparticles in cementitious suspensions under magnetic field, as can be observed from equation (3.7). The influences of particle size of nano-Fe_3O_4 particles on the evolution of storage modulus and loss modulus of cementitious pastes with nanoparticles concentration of 1 wt.% and 3 wt.% are depicted in Figure 3.19. The physical properties of the three nano-Fe_3O_4 particles are shown in Table 3.2. It can be clearly observed from Figure 3.19 (a–b) that the addition of nano-Fe_3O_4 particles strengthens the structural build-up of the cementitious paste in the absence of magnetic field, irrespective of the particle size. The extent of variation of the storage modulus relies on the concentration of nano-Fe_3O_4 particles. In the case of nano-Fe_3O_4 of 1 wt.%, the particle size of the nano-Fe_3O_4 particles seems to have little influence on the evolution of the storage modulus. However, at the nano-Fe_3O_4 content of 3 wt.%, the cementitious paste containing larger nanoparticles shows lower magnitude of storage modulus. This can be explained by the water film thickness theory [35, 74]. After adding nanoparticles with ultra-high specific surface area, more water is required to coat the nanoparticles, leading to a relatively low water film thickness of solid particles. Moreover, the number of particles increases with the decrease of particle size under the same weight. This means that smaller nano-Fe_3O_4 particles could create more contacts and frictional interactions between solid particles, exhibiting an increased effect on the structural build-up. Therefore, at the same addition content, higher stiffness can be obtained for the cementitious paste containing nanoparticles with smaller particle size.

In the presence of an external magnetic field, the evolution of the storage modulus at very early age is less dependent on the particle size of the nanoparticles. However, after exposing to the external magnetic field for a longer period, the storage modulus of the cementitious pastes with MNP2 and MNP3 is far higher than that of the cementitious paste containing

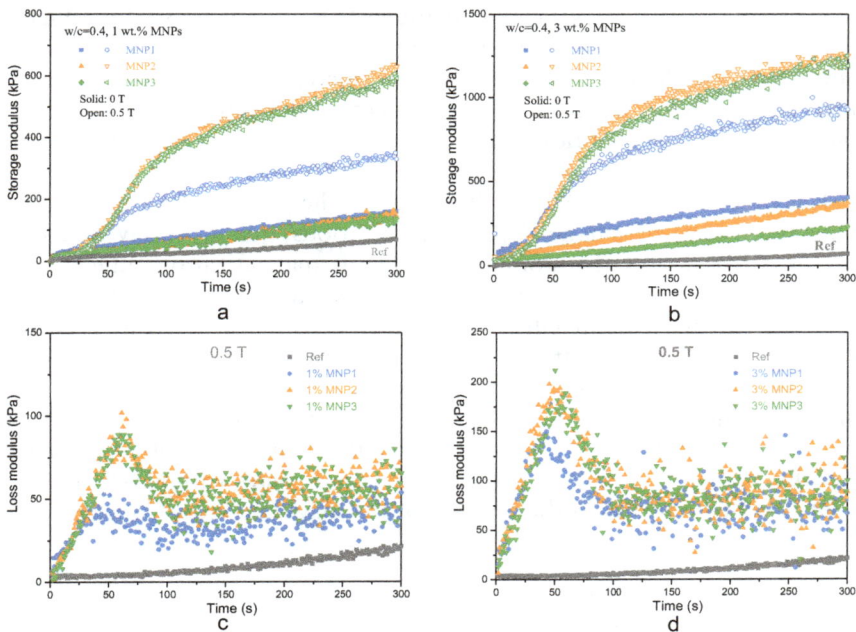

Figure 3.19 Influence of particle size of nano-Fe$_3$O$_4$ particles on the evolution of storage modulus and loss modulus of cementitious pastes (w/c=0.4). (a) Storage modulus (1 wt.%); (b) storage modulus (3 wt.%); (c) loss modulus (1 wt.%); and (d) loss modulus (3 wt.%). Source: after [63].

Table 3.2 Physical properties of the nano-Fe3O4 particles used in Figure 3.19

Material	Particle size (nm)	Saturation magnetization (Ms, emu/g)	Remanent magnetization (Mr, emu/g)	Coercive field (Hc, Oe)
MNP1	20–30	49.48	5.13	48.88
MNP2	100	77.29	14.81	134.01
MNP3	200	77.56	10.23	108.27

MNP1. Distinct trends can also be observed from the evolution of the loss modulus in Figure 3.19 (c–d). More specifically, the increase rate of the loss modulus at early age is mainly determined by the concentration of the nanoparticles, instead of the particle size. By contrast, the peak values of the loss modulus and the corresponding peak time of cementitious pastes containing MNP2 and MNP3 are higher than that of cementitious paste

with MNP1, regardless of the nano-Fe_3O_4 concentration. The results indicate that the cementitious pastes containing MNP2 or MNP3 exhibit higher rheological response to the external magnetic field compared to the cementitious paste with MNP1. Most interestingly, increasing the particle size from 100 nm to 200 nm has negligible influences on the stiffness and structural evolution of the cementitious paste in the presence of an external magnetic field with 0.5 T.

The effect of particle size on the magneto-rheological effect of cementitious paste at various nano-Fe_3O_4 concentrations, quantitatively described by the difference of the storage modulus at magnetization time of 300 s between 0.5 T and 0 T (D-value), is shown in Figure 3.20. Generally, the magneto-rheological effect is considered to show a positive correlation with the concentration of the nano-Fe_3O_4 particles. The cementitious pastes containing coarser nanoparticles (MNP2 and MNP3) exhibit higher magneto-rheological effect than that of the cementitious pastes with finer nanoparticles (MNP1) due to the relatively high magnetic properties. Another reason is the fact that the cementitious pastes with MNP1 possess relatively high rigidity in the absence of the magnetic field. Concentration

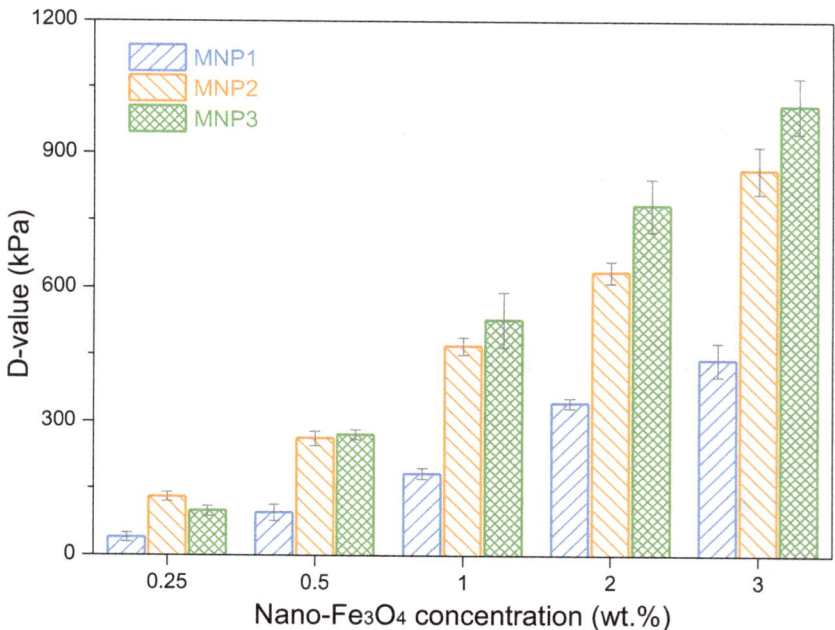

Figure 3.20 Effect of particle size on magnetorheological effect of cementitious paste (w/c=0.4). Error bar indicates the standard deviation. Source: after [63].

plays a significant role in the influence of MNP2 and MNP3 on the magneto-rheological effect. At relatively low nanoparticles concentration, the cementitious pastes with MNP2 show approximately similar magneto-rheological effect to the pastes with MNP3. However, at the nanoparticles content higher than 1 wt.%, the magneto-rheological effect of the pastes with MNP3 is slightly higher than that of MNP2-containing pastes. This can be attributed to the difference of viscoelastic properties of the cementitious pastes without magnetic field. The storage modulus shows less difference under low nanoparticles additions, whereas larger particles result in smaller storage modulus in the case of relatively high nanoparticles concentration. Accordingly, a concentration-dependent magneto-rheological effect is observed.

The relationship between measured magneto-rheological properties and calculated theoretical parameters is established, as depicted in Figure 3.21 (a). It can be seen that the *D-value* shows a good linear relationship with the estimated magnetic yield parameter for each batch of cementitious pastes. Cementitious pastes containing nano-Fe_3O_4 particles with larger size and higher magnetic properties generally have higher magnetic yield parameters and *D-values*. The linear correlation indicates that the calculated magnetic yield parameter can be used to predict the magneto-rheological effect of cementitious pastes with various contents of specific magnetic nanoparticles. In other words, the influence of particle concentration and viscoelastic properties (which is concentration-dependent) of pastes containing specific nano-Fe_3O_4 particles on the magneto-rheological properties can be converted into one factor, i.e., magnetic yield parameter.

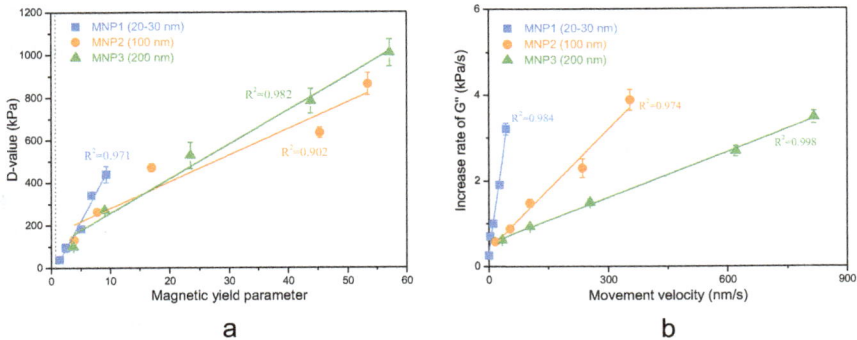

a   b

Figure 3.21   Relationship between measured magneto-rheological properties and calculated theoretical parameters. (a) D-value versus magnetic yield parameter, and (b) early increase rate of loss modulus versus movement velocity of nanoparticles. Error bar indicates the standard deviation. Source: after [63].

The increase of the loss modulus of cementitious paste after applying a magnetic field, describing the improvement of liquid-like properties, is a result of nanoparticles rearrangement and displacement. From Figure 3.21 (b), it can be seen that the increase rate of the loss modulus has an apparent linear correlation with the movement velocity for specific nano-Fe_3O_4 particles. For specific nano-Fe_3O_4 particles, higher concentration means higher movement velocity of nanoparticles and more destroyed early hydration products bridges and flocculation structures between cement particles. Accordingly, higher early increase rate of the loss modulus and more liquid-like behaviour can be achieved. At the same particle concentration, a larger particle size results in a significantly higher movement velocity. However, the increase rates of the loss modulus are almost in the same order of magnitude, which can be clearly observed from Figure 3.19 (c–d). The formation of magnetic clusters after applying a magnetic field is quite fast. The variation of the monitored loss modulus is a post-effect of the displacement of nanoparticles. Therefore, the increase rate of the loss modulus at early age seems to be mainly controlled by the particle concentration, rather than the particle size or magnetic properties. This can be observed from the inserted relationship between the increase rate of the loss modulus and the particle concentration in Figure 3.21 (b). Nevertheless, the conceptual movement velocity of nanoparticles is an effective parameter to describe the improvement of liquid-like behaviour of cementitious pastes with specific nano-Fe_3O_4 particles.

Overall, the particle size plays a major role in dominating the viscoelastic properties of cementitious paste without magnetic field, and the response of the cementitious paste to an external magnetic field is mainly determined by the magnetic properties of the nanoparticles [63]. It can be concluded that a cementitious paste containing nano-Fe_3O_4 particles with larger particle size and higher magnetic properties exhibits increased magneto-rheological effect.

3.3.6   Influence of magnetic field types

Beyond the mix proportion parameters, the external magnetic field itself also exerts a significant impact on the magneto-rheological responses of cement-based MR suspensions. The evolution of storage modulus of cementitious paste under various constant magnetic fields is shown in Figure 3.22. Weak magnetic fields such as 0.08 T and 0.16 T have little effects on the magnitude of storage modulus. Under relatively high magnetic fields such as 0.24 T and 0.5 T, however, the storage modulus increases by 1.4 and 2 times after applying the magnetic field for 180 s, respectively. In other words, the magneto-rheological response is only obvious at sufficiently high magnetic field strength. Furthermore, the increase rate (slope) of the storage modulus at steady state slightly increases with the increase of strengths of

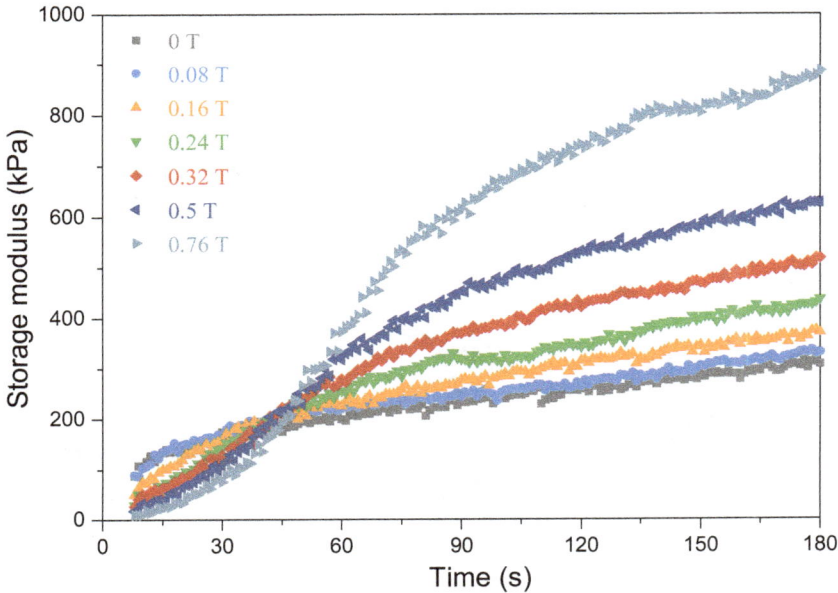

Figure 3.22 Evolution of storage modulus of cementitious paste (w/c is 0.4, and nano-Fe$_3$O$_4$ (20–30 nm) concentration is 3%) under various constant magnetic fields. Source: after [75].

the constant magnetic fields. With increasing magnetic field strength, the number of effective nanoparticles contributing to the clusters increases, resulting in an amplification of the magnetic clusters. Therefore, a higher magnetic field strength corresponds to a greater bonding between magnetic nanoparticles and thus a more obvious magneto-rheological response. Once the solid particles reach their final positions, the storage modulus increases due to the generation of C-S-H bridges between cement particles as well as the continuous magnetic connections between nanoparticles. This possibly can be used to explain the slight increase in the increase rate of the storage modulus after longer magnetization period.

Figure 3.23 presents the evolution of storage modulus of cementitious paste under step-increased magnetic field. The experimental results of reference cement paste (0 T) and the cementitious paste under constant magnetic fields (0 T, 0.25 T and 0.5 T) are also presented. When the magnetic field strength increases from zero to higher value (0.25 T or 0.5 T), the storage modulus experiences an abrupt decrease to a value close to that of the reference cement paste. This means that the internal microstructure of the cementitious paste is disrupted. Afterwards, a marked increase in storage

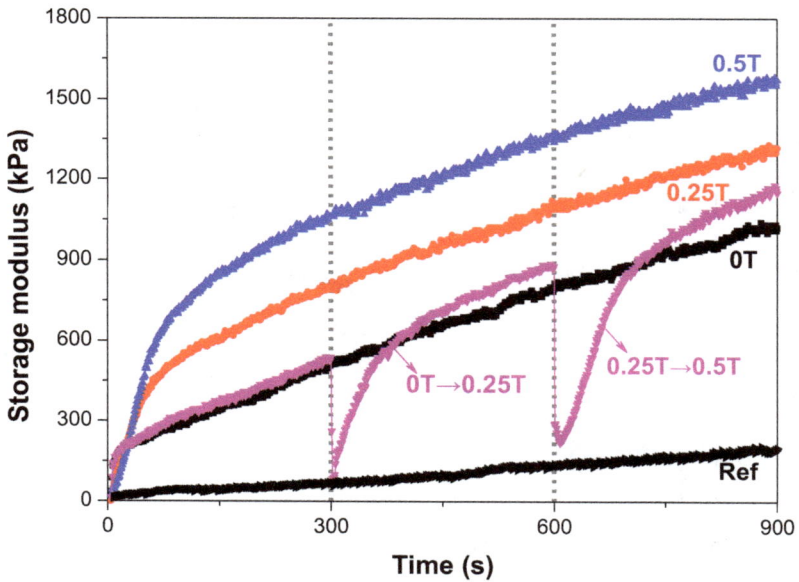

Figure 3.23 Evolution of storage modulus of cementitious paste (w/c is 0.4, and nano-Fe$_3$O$_4$ (20–30 nm) concentration is 5%) under step-increased magnetic field. Source: after [76].

modulus is observed, indicating higher structural build-up with magnetization time. This behaviour is due to the movements and distributions of nano-Fe$_3$O$_4$ particles in the cement paste under magnetic field [76]. At 600 s when increasing the magnetic field from 0.25 T to 0.5 T, the storage modulus also shows a sharp decrease. This is explained by the fact that the nanoparticles which are already in the chains or clusters structures will agitate and redistribute to form stronger structures when the strength of the magnetic field increases. Their movement will break down the links between clogged particles and lead to a decrease in the storage modulus. Another explanation could be that the magnetic field of 0.25 T is not strong enough to insert all the nanoparticles in the chain-like structure. When the magnetic field increases to 0.5 T, the remaining nanoparticles will contribute to the formation of the chains or clusters. Their movement leads to a decrease in the storage modulus of the cementitious paste. Besides, the abrupt decrease in the storage modulus at 600 s is larger than that at 300 s, but a higher storage modulus is still observed at 600 s. This behaviour is possibly due to the strengthened connections between particles in the cement paste medium with the elapsed time. More experimental results and underlying mechanisms about the structural evolution of cementitious paste under time-varying magnetic field can be found in literature [76].

3.3.7   Magneto-responsive flowing behaviour

Rheological properties, with yield stress and plastic viscosity as two intrinsic physical parameters, are of great importance in describing flowability and workability, evaluating pumpability, and predicting formwork filling of cement-based materials [35]. The measured shear stress of cementitious paste with w/c of 0.4 and nano-Fe_3O_4 (20–30 nm) content of 3 wt.% under a magnetic field of 0.5 T is even lower than that obtained without magnetic field, due to the obvious nanoparticle agglomeration and bleeding in the interface between the cementitious paste and the upper rotating plate [77]. By examining the magnetic properties of the hardened cementitious paste powders in the plate, it is revealed that the nanoparticle agglomeration is contributed by the combined effect of high magnetic field and high-rate shearing. Care should be taken when studying the dynamic flow properties of cementitious paste under magnetic field in a parallel plate rheometry.

In addition to shear curve, the influence of magnetic field on flow behaviour of cementitious-like suspension containing nano-Fe_3O_4 particles at constant low shear rate is revealed. Figure 3.24 shows the evolution of apparent viscosity of limestone powder suspension (W/LP of 0.35 and nano-Fe_3O_4 (20–30 nm) of 3 wt.%) at constant shear rate of 10 s^{-1} under various constant magnetic fields. The magnetic fields were exerted on the suspension at the beginning of the shearing. In the absence of magnetic field, the suspension exhibits a significant reduction in the apparent viscosity immediately after initiation of the shearing. There is an obvious decay in the apparent viscosity followed by an approximate steady-state after a longer period of shearing.

In the presence of an external magnetic field, the development of apparent viscosity with time shows distinct behaviour. On the one side, it can be seen from Figure 3.24 that the initial apparent viscosity of the suspension under an external magnetic field, irrespective of the magnetic field strength, is slightly higher than that obtained without magnetic field. This implies that the magnetic clusters of nanoparticles are formed immediately after applying an external magnetic field. On the other side, under constant shearing, the hydrodynamic forces tending to destroy the internal structures compete with the interactions facilitating the build-up of colloidal structures and magnetic clusters. Therefore, the apparent viscosity at constant shear rate tends to evolve differently, depending on the strength of the magnetic field. A higher magnetic field strength corresponds to a more rapid response of the suspension under constant shearing at 10 s^{-1}. For example, the suspension starts to respond at shearing time around 230 s under a magnetic field of 0.25 T. However, in the presence of a magnetic field of 0.75 T,

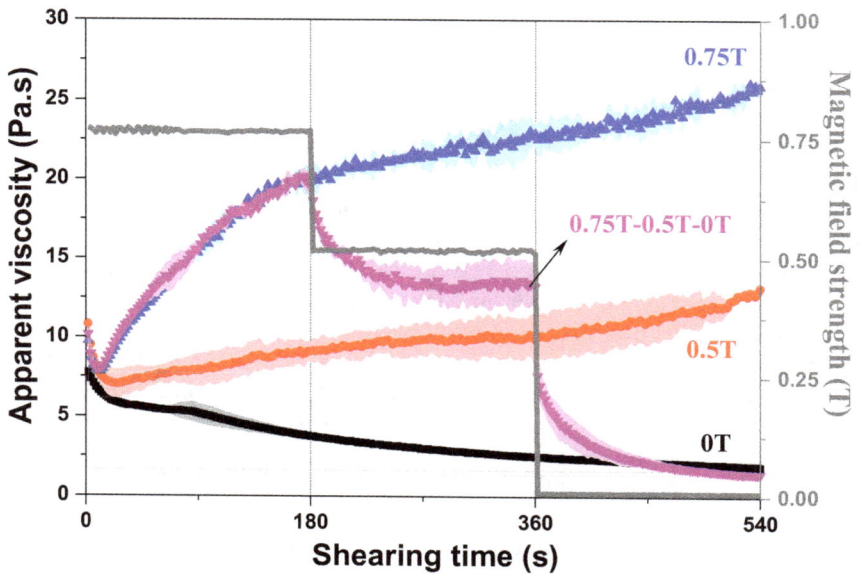

Figure 3.24 Evolution of apparent viscosity of limestone powder suspension (W/LP of 0.35 and nano-Fe_3O_4 (20–30 nm) of 3 wt.%) under various constant magnetic fields (0 T, 0.25 T, 0.5 T and 0.75 T), at constant shear rate of 10 s^{-1}. Source: after [78].

the apparent viscosity of the suspension exhibits a significant increase at around 10 s. Besides, the apparent viscosity at the shearing time of 540 s rises from 1.9 Pa.s to 25.8 Pa.s with increasing magnetic field strength from 0 T to 0.75 T. The non-linear increase of the apparent viscosity with magnetic field strength can possibly be attributed to the viscoelastic properties of the suspension [38]. Nevertheless, applying a constant magnetic field increases the viscosity of the cementitious-like suspension containing nano-Fe_3O_4 particles under constant shearing state.

3.4 MAGNETO-RHEOLOGICAL BEHAVIOUR OF FLY ASH-CEMENT PASTE

From theoretical and lab-scale points of view, ferromagnetic particles with high magnetic properties such as carbonyl iron particles and nano-Fe_3O_4 particles can be used as magnetizable additives in cementitious materials, as previously described. Due to the high costs, however, the extensive usage of ferromagnetic particles in engineering applications of cementitious materials is probably not the first option, especially for mortar and concrete.

In this case, traditionally available mineral additives or waste materials with magnetic properties might provide better alternatives to achieve magneto-rheology control of cementitious materials.

3.4.1   Magnetic properties of fly ash

As the main by-product of coal-fired power plants, fly ash is an aluminosilicate material widely used in cement and concrete with economic, technical and ecological benefits [79–82]. From the perspective of chemical composition, fly ash typically consists of large amounts of quartz, mullite and anhydrite, as well as a small amount of hematite, magnetite and maghemite [83, 84]. Depending on the nature of the coal source, the iron oxides in fly ash typically range from 2% to 20%. The iron oxide content of a fly ash mainly depends on the amount of pyrite (FeS_2) in the burnt coal. During the burning of the pulverized coal at high temperatures, the pyrite is decomposed and spherical iron particles and sulphur gas are formed and further oxidized to magnetite (Fe_3O_4) and sulphur dioxide [85]. The course of coal combustion and the amount of oxygen in the combustion chamber determine the degree of iron oxidation and influence the amount of weakly magnetic iron oxide hematite (Fe_2O_3) and strongly magnetite (Fe_3O_4). In addition to the magnetic crystalline phase magnetite (Fe_3O_4), α-Fe_2O_3 hematite, γ-Fe_2O_3 maghemite, Al-Mg spinels, wüstite and pyrrhotite are available and they are often trapped in aluminosilicate and calcium-aluminosilicate basic glass [86]. Due to the presence of the ferromagnetic spinel structures, i.e. magnetite and maghemite, fly ash particles are highly magnetic. It is well recognized that the magnetic properties of fly ash (such as magnetic susceptibility and saturation magnetization) show a linear correlation with the content of iron oxides [87–90]. Note that the Fe in the spinel structures could be substituted by Al, Mg, Mn or Ti, resulting in a decrease in the magnetic properties of fly ash [91, 92]. The magnetic properties of fly ash have attracted some attention in cement and concrete materials [93–95].

The magnetization curves of four representative fly ashes obtained from VSM measurement are shown in Figure 3.25 [96]. The magnetization curve of a Portland cement is also presented. The corresponding magnetic parameters are summarized in Table 3.3. It can be seen that the saturation magnetization of all the particles is achieved at magnetic field strength around 0.2 T. All fly ashes exhibit higher saturation magnetization than the Portland cement due to the presence of ferromagnetic structures such as magnetite and maghemite. All fly ash particles exhibit slightly higher remnant magnetization and coercivity than the Portland cement particles. This indicates that some residual magnetic clusters or structures formed under magnetic field might be present in cementitious suspensions after removal of the magnetic field.

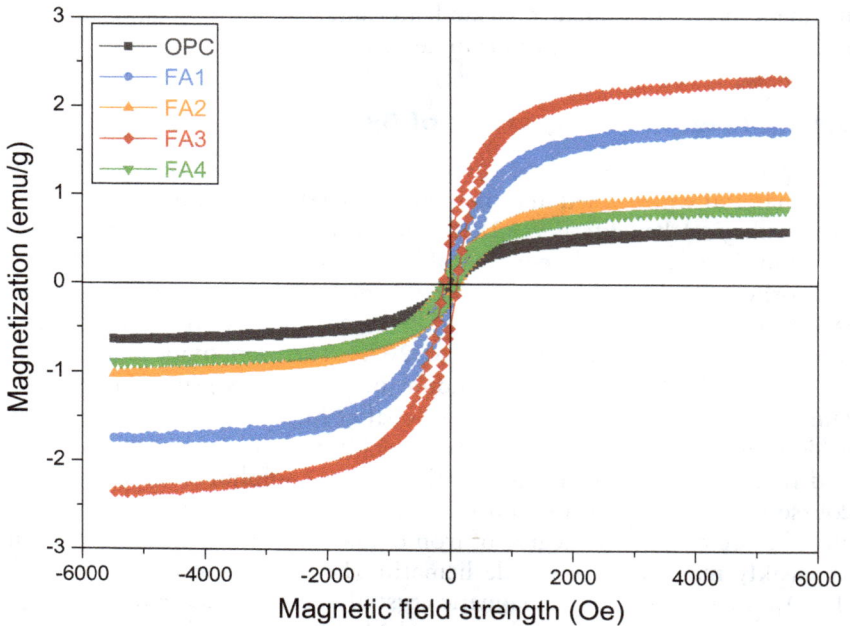

Figure 3.25 Magnetization loop curves of representative fly ashes. Source: after [96].

Table 3.3 Magnetic properties of four different fly ashes

Material	Particle size (μm)	Magnetic fraction (wt.%)	Saturation magnetization (Ms, emu/g)	Remanent magnetization (Mr, emu/g)	Coercive field (Hc, Oe)
OPC	9.46	-	0.59	0.05	70.72
FA1	8.01	11.2	1.74	0.23	101.95
FA2	10.17	4.6	0.99	0.13	109.02
FA3	13.32	33.9	2.31	0.48	111.58
FA4	15.39	5.8	0.85	0.14	104.80

Source: [96].

Owing to the magnetic properties, fly ash can be separated into non-magnetic and magnetic fractions by passing through an external magnetic field [88, 93, 97, 98]. Dry and wet separation methods are generally utilized. The dry separation method is less effective than the wet separation method, because of the strong electrostatic attractions between solid particles [88]. The wet separated magnetic fraction by mass for the above-mentioned four

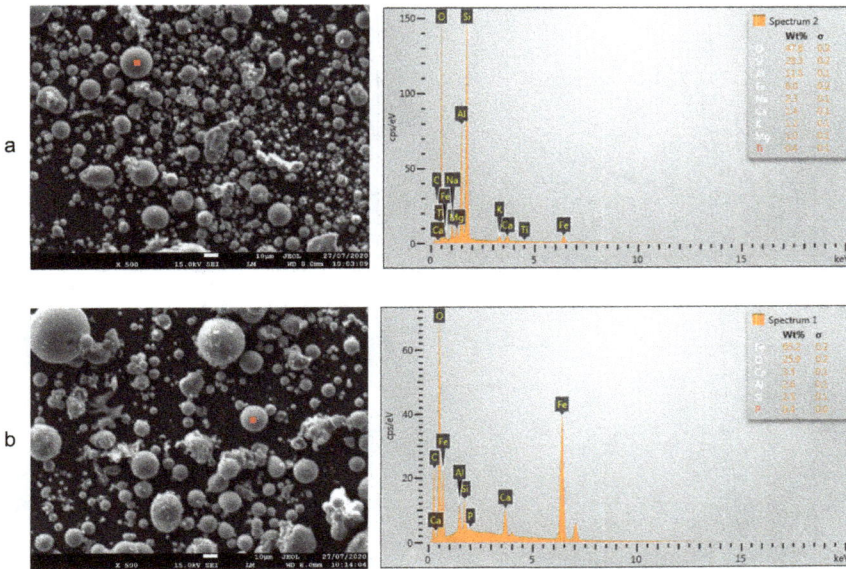

Figure 3.26 Typical SEM images and corresponding EDX spectra of fly ash. (a) non-magnetic part, and (b) magnetic part. Red spots are the area where EDX spot analysis was taken. Source: after [96].

fly ashes is presented in Table 3.3. Generally, fly ash with higher saturation magnetization has higher magnetic fraction. In addition, typical SEM micrographs as well as EDX spot spectra of fly ash are shown in Figure 3.26. We can see that the Fe element amount in the magnetic fraction is larger than that in the non-magnetic fraction [96].

3.4.2   Influence of fly ash volume fraction

The influence of fly ash volume fraction on the magneto-responsive structural build-up of cement pastes under external magnetic field is presented in Figure 3.27. The increase rate of the storage modulus at time scale of 200–300 s (ΔG') is calculated as an indicator to describe the structural build-up rate of cement paste at steady state. In the absence of magnetic field, the volumetric replacement of cement with fly ash reduces the intensity and rate of structural build-up of the cement pastes, especially at higher fly ash volume fraction. The storage modulus at 300 s and the structural build-up rate at steady state are 95 kPa, 83 kPa, 30 kPa and 0.24 kPa/s, 0.21 kPa/s,

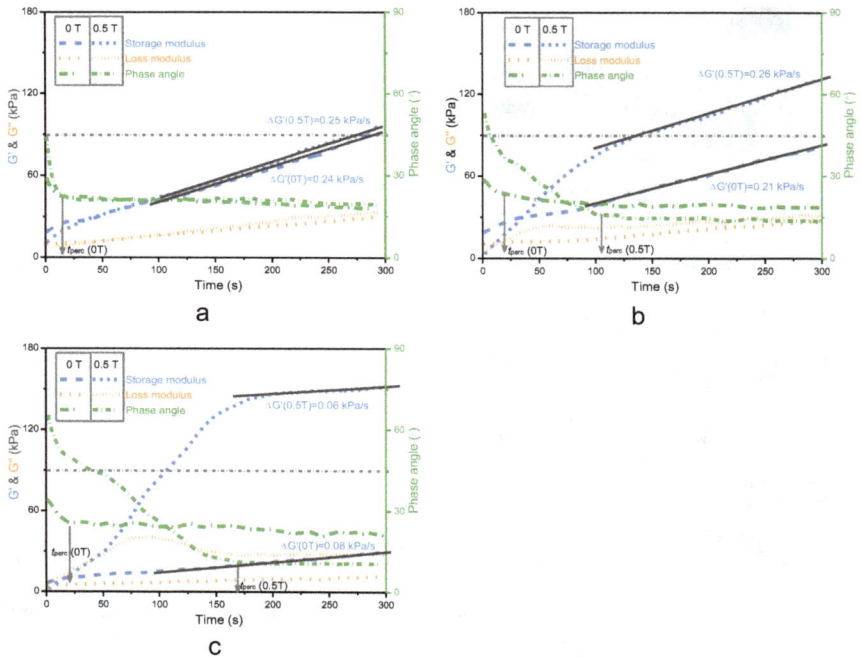

Figure 3.27 Influence of fly ash volume fraction on the structural evolution of cement paste under external magnetic field. (a) 0 vol.% FA; (b) 25 vol.% FA; and (c) 50 vol.% FA. Source: after [96].

0.08 kPa/s for the cement pastes with fly ash volume fraction of 0%, 25% and 50%, respectively.

Under the magnetic field of 0.5 T, the fly ash-incorporated cement pastes exhibit obvious magneto-rheological responses. The rheological responses of the fly ash-incorporated cement pastes to an external magnetic field are generally similar to that of cementitious paste containing nano-Fe_3O_4 particles [36]. The obvious rheological response of cement paste with fly ash to an external magnetic field can be attributed to the magnetic properties of the fly ash particles, where the saturation magnetization of the cement and fly ash is 0.59 emu/g and 1.74 emu/g, respectively. The magneto-rheological effect amplifies with the increase of fly ash volume fraction due to the increased concentration of the magnetic fly ash particles. For example, the difference of the storage modulus at 300 s between 0 T and 0.5 T (D-value) is 49 kPa and 122 kPa for the cement paste with 25% and 50% fly ash, respectively. Besides, the liquid-like properties dominating duration ($t_{G'=G''}$) and the percolation time (t_{perc}) increase from 6 s to 40 s and from 105 s to 170 s with increasing fly ash replacement from 25% to 50%, respectively.

3.4.3   Influence of fly ash type

At the same volumetric replacement of 50%, the influence of fly ash type on the magneto-rheological response of cement paste is presented in Figure 3.28. The magnetic properties of the fly ashes are presented in Table 3.3. The structural evolution of the cement pastes depends on the type of fly ash, regardless of the application of the magnetic field. Without the external magnetic field, 50%FA1 and 50%FA3 have similar structural build-up, while 50%FA2 and 50%FA4 show relatively higher intensity and rate of structural evolution. Indeed, the storage modulus at magnetization of 300 s is around 30 kPa and 58 kPa for the mixtures of 50%FA1 and 50%FA4, respectively. The corresponding increase rate of the storage modulus at steady state is 0.08 kPa/s and 0.15 kPa/s for these two mixtures, respectively. This can be explained by the morphology and chemical activity of the fly ash particles. The particles of FA1, FA2 and FA3 are spherical, while FA4 has irregular shape. The chemical activity of FA4 is higher than that

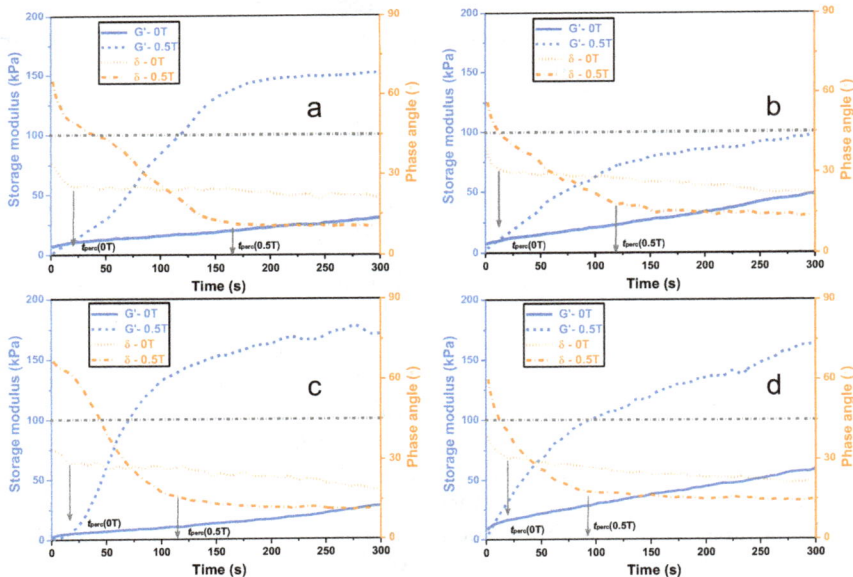

Figure 3.28 Effect of fly ash types on magnetorheological response of cement paste. (a) 50% FA1; (b) 50% FA2; (c) 50% FA3 and (d) 50% FA4. Source: after [96].

of FA1, FA2 and FA3. Consequently, 50%FA4 shows slightly higher structural build-up rate in comparison to 50%FA1 and 50%FA3. The higher structural build-up of the cement paste with FA2 is possibly due to the high specific surface area of the fly ash.

In the presence of an external magnetic field of 0.5 T, 50%FA2 exhibits lower magneto-rheological response than 50%FA1. More specifically, 50%FA2 shows significantly shorter liquid-like properties dominating duration and percolation time than 50%FA1. The D-value of G' between 0.5 T and 0 T is 122 kPa/s and 50 kPa/s for the mixtures of 50%FA1 and 50%FA2, respectively. This is in good agreement with the relatively low saturation magnetization and magnetic fraction of FA2 as compared to FA1. In comparison to 50%FA1, 50%FA3 exhibits more obvious magneto-rheological response, characterized by the faster structural build-up (i.e. higher D-value and ΔG'), which can be attributed to the high magnetic properties of FA3 and thus the high inter-particle connections between magnetic particles. In spite of the higher magneto-rheological response, 50%FA3 has similar liquid-like properties domination duration and slightly shorter percolation time than 50%FA1. This indicates that the solid particles in the suspension of 50%FA3 can easily reach their equilibrium positions after applying the external magnetic field. This is probably because of the coarser particles of FA3 compared to FA1. In the case of 50%FA4, the suspension shows a relatively high structural build-up rate compared with other suspensions, possibly due to the higher chemical hydration. Interestingly, 50%FA4 has quite higher magneto-rheological response than 50%FA2. This can be attributed to the morphological effect of the magnetic particles, which increases the friction and connection between magnetic clusters [99, 100].

3.4.4 Relationship between magneto-rheological effect and magnetic properties of FA

To correlate the magneto-rheological effect of cement pastes with the magnetic properties of fly ash, the D-value of storage modulus at 300 s is used to quantitatively describe the magneto-rheological effect, and the saturation magnetization of unseparated fly ash is selected to characterize the magnetic properties of fly ash. The points of D-value versus saturation magnetization of original fly ash are plotted in Figure 3.29. It can be seen that at the same volumetric replacement, the magneto-rheological effect has a considerable linear proportional relationship with the saturation magnetization of the original fly ash, with the coefficient of determination R^2 higher than 0.97, except for the mixture of 50%FA4. The results infer that the saturation magnetization of original fly ash could be used as an effective parameter to evaluate the magneto-rheological properties of fly ash-incorporated cement pastes. The aberrant point of 50%FA4 is possibly due to the surface texture, irregular shape and high chemical reactivity of the fly ash,

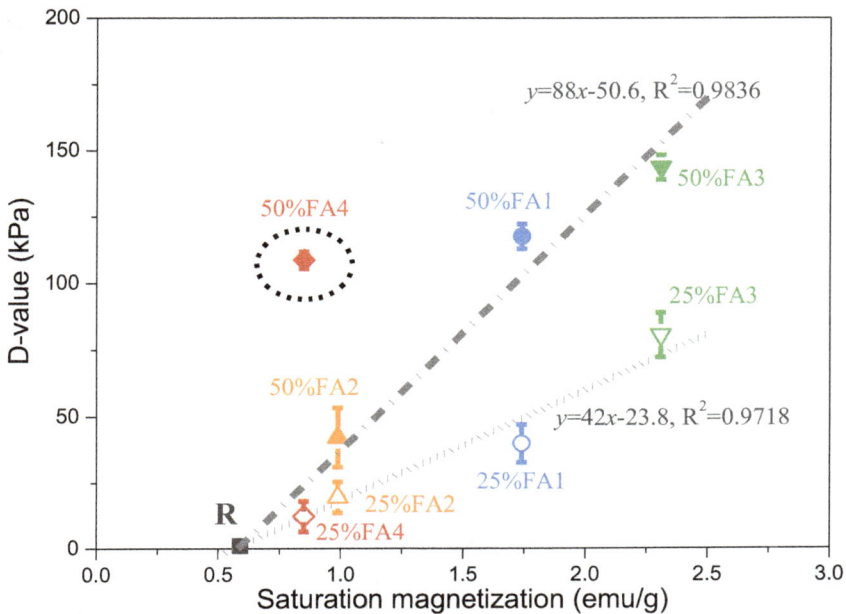

Figure 3.29 Relationship between D-value and saturation magnetization of original fly ash. The point of 50%FA4 is not included in the fitted curve. Error bar indicates the standard deviation. Source: after [96].

increasing the inter-particle frictions between magnetic fly ash clusters. In the case of cementitious paste with nano-Fe_3O_4 particles, the particle size of the nanoparticles plays a significant role in dominating the viscoelastic properties without magnetic field, while the response of the cementitious paste to an external magnetic field is mainly determined by the magnetic properties and crystalline structures of the nanoparticles [40, 63]. For the fly ash-incorporated cement pastes, the median size of the used fly ash only varies from 8 μm to 15 μm. In the presence of a magnetic field, the structural evolution of fly ash-incorporated cement pastes is mainly determined by the magnetic properties of the fly ash, rather than the particle size. Although more data is required, it is reasonable to correlate the magneto-rheological effect with the saturation magnetization of original fly ash.

3.5   OTHER POTENTIAL MATERIALS

As explained in the previous sections, magneto-responsiveness can be introduced to Portland cement-based composites by the incorporation of nano, sub-micron or micron size ferromagnetic particles. The source of ferromagnetic particles with sufficient concentration in a Portland cement-based mixture can be commercially available nano-magnetite particles added to the concrete mixture or some crystalline phases from already available mineral admixtures [101], such as fly ash or different types of slags and even some special type of cements. Except for these two, the other types of additives naturally containing magnetic phases in their microstructure can also introduce the magneto-responsive property to Portland cement systems when they are used as an additive.

3.5.1   Steel fibres, metal scraps and chips

As well known, man-made metallic discrete fibres (steel fibres) are generally used to enhance the mechanical properties of hardened concrete under static and dynamic loading conditions. Fibres may increase the strength of the composite by transferring stress across cracks and increase the toughness of the composite by providing energy absorbing mechanism during debonding and pull-out process of the fibres bridging the cracks [102]. Ideally, the volume distribution of discrete steel fibres should be uniform and the fibres should be randomly oriented in the matrix for ideal isotropic behaviour. But in reality, due to the mixing and consolidation conditions and the geometrical constraints related to the formworks, fibre distribution is not uniform and hence the expected engineering properties of fibre-reinforced cementitious composite could be lower than the expected performance level based on the homogeneous fibre distribution in random orientation. Therefore, there has been an increasing interest for years in the development of a reliable non-destructive method to determine the fibre content and distribution in fibre-reinforced concrete sections based on electromagnetic and dielectric properties of fibres [103–109]. Although the details of the mentioned non-destructive methods for determining fibre distribution and orientation are far beyond the subject of this book, the big potential of the magnetic field to provide better compaction and more favorable fibre orientation in fibre-reinforced composites is worth mentioning here.

As known, the main difficulty in steel fibre-reinforced cementitious composite production with superior mechanical performance is to be able to introduce high aspect ratio (fibre length to diameter ratio) fibres in uniformly dispersed high volume while keeping the workability of the composite sufficient. This is generally not that easy because the presence of high aspect ratio fibres in high volume adversely affects the workability and hence mixing, placing and finishing become more difficult. As well known, after the completion of the mixing period in Portland cement systems, there is no way to increase the workability to reach a certain consolidation except for an internal and external vibration. Although there is an assumption that the fibres are uniformly distributed throughout the matrix, in reality, this assumption is not correct for a fibre-reinforced concrete that has been placed and compacted by vibrators [102], [110]. On the other hand, the previous theoretical and experimental studies showed that if the orientation of steel fibres, which could be characterized by the fibre orientation factor, in a concrete element could be aligned in the direction of the principal tensile stress, the tensile performance of fibre-reinforced element could increase significantly [102, 111]. In some of the real fibre-reinforced concrete casting operations, the preferred fibre alignment is governed by two physical effects. The first is the shear-induced orientation, which is observed when the flow is dominated by shear stress, and hence a parabolic velocity profile is formed. In this case, the torque acting on the fibre due to the different fluid velocities along with the fibre during flow makes the fibre aligned naturally parallel to the flow direction such that the torque becomes minimum on the aligned fibre. However, if there is a plug region in the flow, where the shear stress is lower than the yield stress of the material, the fibres do not align due to the constant velocity profile in the plug region. The second is a well-known phenomenon called wall effect: there are no fibres at all perpendicular to a wall at a distance lower than half of the fibre [112]. The extent of the preferred alignment of fibres depends on many factors, mainly including the rheology of the concrete and geometric constraints related to fibres and the mould dimensions, and currently cannot be achieved in a controlled manner during casting. Therefore, magnetically aligned steel fibre composite production, based on providing the alignment of metallic fibres after casting, has been proposed and first patented in the late 1980s by Bayer AG [113] and this idea was followed by Emami and Nielsen in 1997 for concrete [114]. Recently some new approaches for better compaction through the micro-vibration of steel fibres and preferable orientation in fibre-reinforced concrete have been introduced based on the ferromagnetic nature of common steel fibres used in modern concrete technology. Michels and Gams [115] introduced the orientation of randomly distributed steel fibres in cement mortar when the steel fibre-reinforced cement mortar samples in the moulds are placed in a helical coil and exposed to electromagnetic fields. The authors have proven the orientation of commercially available hooked-end steel fibres (30 mm length, 0.6 mm diameter and 50 aspect ratio) along the magnetic field by X-ray photos. They also reported that the flexural performance of aligned fibre-reinforced cementitious mortars was better than randomly distributed fibre-reinforced mortars. This study was followed by several other studies documenting the effectiveness of the magnetic field-based fibre alignment method in aligned steel fibre cement mortar production [8], [111], [116], [117]. As shown by Villa and Medina [117], hooked-end steel fibres in the fresh concrete mixture can be aligned if the generated magnetic torque on steel fibres is higher than the yield torque of the surrounding cementitious matrix. In addition, an example of applying alternating magnetic to improve the compaction of freshly cast concrete incorporating steel chips has been shown in a study by Abavisani et al. [7]. The authors showed that the alternating magnetic field application to freshly cast concrete incorporating steel chips could provide better compaction of steel-chip incorporated concretes and could provide better mechanical properties due to the internal micro-vibration effect of steel chips under alternating magnetic field; however, the mechanical performance of the hardened composite is dependent on the exposure direction.

3.5.2 Specific types of slags

The magnetism in copper slag (CS) has first been reported by Edmund Mondy in a letter to the editor in one of the issues of *Nature* published in 1872 [118]. He mentioned that a specimen of ore furnace slag from copper smelting was strongly polar magnetic. Today it is known that different types of slags, which were obtained as a waste of different metal industries, present different magnetic properties. Figure 3.30 shows an example of the magnetization curves (obtained by the vibrating sample magnetometer – VSM at room temperature in an applied field up to 10 kOe) of ground granulated blast furnace slag (GGBFS), copper slag (CS) and stainless steel slag (SSS) samples together with a Portland cement sample. As shown in Figure 3.30, different types of slags present different magnetization properties. The magnetization curves of all presented slags in Figure 3.30 display a hysteresis loop due to the domains that cannot return to their original orientation when the external magnetic field is decreased after the saturation

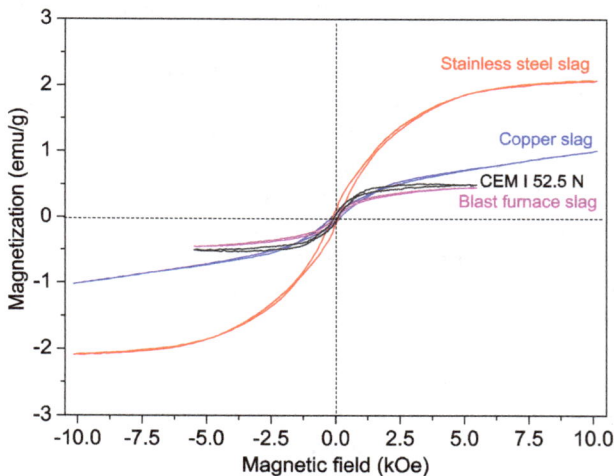

Figure 3.30 Magnetization curves of different slags together with CEM I 52.5 N cement.

value is reached. In addition, remanence and coercivity values, slopes of the magnetization loops and saturation magnetization values of different slags are quite different due to the differences in their atomic structures, causing a difference in their susceptibility and permeability. This type of magnetic hysteresis curve is typical for ferromagnetic materials, which tend to magnetize in the direction of the magnetic field and some of them are retained after the removal of the external magnetic field [12]. The type of ferrites in a material also plays an important role in the magnetization behaviour. Hematite is weakly ferromagnetic at room temperature and its magnetic behaviour is dependent on its crystallinity, particle size and on the extent of cation substitution. Maghemite and magnetite are ferromagnetic and ferrimagnetic at room temperature [119]. Such differences are probably the case for CEM I 52.5 N cement and GGBFS which present very close magnetization despite their very different Fe_2O_3 contents (Table 3.4). A more dramatic example of the importance of the type of ferrites in the material is the copper slag case. As can be seen from Figure 3.30, despite its very high Fe_2O_3 content (45.4% in Table 3.4), the magnetization of copper slag is not as high as the other materials with much less Fe_2O_3 contents (Table 3.4). In addition, it never reaches to a saturation magnetization value and it presents a somewhat combination of ferromagnetic and paramagnetic material responses with a gradually increasing magnetization and presenting a remanent magnetization

Table 3.4 Chemical compositions of different slags and a Portland cement sample in Figure 3.30

Element	Blast furnace slag	Copper Slag	Stainless steel slag	CEM I 52.5 N
SiO_2	31.1	29.4	28.0	18.6
CaO	40.9	2.60	44.2	64.1
MgO	9.16	1.05	4.41	1.20
Al_2O_3	13.7	9.82	6.30	5.2
Fe_2O_3	0.401	45.4	2.04	4.9
SO_3	2.31	0.845	0.491	3.5
Na_2O	-	3.66	-	0.64 (eq.)
K_2O	0.685	0.188	0.044	
P_2O_5	-	0.504	-	N.A.
TiO_2	1.26	0.294	1.64	
Cr_2O_3	-	1.36	9.90	
MnO	0.310	0.653	2.65	
ZrO_2	0.119	0.237	0.185	
Nb_2O_5	-	-	0.069	
NiO	-	0.0178	0.054	
SrO	0.0741	0.0456	0.033	
CuO	0.0054	0.547	0.014	
ZnO	0.0030	3.32	0.011	

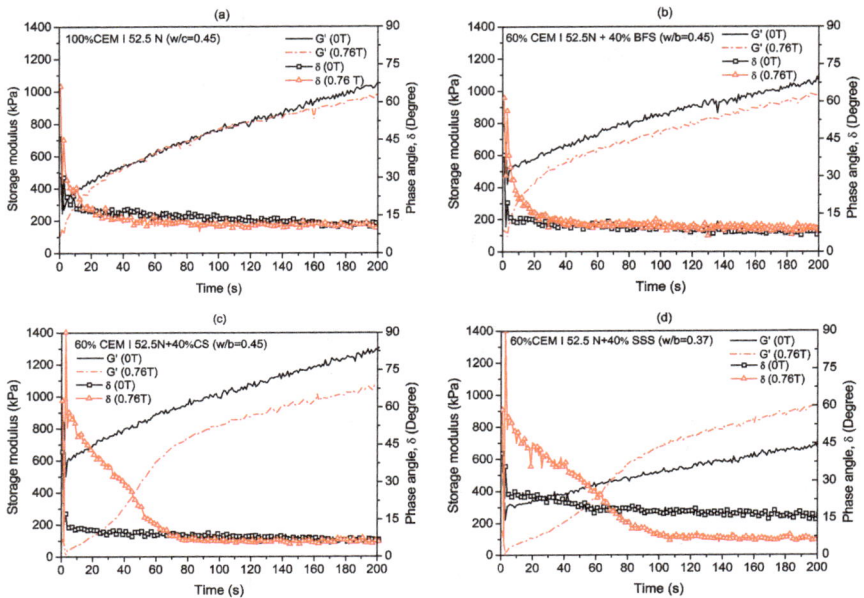

Figure 3.31 Storage modulus and phase angle evolution of Portland cement paste (a), blast furnace slag (b), copper slag (c) and stainless steel slag (d) incorporated pastes without (0 T) and with magnetic field of 0.76 T.

when the external magnetic field is decreased to zero. Its magnetization is higher than GGBFS but much smaller than SSS. SSS presents significantly higher saturation magnetization as compared to CS and GGBFS.

The effect of applying a constant magnetic field on the viscoelastic properties of Portland cement paste and blast furnace slag (BFS), SSS and CS incorporated Portland cement pastes is shown in Figure 3.31 (a, b, c and d), respectively. The viscoelastic properties presented here as the storage modulus (G') and loss modulus (G") are the parameters measured by the small-amplitude oscillation test (SAOS) performed in an oscillation rheometer with parallel plate geometry under a constant magnetic field applied perpendicular to the plate plane during the measurement.

As shown in Figure 3.31 a and b, applying a constant magnetic field of 0.76T does not cause any significant change on the storage modulus (structural build-up) and phase angle evolutions of the paste mixtures with 100% CEM I 52.5 N cement and the blend of 60% CEM I 52.5 N and 40% BFS (by weight of cement). This can be attributed to the low magnetization properties of Portland cement and BFS, as observed in their magnetization curves given in Figure 3.30. On the other hand, despite their low magnetization capacity, it is obvious that these materials contain some magnetic phases responding to magnetic field applications. One could notice from Figure 3.31 a and b that the storage modulus under a magnetic field shows a very sudden decrease in the first few couples of seconds after the magnetic field application and the storage modulus shows a gradual increase for the first 20 seconds. This is probably due to the very initial micro-agitation effect of the magnetizable particles, which are first trying to overcome the drag forces of the surrounding pore solution to be aligned with the magnetic field lines during the first couple of seconds of the magnetic field application and the further accumulation of more magnetic particles contributing to the columnar formation of magnetic particles. On the other hand, it seems that the amount of magnetic particles, or the level of their magnetic properties, is not enough for the formation of a strong columnar arrangement of magnetic particles perpendicular to the oscillation strain plane as they do not show a higher storage modulus compared to the no magnetic field case for these materials. On the other hand, the higher phase angle recorded at the beginning of the magnetic field application could also support the hypothesis that the magnetic field applied to a cement paste containing magnetizable particles could cause a more liquid-like behaviour at the beginning of the magnetic field application. This phenomenon seems to be more obvious in CS and SSS cases, as discussed below.

Before discussing the magneto-rheological behaviour of CS- and SSS- incorporated cement pastes, several important points that were observed in the magnetization curves of CS and GGBFS (Figure 3.30) should be summarized. As shwon in the figure, first, the SSS has significantly higher magnetization than the CS under the same magnetic field levels. Second, while the SSS reaches its saturation magnetization in relatively higher magnetic field levels compared to fly ash or Portland cement, the CS presents gradually increasing magnetization without saturation up to the 10 kOe magnetization level. This shows that the domains of CS are not fully aligned to the magnetic field lines, even at very high magnetic field levels. The reflections of the mentioned magnetic properties of CS and SSS materials are somewhat observed in the magneto-rheological responses of CS- and SSS-incorporated cementitious pastes. As shown in Figure 3.31c, when there is no magnetic field, the storage modulus of 40% CS-incorporated paste started from around 600 kPa, indicating less flowability compared to the mixture without CS presenting a lower initial structuration as shown in Figure 3.31a, and showed a continuous increase for the entire test duration of 200 s. Applying a constant magnetic field of 0.76T caused a sudden decrease in its storage modulus and a significant increase in the phase angle, indicating more liquid-like behaviour of 40% CS-incorporated mixture under a magnetic field. It should be noted here that a higher phase angle shows more liquid-like behaviour of a given mixture. As the magnetic field was kept on the material, the storage modulus showed a higher increasing rate up to around 70 s compared to the increase rate in storage modulus without a magnetic field. During this time, the phase angle continuously decreased until reaching its level for no magnetic field case around 70 s. The level of storage modulus was lower than that of no magnetic field case, but the increase rate of storage modulus was similar without and with a magnetic field. Similar behaviour has also been observed in SSS case. Similar to the CS case, applying a constant magnetic field to SSS-incorporated mixture provided more deflocculated microstructure at the beginning of the magnetic field application and it was followed by a rapid increase in structuration until around 80 s. During this time, as similar to the CS-incorporated mixture case, the phase angle was significantly higher than no magnetic field case, but has a decreasing trend, indicating a relative increase in solid-like behaviour. After around 80 s, the increased rate of storage modulus was reduced to its increased level without a magnetic field. During this regime, opposite the CS-incorporated mixture, the extent of the storage modulus was higher than its no magnetic field case, indicating the formation of more columnar magnetic particle formations perpendicular to the oscillation plane.

The given example in Figure 3.31 shows the potential usage of different types of slags in the development of magneto-rheological cementitious materials. When considering the magnetization curves of different types of slags (Figure 3.30) and the magneto-rheological response of those slag-incorporated cement pastes (Figure 3.31) together, it could be concluded that the magnetization properties of the slags define the magneto-rheological response of cementitious mixtures containing different types of slags. It should be mentioned here that the utilized amount of CS and SSS in Figure 3.31c and d do not suggest the typical usage amount in cement and concrete technology.

3.6   SHORT SUMMARY OF STATE OF THE ART AND OUTLOOK

Active rheology control by means of magnetic additives and magnetic fields is an essential way to advance construction technology and make pumping and casting processes more reliable. For this end, the fundamental understanding of the rheological responses of cementitious paste containing magnetic particles has been given. The main conclusions can be summarized as follows:

Applying a magnetic field promotes magnetic particles to move to agglomerate and form clusters. The corresponding cementitious paste shows improved liquid-like behaviour at very early age and increased stiffness after longer period of magnetization. The intensity of magneto-rheological responses is related to the viscoelastic and rheological properties of the suspension. At relatively low w/c, nano-Fe_3O_4 incorporated cement pastes with low PCE additions show negligible magneto-rheological responses due to the high viscoelastic yield stress and viscosity of the suspension. For the paste with high PCE dosage, however, slight magneto-rheological responses can be observed after longer period of magnetization. Increasing nano-Fe_3O_4 particles concentration increases the intensity of liquid-like properties immediately after initiation of the magnetic field and enhances the stiffness of cementitious paste after undergoing longer magnetization due to a large number of available nanoparticles contributing to the formation of magnetic clusters. Particle size of nano-Fe_3O_4 plays a considerable role in dominating the viscoelastic properties of cementitious paste without magnetic field. The responses of the cementitious paste to an external magnetic field, however, are mainly determined by the crystalline structures and magnetic properties of the nano-Fe_3O_4 particles.

The rheological response of cementitious paste is dependent on the applied magnetic field. Weak magnetic fields have little effects on the structural evolution, while at sufficiently high magnetic fields, the magneto-rheological effect exhibits a linear correlation with the magnetic field strength. When suddenly applying an external magnetic field to a resting cementitious paste, an abrupt decrease in storage modulus indicating the breakdown of C-S-H bridges and flocculated networks by the movement of magnetic nanoparticles is observed. With the magnetic field changing from low (not zero) to higher value, the nanoparticles in the existing clusters will redistribute and the nanoparticles remaining in the gap between cement particles start to move to form stronger and larger clusters, resulting in a sharp decrease in the storage modulus. Step-decreasing magnetic field results in the disintegration of cluster structures, so that a reduction in storage modulus is observed. After removing the magnetic field, the cementitious paste shows a higher structural build-up rate than in the situation without experiencing any magnetic field. The structural build-up of the cementitious paste under linear magnetic field sweep mode is history-dependent.

Conceptual equations about magneto-dynamic force and movement velocity of nanoparticles in cement-based suspensions are derived. The estimated magnetic yield parameter, describing the competition between the magnetic force and the viscoelastic stress of the suspension, and the predicted movement velocity of nanoparticles can be used as relevant indicators to characterize the magneto-rheological responses of cementitious paste containing magnetic nanoparticles. Specifically, the estimated magnetic yield parameter shows a linear relationship with the magneto-rheological effect of the cementitious paste. The conceptual movement velocity of nanoparticles is proportional to the increase rate of loss modulus at very early age.

In the case of the dynamic rheological properties of nano-Fe_3O_4 incorporated cementitious paste under external magnetic field, bleeding and nanoparticles agglomeration on the interface between cementitious paste and upper rotating plate can be observed after shear curve test. This is due to the magnetic cluster migrations induced by magnetic force gradient and shear rate gradient. Under constant shearing with low shear rate, limestone powder suspension with nanoparticles exhibits obvious rheological responses to an external magnetic field. A higher magnetic field strength corresponds to a faster and more pronounced response. Upon applying a step-increased magnetic field, the suspension shows an immediate enhancement in stiffness. Step-decrease magnetic field results in a drop decrease in shear viscosity, followed by a gradual reduction behaviour. Linearly increasing magnetic field from 0 T to approx. 0.3 T shows less influences on the shear viscosity, and at the magnetic field higher than 0.3 T, the shear viscosity gradually increases and the increase rate is similar to that obtained under constant magnetic field of 0.75 T. When the magnetic field suddenly changes from 0.75 T to 0 T, the shear viscosity exhibits a sharp reduction. If the magnetic field linearly decreases from 0.75 T to 0 T, however, the shear viscosity continuously increases at a decreasing increase rate within the first few tens of seconds, and after reaching a peak, the shear viscosity gradually decreases.

The pioneering tests on the magneto-rheological properties of fly ash-incorporated cement pastes are promising. For fly ashes with higher saturation magnetization than Portland cement particles, the prepared fly ash-cement pastes exhibit obvious magneto-rheological responses, and the magneto-rheological effect increases with the volume fraction of fly ash. The magneto-rheological response depends on the characteristics of fly ash. Under the same volumetric replacement, the magneto-rheological effect generally has a considerable linear proportional relationship with the saturation magnetization of the original fly ash. Similar to fly ash, some types of metallurgical slags, such as copper slag and stainless steel slag, seem to be promising materials for active rheology control in cementitious materials.

Despite the achievements in the fundamental study of active rheology control, further studies should be considered in future research. For example, the particle size of solid grains in cement-based suspensions with nanoparticles varies from few nanometres to hundreds of micrometres. Under an external magnetic field, an additional magnetic force is applied to the magnetic particles, and the complex solid systems make it more difficult to accurately calculate the numerical values of each inter-particle force. More theoretical work is needed to fully understand the magnetic force between magnetic particles in cement-based suspensions concerning magnetic clustering, mutual effect between particles and changing of the surrounding solutions. The temperature of cement-based materials is always changing in the process of pumping and formwork casting. Furthermore, applying an external magnetic field will create additional electromagnetic thermal heat to the cementitious materials. Therefore, further research about the influence of temperature on the magneto-rheological behaviour of cementitious paste is necessary. Last but not least, more available commercial mineral additives with magnetic properties should be investigated more comprehensively from the viewpoints of mortar and concrete, and advanced setups need to be developed to simulate the flowing behaviour of cementitious pastes in a pipe or formwork under an external magnetic field.

REFERENCES

[1] Lesage, K. and G. De Schutter, *Admixture for a cementitious material to influence the rheology properties of the cementitious material.* patent application, EP3873869A1, US20210317039A1, 2019.

[2] De Schutter, G., et al., Vision of 3D printing with concrete - Technical, economic and environmental potentials. *Cement and Concrete Research*, 2018. 112: p. 25–36.

[3] Khorshidi, N., M. Ansari, and M. Bayat, An investigation of water magnetization and its influence on some concrete specificities like fluidity and compressive strength. *Computers and Concrete*, 2014. 13(5): p. 649–657.

[4] Su, N. and C.-F. Wu, Effect of magnetic field treated water on mortar and concrete containing fly ash. *Cement and Concrete Composites*, 2003. 25(7): p. 681–688.

[5] Wei, H., Y. Wang, and J. Luo, Influence of magnetic water on early-age shrinkage cracking of concrete. *Construction and Building Materials*, 2017. 147: p. 91–100.

[6] De Schutter, G. and K. Lesage, Active control of properties of concrete: a (p)review. *Materials and Structures*, 2018. 51(5): p. 123.

[7] Abavisani, I., O. Rezaifar, and A. Kheyroddin, Alternating magnetic field effect on fine-aggregate steel chip–Reinforced concrete properties. *Journal of Materials in Civil Engineering*, 2018. 30(6): p. 04018087.

[8] Mu, R., et al., Aligning steel fibres in cement mortar using electro-magnetic field. *Construction and Building Materials*, 2017. 131: p. 309–316.

[9] Hajforoush, M., et al., The effects of uniform magnetic field on the mechanical and microstructural properties of concrete incorporating steel fibres. *Scientia Iranica*, 2021. 28(5): p. 2557–2567.

[10] Choi, M.S., et al., Effects of an externally imposed electromagnetic field on the formation of a lubrication layer in concrete pumping. *Construction and Building Materials*, 2014. 61: p. 18–23.

[11] Soto-Bernal, J.J., et al., Effects of static magnetic fields on the physical, mechanical, and microstructural properties of cement pastes. *Advances in Materials Science and Engineering*, 2015. 2015: p. 1–9.

[12] Ida, N., *Engineering electromagnetics.* third ed. 2015. Springer International Publishing-Switzerland.

[13] Getzlaff, M., *Fundamentals of magnetism.* 2008. Springer-Verlag Berlin Heidelberg.

[14] Carmichael, R.S., *Practical handbook of physical properties of rocks and minerals.* 1989. CRC Press Taylor&Francis Group.

[15] Schwerdt, J.I., et al., Magnetic field-assisted gene delivery: Achievements and therapeutic potential. *Current Gene Therapy*, 2012. 12(2): p. 116–126.

[16] Liu, T. and Y. Xu, Magnetorheological elastomers: materials and applications. In *Smart and Functional Soft Materials*. 2019, IntechOpen London, UK. p. 147–180.

[17] Sista, K.S., et al., Carbonyl iron powders as absorption material for microwave interference shielding: A review. *Journal of Alloys and Compounds*, 2021. 853(2021): p. 157251.

[18] Nair, S.D. and R.D. Ferron, Set-on-demand concrete. *Cement and Concrete Research*, 2014. 57: p. 13–27.

[19] Rashad, A.M., A synopsis about the effect of nano-Al2O3, nano-Fe2O3, nano-Fe3O4 and nano-clay on some properties of cementitious materials – A short guide for Civil Engineer. *Materials & Design (1980–2015)*, 2013. 52: p. 143–157.

[20] Li, H., et al., Microstructure of cement mortar with nano-particles. *Composites Part B: Engineering*, 2004. 35(2): p. 185–189.

[21] Amin, M.S., S.M.A. El-Gamal, and F.S. Hashem, Effect of addition of nano-magnetite on the hydration characteristics of hardened Portland cement and high slag cement pastes. *Journal of Thermal Analysis and Calorimetry*, 2012. 112(3): p. 1253–1259.

[22] Shekari, A.H. and M.S. Razzaghi, Influence of nano particles on durability and mechanical properties of high performance concrete. *Procedia Engineering*, 2011. 14: p. 3036–3041.

[23] El-Diasty, F., et al., Complex susceptibility analysis of magneto-fluids: Optical band gap and surface studies on the nanomagnetite-based particles. *Current Opinion in Solid State and Materials Science*, 2009. 13(1–2): p. 28–34.

[24] Bragança, M.O.G.P., et al., Performance of Portland cement concretes with 1% nano-Fe3O4 addition: Electrochemical stability under chloride and sulfate environments. *Construction and Building Materials*, 2016. 117: p. 152–162.

[25] Mansouri, I., et al., Effect of magnetite nanoparticles (ferroferric oxide) on discrete concrete properties. *Proceedings of the Institution of Civil Engineers-Construction Materials*, 2018: p. 1–8.

[26] Rabinow, J., The magnetic fluid clutch. *Electrical Engineering*, 1948. 67(12): p. 1167–1167.

[27] Park, J.H., B.D. Chin, and O.O. Park, Rheological properties and stabilization of magnetorheological fluids in a water-in-oil emulsion. *Journal of Colloid and Interface Science*, 2001. 240(1): p. 349–354.

[28] Wereley, N.M., et al., Bidisperse magnetorheological fluids using fe particles at nanometer and micron scale. *Journal of Intelligent Material Systems and Structures*, 2006. 17(5): p. 393–401.

[29] Vinod, S., R. John, and J. Philip, Magnetorheological properties of sodium sulphonate capped electrolytic iron based MR fluid: A comparison with CI based MR fluid. *Smart Materials and Structures*, 2017. 26: p. 025003.

[30] Genç, S. and P.P. Phulé, Rheological properties of magnetorheological fluids. *Smart Materials and Structures*, 2002. 11(1): p. 140–146.

[31] Bica, I., Y.D. Liu, and H.J. Choi, Physical characteristics of magnetorheological suspensions and their applications. *Journal of Industrial and Engineering Chemistry*, 2013. 19(2): p. 394–406.

[32] Ahamed, R., S.-B. Choi, and M.M. Ferdaus, A state of art on magneto-rheological materials and their potential applications. *Journal of Intelligent Material Systems and Structures*, 2018. 29(10): p. 2051–2095.

[33] Tattersall, G.H. and P.F. Banfill, *The rheology of fresh concrete*. 1983.

[34] Schultz, M.A. and L.J. Struble, Use of oscillatory shear to study flow behaviour of fresh cement paste. *Cement and Concrete Research*, 1993. 23(2): p. 273–282.

[35] Jiao, D., et al., Effect of constituents on rheological properties of fresh concrete-A review. *Cement and Concrete Composites*, 2017. 83: p. 146–159.

[36] Jiao, D., et al., Rheological behaviour of cement paste with nano-Fe3O4 under magnetic field: Magneto-rheological responses and conceptual calculations. *Cement and Concrete Composites*, 2021. 120 : p. 104035.

[37] Rich, J.P., P.S. Doyle, and G.H. McKinley, Magnetorheology in an aging, yield stress matrix fluid. *Rheologica Acta*, 2012. 51(7): p. 579–593.

[38] Rankin, P.J., A.T. Horvath, and D.J. Klingenberg, Magnetorheology in viscoplastic media. *Rheologica Acta*, 1999. 38(5): p. 471–477.

[39] He, Y.B., J.S. Laskowski, and B. Klein, Particle movement in non-Newtonian slurries: the effect of yield stress on dense medium separation. *Chemical Engineering Science*, 2001. 56 : p. 2991–2998.

[40] Jiao, D., *Active rheology control of cementitious materials using magnetic field*. 2021, Ghent University.

[41] Ngatu, G.T., et al., Dimorphic magnetorheological fluids: exploiting partial substitution of microspheres by nanowires. *Smart Materials and Structures*, 2008. 17(4): p. 045022.

[42] Chung, K., U. Jeong, and J.-E. Oh, Effects of magnetic field input cycle and peptizer on the MR effect of magneto-rheological elastomer based on natural rubber. *Polymer Engineering & Science*, 2015. 55(11): p. 2669–2675.

[43] Jiao, D., et al., Quantitative assessment of the influence of external magnetic field on clustering of nano-Fe3O4 particles in cementitious paste. *Cement and Concrete Research*, 2021. 142 : p. 106345.

[44] Bayerl, T., R. Schledjewski, and P. Mitschang, Induction heating of thermoplastic materials by particulate heating promoters. *Polymers & Polymer Composites*, 2012. 20(4): p. 333–341.

[45] Naud, C., et al., Cancer treatment by magneto-mechanical effect of particles, a review. *Nanoscale Advances*, 2020. 2(9): p. 3632–3655.

[46] Kulkarni, S.A., et al., Vibrational response of clusters of Fe3O4 nanoparticles patterned on glass surfaces investigated with magnetic sample modulation AFM. *Nanoscale*, 2018. 10(43): p. 20426–20434.

[47] Yamaguchi-Sekino, S., M. Sekino, and S. Ueno, Biological effects of electromagnetic fields and recently updated safety guidelines for strong static magnetic fields. *Magnetic Resonance in Medical Sciences*, 2011. 10(1): p. 1–10.

[48] Coussot, P., *Rheometry of pastes, suspensions, and granular materials: Applications in industry and environment*. 2005: John Wiley & Sons, Inc.

[49] Connord, V., et al., Real-time analysis of magnetic hyperthermia experiments on living cells under a confocal microscope. *Small*, 2015. 11(20): p. 2437–2445.

[50]   Bueno, M., M. Arraigada, and M.N. Partl, Damage detection and artificial healing of asphalt concrete after trafficking with a load simulator. *Mechanics of Time-Dependent Materials*, 2016. 20(3): p. 265–279.

[51]   Jeoffroy, E., et al., Iron-based particles for the magnetically-triggered crack healing of bituminous materials. *Construction and Building Materials*, 2018. 164: p. 775–782.

[52]   Tarbozagh, A.S., O. Rezaifar, and M. Gholhaki, Electromagnetism in taking concrete behaviour on demand. *Structures*, 2020. 27 : p. 1057–1065.

[53]   Jorge, G.A., M. Llera, and V. Bekeris, Magnetic particles guided by ellipsoidal AC magnetic fields in a shallow viscous fluid: Controlling trajectories and chain lengths. *Journal of Magnetism and Magnetic Materials*, 2017. 444: p. 467–471.

[54]   Melle, S., G.G. Fuller, and M.A. Rubio, Structure and dynamics of magnetorheological fluids in rotating magnetic fields *Physical Review E*, 2000. 61(4): 4111.

[55]   Hill, R.J., J.R. Craig, and G. Gibbs, Systematics of the spinel structure type. *Physics and Chemistry of Minerals*, 1979. 4(4): p. 317–339.

[56]   Schwertmann, U. and R.M. Cornell, *Iron oxides in the laboratory: preparation and characterization*. 2008: John Wiley & Sons.

[57]   Blaney, L., *Magnetite (Fe3O4): Properties synthesis and applications*. 2007.

[58]   Chen, Y., et al., Micro-sized and nano-sized Fe3O4 particles as anode materials for lithium-ion batteries. *Journal of Materials Science & Technology*, 2011. 27(1): p. 41–45.

[59]   Akbarzadeh, A., M. Samiei, and S. Davaran, Magnetic nanoparticles: preparation, physical properties, and applications in biomedicine. *Nanoscale Res Lett*, 2012. 7(1): p. 144.

[60]   Li, Q., et al., Correlation between particle size/domain structure and magnetic properties of highly crystalline Fe3O4 nanoparticles. *Scientific Reports*, 2017. 7(1): p. 9894.

[61]   Goya, G.F., et al., Static and dynamic magnetic properties of spherical magnetite nanoparticles. *Journal of Applied Physics*, 2003. 94(5): p. 3520–3528.

[62]   Sung Lee, J., et al., Magnetic multi-granule nanoclusters: A model system that exhibits universal size effect of magnetic coercivity. *Scientific Reports*, 2015. 5: p. 12135.

[63]   Jiao, D., et al., Structural evolution of cement paste with nano-Fe3O4 under magnetic field - Effect of concentration and particle size of nano-Fe3O4. *Cement and Concrete Composites*, 2021. 120: p. 104036.

[64]   Jiao, D., et al., Thixotropic structural build-up of cement-based materials: A state-of-the-art review. *Cement and Concrete Composites*, 2021. 122 : p. 104152.

[65]   Roussel, N., et al., The origins of thixotropy of fresh cement pastes. *Cement and Concrete Research*, 2012. 42(1): p. 148–157.

[66]   Yuan, Q., et al., Small amplitude oscillatory shear technique to evaluate structural build-up of cement paste. *Materials and Structures*, 2017. 50(2): p. 112.

[67]   Jiang, S.P., J.C. Mutin, and A. Nonat, Studies on mechanism and physicochemical parameters at the origin of the cement setting. I. The fundamental

processes involved during the cement setting. *Cement and Concrete Research*, 1995. 25(4): p. 779–789.

[68]   Mostafa, A.M. and A. Yahia, New approach to assess build-up of cement-based suspensions. *Cement and Concrete Research*, 2016. 85: p. 174–182.

[69]   Nair, S.D. and R.D. Ferron, Real time control of fresh cement paste stiffening: Smart cement-based materials via a magnetorheological approach. *Rheologica Acta*, 2016. 55(7): p. 571–579.

[70]   Jiao, D., et al., Effect of solid volume fraction on the magneto-rheological response of nano-Fe3O4 incorporated cementitious paste. *Journal of Materials in Civil Engineering*, 2021. Under review.

[71]   Sha, S., et al., Influence of the structures of polycarboxylate superplasticizer on its performance in cement-based materials-A review. *Construction and Building Materials*, 2020. 233: p. 117257.

[72]   Yazdi, N.A., et al., To study the effect of adding Fe2O3 nanoparticles on the morphology properties and microstructure of cement mortar. *Life Science Journal-Acta Zhengzhou University Overseas Edition*, 2011. 8(4): p. 550–554.

[73]   Sikora, P., et al., The influence of nano-fe3o4 on the microstructure and mechanical properties of cementitious composites. *Nanoscale Research Letters*, 2016. 11(1): p. 182.

[74]   Hu, J., *A study of effects of aggregate on concrete rheology*. 2005. Iowa State University.

[75]   Jiao, D., et al., Structural build-up of cementitious paste under external magnetic fields. In *Rheology and processing of construction materials*. 2020. Springer. p. 36–42.

[76]   Jiao, D., et al., Structural build-up of cementitious paste with nano-Fe3O4 under time-varying magnetic fields. *Cement and Concrete Research*, 2019. 124 : p. 105857.

[77]   Jiao, D., et al., Rheological properties of cement paste with nano-fe3o4 under magnetic field: flow curve and nanoparticle agglomeration. *Materials*, 2020. 13(22): p. 5164.

[78]   Jiao, D., et al., Flow behaviour of cementitious-like suspension with nano-Fe3O4 particles under external magnetic field. *Materials and Structures*, 2021. 54 : p. 209.

[79]   Bentz, D.P. and C.F. Ferraris, Rheology and setting of high volume fly ash mixtures. *Cement and Concrete Composites*, 2010. 32(4): p. 265–270.

[80]   Jalal, M., M. Fathi, and M. Farzad, Effects of fly ash and TiO2 nanoparticles on rheological, mechanical, microstructural and thermal properties of high strength self compacting concrete. *Mechanics of Materials*, 2013. 61 : p. 11–27.

[81]   Jiao, D., et al., Mixture design of concrete using simplex centroid design method. *Cement and Concrete Composites*, 2018. 89: p. 76–88.

[82]   Giergiczny, Z., Fly ash and slag. *Cement and Concrete Research*, 2019. 124.

[83]   Hansen, L.D. and G.L. Fisher, Elemental distribution in coal fly ash particles. *Environmental Science & Technology*, 1980. 14(9): p. 1111–1117.

[84]   Mccarthy, G.J., et al., Use of a database of chemical, mineralogical and physical properties of North American fly ash to study the nature of fly

ash and its utilization as a mineral admixture in concrete. *MRS Online Proceedings Library Archive*, 1989. 178: p. 3–33.

[85]   Flanders, P.J., Collection, measurement, and analysis of airborne magnetic particulates from pollution in the environment. *Journal of Applied Physics*, 1994. 75(10): p. 5931–5936.

[86]   Strzalkowska, E., Morphology, chemical and mineralogical composition of magnetic fraction of coal fly ash. *International Journal of Coal Geology*, 2021. 240: p. 103746.

[87]   Veneva, L., et al., Rock magnetic, mineralogical and microstructural characterization of fly ashes from Bulgarian power plants and the nearby anthropogenic soils. *Physics and Chemistry of the Earth, Parts A/B/C*, 2004. 29(13-14): p. 1011–1023.

[88]   Kukier, U., et al., Composition and element solubility of magnetic and non-magnetic fly ash fractions. *Environmental Pollution*, 2003. 123(2): p. 255–266.

[89]   Liu, D., et al., Experimental study on the magnetic characteristics of coal fly ash at different combustion temperatures. *Environmental Technology*, 2018. 39(15): p. 1967–1975.

[90]   Shoumkova, A.S., Magnetic separation of coal fly ash from Bulgarian power plants. *Waste Management Research*, 2011. 29(10): p. 1078–89.

[91]   Dekkers, M.J. and H.S. Pietersen, Magnetic properties of low-ca fly ash: A rapid tool for fe-assessment and a survey for potentially hazardous elements. *MRS Online Proceedings Library Archive*, 1991. 245: p. 37–47.

[92]   Gomes, S., et al., Characterization of magnetite in silico-aluminous fly ash by SEM, TEM, XRD, magnetic susceptibility, and Mössbauer spectroscopy. *Cement and Concrete Research*, 1999. 29: p. 1705–1711.

[93]   Payá, J., et al., Comparisons among magnetic and non-magnetic fly ash fractions: strength development of cement-fly ash mortars. *Waste Management*, 1996. 16: p. 119–124.

[94]   Presuel-Moreno, F.J. and A.A. Sagüés, Bulk magnetic susceptibility measurements for determination of fly ash presence in concrete. *Cement and Concrete Research*, 2009. 39(2): p. 95–101.

[95]   Gopalakrishnan, R., S. Barathan, and D. Govindarajan, Magnetic susceptibility measurements on fly ash admixtured cement hydrated with groundwater and seawater. *American Journal of Materials Science*, 2012. 2(1): p. 32–36.

[96]   Jiao, D., et al., Possibilities of fly ash as responsive additive in magnetorheology control of cementitious materials. *Construction and Building Materials*, 2021. 296: p. 123656.

[97]   Prakash, S., et al., Characterisation and removal of iron from fly ash of Talcher area, Orissa, India. *Minerals Engineering*, 2001. 14(1): p. 123–126.

[98]   Garcés, P., et al., The effect of processed fly ashes on the durability and the corrosion of steel rebars embedded in cement–modified fly ash mortars. *Cement and Concrete Composites*, 2010. 32(3): p. 204–210.

[99]   Arief, I. and P.K. Mukhopadhyay, Preparation of spherical and cubic Fe 55 Co 45 microstructures for studying the role of particle morphology in magnetorheological suspensions. *Journal of Magnetism and Magnetic Materials*, 2014. 360: p. 104–108.

[100] Lee, J.Y., S.H. Kwon, and H.J. Choi, Magnetorheological characteristics of carbonyl iron microparticles with different shapes. *Korea-Australia Rheology Journal*, 2019. 31(1): p. 41–47.

[101] Tokyay, M., *Cement and concrete mineral admixtures*. 2016. CRC Press Taylor & Francis Group.

[102] Bentur, A. and S. Mindess, *Fibre reinforced cementitious composites*. 2nd ed. 2007. Taylor & Francis

[103] Faifer, M., et al., Steel fibre reinforced concrete characterization based on a magnetic probe. *2010 IEEE International Instrumentation and Measurement Technology Conference I2mtc 2010, Proceedings*, 2010. p. 157–162.

[104] Faifer, M., et al., Nondestructive Testing of Steel-Fibre-Reinforced Concrete Using a Magnetic Approach. *IEEE Transactions on Instrumentation and Measurement*, 2011. 60(5): p. 1709–1717.

[105] Ferrara, L., M. Faifer, and S. Toscani, A magnetic method for non destructive monitoring of fibre dispersion and orientation in steel fibre reinforced cementitious composites-part 1: method calibration. Materials and Structures, 2012. 45(4): p. 575–589.

[106] Al-Mattarneh, H., *Electromagnetic quality control of steel fibre concrete*. Construction and Building Materials, 2014. 73: p. 350–356.

[107] Li, L.F., et al., Fibre distribution characterization of ultra-high perform-ance fibre-reinforced concrete (UHPFRC) plates using magnetic probes. *Materials*, 2020. 13(13): p. 5064.

[108] Ozyurt, N., et al., Monitoring fibre dispersion in fibre-reinforced cemen-titious materials: Comparison of AC-impedance spectroscopy and image analysis. Paper by Nilufer Ozyurt, Leta Y. Woo, Thomas O. Mason, and Surendra P. Shah - Authors' closure. *ACI Materials Journal*, 2007. 104(4): p. 436–437.

[109] Torrents, J.M., et al., Inductive method for assessing the amount and orientation of steel fibres in concrete. *Materials and Structures*, 2012. 45(10): p. 1577–1592.

[110] Gettu, R., et al., Study of the distribution and orientation of fibres in SFRC specimens. *Materials and Structures*, 2005. 38(275): p. 31–37.

[111] Abrishambaf, A., M. Pimentel, and S. Nunes, Influence of fibre orienta-tion on the tensile behaviour of ultra-high performance fibre reinforced cementitious composites. *Cement and Concrete Research*, 2017. 97 : p. 28–40.

[112] Martinie, L. and N. Roussel, Simple tools for fibre orientation predic-tion in industrial practice. *Cement and Concrete Research*, 2011. 41(10): p. 993–1000.

[113] Bayer, A., *Process for producing a prepreg with aligned short fibres. Patent No. DE3641828A1*. 1986. Germany.

[114] Emami, A. and K. Nielsen, *Building material, especially concrete or mortar, contains magnetically or electrically aligned parallel fibres. Patent No. DE19750746A1*. 1997. Germany.

[115] Michels, J. and M. Gams, Preliminary study on the influence of fibre orientation in fibre reinforced mortars. *GRAĐEVINAR*, 2016. 68(8): p. 645–655.

[116] Mu, R., et al., Preparation of aligned steel fibre reinforced cementitious composite and its flexural behaviour. *Jove-Journal of Visualized Experiments*, 2018(136): p. 56307.

[117] Villa, V.P. and N.F. Medina, Alignment of hooked-end fibres in matrices with similar rheological behaviour to cementitious composites through homogeneous magnetic fields. *Construction and Building Materials*, 2018. 163: p. 256–266.

[118] Mondy, E., Magnetism in copper slags. *Nature*, 1872. 6(43).

[119] Teja, A.S. and P.Y. Koh, Synthesis, properties, and applications of magnetic iron oxide nanoparticles. *Progress in Crystal Growth and Characterization of Materials*, 2009. 55(1–2): p. 22–45.

Active rheology control of cementitious materials with responsive polymers

Metwally Ezzat, Richard Hoogenboom, Karel Lesage, and
Geert De Schutter

CONTENTS

4.1   General introduction   138
4.2   Stimuli-responsive polymers   139
      4.2.1   Electro-responsive polymers   139
      4.2.2   Magneto-responsive polymer composites   140
4.3   Stimuli-responsive polymers for active rheology control (ARC) 141
      4.3.1   Redox-responsive polymers for ARC   141
              4.3.1.1   Synthesis of the redox-responsive
                        superplasticizers   141
              4.3.1.2   Redox switchability behaviour   146
      4.3.2   Magneto-responsive polymer composites for ARC   149
              4.3.2.1   Introduction   149
              4.3.2.2   Synthesis of poly(MAA-co-PEGMA$_{500}$-co-
                        DopMA) terpolymers   150
              4.3.2.3   Preparation and magnetic properties of
                        polymer-coated Fe_3O_4 NPs   152
              4.3.2.4   Effect of poly(MAA-co-PEGMA$_{500}$-co-
                        DopaMA) on cement hydration   153
              4.3.2.5   Dispersing performance of poly(MAA-co-
                        PEGMA$_{500}$-co-DopaMA)   154
              4.3.2.6   Evaluation of magneto-responsive
                        rheological behaviour of cement paste   156
              4.3.2.7   Rheological properties of reference cement
                        paste   157
              4.3.2.8   Magneto-rheological properties of
                        reference cement paste   158
              4.3.2.9   Magneto-rheological behaviour of cement
                        paste with Fe_3O_4 NPs   158

DOI: 10.1201/9781003289463-4

4.3.2.10   Rheological properties of cement paste
            with poly(MAA-*co*-PEGMA$_{500}$-*co*-DopaMA)   160
4.3.2.11   Magneto-rheological behaviour of cement
            paste containing pDopaMA coated
            Fe$_3$O$_4$ NPs   161
4.4   Summary   163
References   164

4.1   GENERAL INTRODUCTION

Concrete is basically composed of three main constituents: cement, water and aggregates [1]. However, concrete has become more complex in recent decades due to the advent of new ingredients such as filler materials and mineral and chemical admixtures to improve the quality of concrete without losing its properties [2]. The concrete casting process was improved using chemical admixtures that provided control over fresh concrete properties, such as workability and rheological properties [3]. The rheological properties of cementitious materials can be characterized by at least two parameters according to the Bingham model in the equation below [4]:

$$\tau = \tau_o + \mu\dot{\gamma} \qquad\qquad (4.1)$$

where τ is the shear stress (Pa), τ_o is the yield stress (Pa), μ is the plastic viscosity (Pa.s) and $\dot{\gamma}$ is the shear rate (s^{-1}). The rheological parameters that describe the flow behaviour of cement pastes according to the Bingham model are the plastic viscosity (μ) and the yield stress (τ_o). Other models have been employed to study the flow characteristics of cement pastes in certain circumstances where the Bingham equation cannot be applied [5]. Chemical admixtures could significantly influence the rheology of cementitious materials [6].

Oscillatory shear tests have been extensively used to study the viscoelastic properties of cement pastes [7]. This technique could be employed to study the structural build-up evolution during the cement hydration process [8, 9]. Monitoring the storage modulus (G'), the elastic component of the linear viscoelastic region (LVR), can shed light on the structural build-up and elasticity of cement pastes under the ongoing hydration process [8]. The linear increase in G' is related to colloidal interaction forces such as Van der Waals, electrostatic interactions and steric effects in the early time after mixing (a few seconds) and to the C-S-H nucleation at longer periods [10–12].

Concrete superplasticizers such as polycarboxylate ethers (PCEs) are usually added to the cement pastes or fresh concrete to enhance their workability and/or to reduce the water content [13–16]. However, due to the retardation effects of superplasticizers, the structural build-up of cement pastes or concrete could be significantly affected. This behaviour could exert more pressure on the formwork after casting. In modern concrete technology, such as three-dimensional (3D) printing, where the formwork is absent, it is desirable to have a pseudo-solid-like state directly after deposition so that a fast structural build-up can be achieved, which is essential in 3D printing as many layers are required to be printed in a short time [17, 18]. In this perspective, it is essential to develop chemical admixtures that can enhance the workability during mixing and pumping processes and accelerate structuration after casting or deposition. In this work, the concept of smart superplasticizers has been introduced for the first time to enhance the workability of cement pastes before placing and increase the structuration after casting in response to external stimuli.

4.2   STIMULI-RESPONSIVE POLYMERS

Stimuli-responsive polymers or smart polymers are a kind of macromolecules that are sensitive to external stimuli such as temperature, pH, electric or magnetic fields [19, 20]. They can self-assemble or undergo phase transitions or morphology changes via physical or chemical changes in response to such stimuli [19, 21]. These smart polymers can be designed with responsive polymers or polymer-functionalized responsive compounds/materials. In the latter case, the polymer will serve as a template/carrier for the responsive compound [22]. One notable example for that includes polymer-functionalized nanoparticles (NPs) [23, 24]. Some related triggers will be covered in the following section (i.e. temperature, electric and magnetic triggers), and new responsive copolymers based on these stimuli will be introduced in this work.

4.2.1   Electro-responsive polymers

Electroactive polymers (EAPs) are a class of materials that exhibit shape or volume changes in response to various electrical stimuli, capable of transducing electrical energy into mechanical work. EAPs gained high scientific and technological values due to their potential applications in sensors and actuators, robotics and artificial muscles, optical systems, drug delivery, space, ocean and energy harvesting applications [25].

The electro-responsive polymers are classified into current-responsive polymers (ionic polymers and conductive polymers) and voltage-responsive polymers (dielectric gels and elastomers) [26].

Conductive polymers such as polyaniline (PANI), polythiophene (PT) or sulfonated-polystyrene (PSS) are electric-responsive polymers that can swell, shrink or bend in response to an external electric field [20, 22, 27]. PANI is one of the most studied conducting polymers due to its high electronic

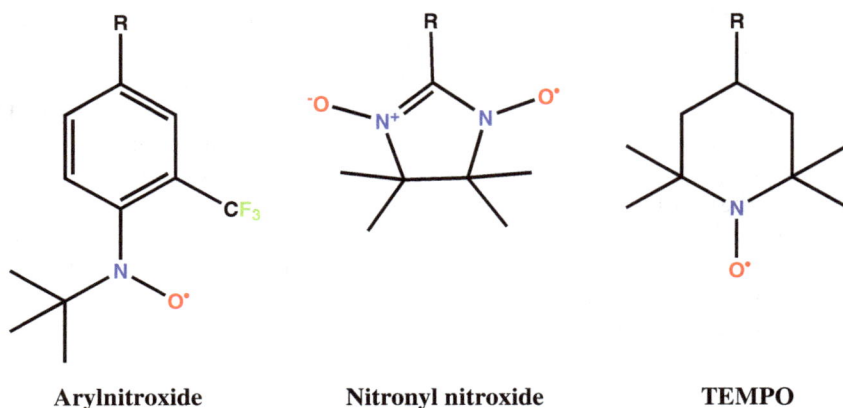

Figure 4.1 Some examples of common nitroxide radicals.

conductivity, redox and ion-exchange properties, excellent environmental stability and ease of preparation from common chemicals.

On the other hand, polymers with redox properties are electroactive macromolecules containing localized sites or groups that can be oxidized (loss of electrons) and reduced (gain of electrons), and this oxidation-reduction (redox) process is reversible [28]. This redox process may result in changes in the polymer properties depending on their oxidation state (oxidized or reduced). As a result, the polymer can exhibit different chemical, electronic, optical or mechanical properties [29]. Redox polymers, including organic chemical groups such as nitroxyl, verdazyl, phenoxyl, carbazol, quinones, viologens or hydrazyl, have been actively pursued by the scientific community over the last few years in many applications, such as the development of batteries, biofuel cells, biosensors, actuators and drug delivery vehicles [30–33]. Nishide and co-workers were pioneers in developing polymers containing stable nitroxide radicals such as 2,2,6,6-tetramethyl-1-piperidinyloxy (TEMPO) radical moieties (Figure 4.1). A variety of redox polymers with nitroxide pendant groups have been synthesized that are good candidates for organic batteries and supercapacitors due to their fast electron transfer kinetics [28].

4.2.2 Magneto-responsive polymer composites

Magneto-responsive polymers are an interesting research topic due to their applications in different fields such as biomedical, coatings and microelectronics [34]. Magneto-responsive polymers have a short response time upon exposure to a magnetic field that could be beneficial when requiring an immediate response. The preparation of highly efficient magnetic responsive materials requires doping polymeric materials with magnetic nanoparticles (MNPs) [34]. MNPs made of superparamagnetic Fe_3O_4, and metallic iron, in addition to hard magnetic materials (e.g. Co, Ni, FeN, or FePd), could be efficiently applied as doping materials for the preparation of magnetic responsive polymer composites [34–37].

4.3 STIMULI-RESPONSIVE POLYMERS FOR ACTIVE RHEOLOGY CONTROL (ARC)

4.3.1 Redox-responsive polymers for ARC

TEMPO radicals are known to possess high stability and fast electron transfer kinetics, enabling a fast response triggering effect [38]. Thus, novel concrete superplasticizers consisting of the redox-responsive nitroxide radical, TEMPO and methacrylic acid (MAA) have been prepared using the RAFT polymerization technique. Moreover, the incorporated bulky TEMPO radical is hypothesized to act as a pendant group, allowing synergistic effects between the steric stabilization induced by the TEMPO pendant groups and electrostatic forces induced by the COO^- groups, allowing the desired features for a typical superplasticizer. While enhancing the cement fluidity before placement, the cement structural evolution after casting could be achieved by applying an external electric potential that induces the oxidation of the TEMPO radical to the corresponding oxoammonium cation, $TEMPO^+$, altering the physical properties of cement pastes by inducing soluble-insoluble and adsorbed-desorbed polymer transitions. In this way, it is hypothesized that we can induce active stiffening control (ASC) for the cement paste or concrete during or after processing using external stimuli.

4.3.1.1 Synthesis of the redox-responsive superplasticizers

The direct radical polymerization of TEMPO-based monomers could not be achieved as it acts as a radical trap [39, 40]. For this reason, a random copolymer having MAA and 2,2,6,6-tetramethyl-4-piperidyl methacrylate (TMPMA) comonomers was first prepared as a polymer precursor using RAFT polymerization (Figure 4.2). ^{1}H-NMR analysis showed the successful copolymerization process and the formation of near-ideal random copolymers as observed from the kinetics plots of both MAA and TMPMA during the copolymerization process (Figure 4.3). The actual chemical composition was further evaluated by acid-base titration of pTMPMA and was found to be in good agreement with the theoretical composition.

The oxidation of the resulting poly(MAA-co-TMPMA) with H_2O_2 in the presence of $Na_2WO_4.2H_2O$/EDTA in methanol yielded the corresponding nitroxide polyradical, poly(MAA-co-TEMPO) (Figure 4.2). This oxidation method is more economical and environmentally friendly than using

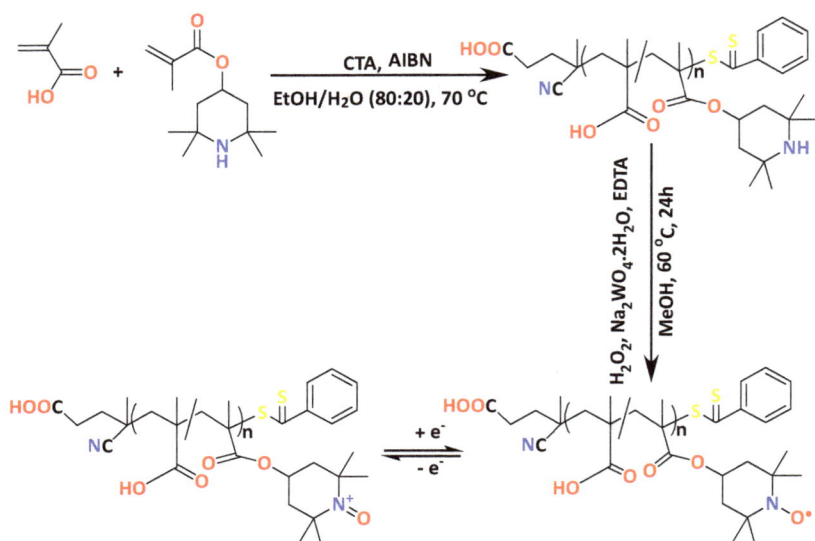

Figure 4.2 Schematic representation for the synthesis of redox-responsive poly(MAA-co-TEMPO).

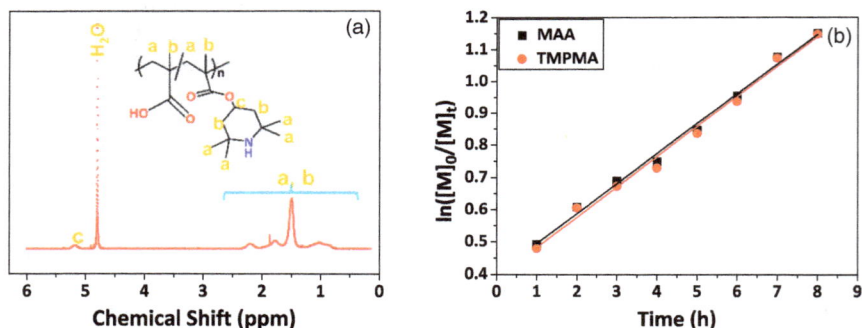

Figure 4.3 ^{1}H-NMR spectra of purified poly(MAA-co-TMPMA) in D_2O (a) and first order kinetic plot for the synthesis of PCA by copolymerization of MAA and TMPMA; [MAA]:[TMPMA]:[CTA]: [AIBN] = 50:50:1:0.2, [M]= 1 M in EtOH/H_2O (80/20) at 70 °C (b).

m-chloroperoxybenzoic acid in an inert solvent such as Dichloromethane (DCM) or CHCl$_3$ [41]. In this work, poly(MAA$_{50}$-co-TMPMA$_{50}$) has been employed in which the subscripts refer to the number of repeating units of MAA and TMPMA units. The further oxidation of poly(MAA-co-TEMPO) to the oxoammonium cation analog, poly(MAA-co-TEMPO$^+$), could be

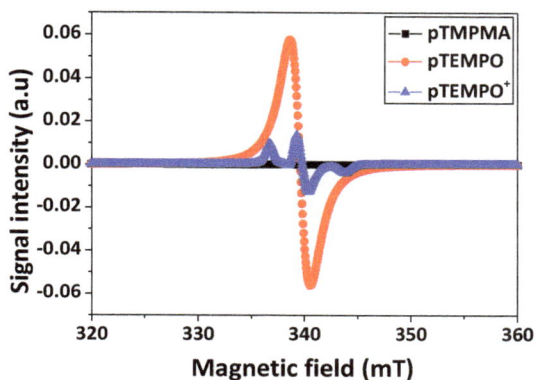

EPR spectra of poly(MAA-co-TMPMA), poly(MAA-co-TEMPO) and poly(MAA-co-TEMPO$^+$). (X-Band) with g=2.005.

achieved by either chemical oxidation or electrochemical oxidation. The chemical oxidation of poly(MAA-co-TEMPO) was performed using NaOCl as an oxidant in water.

Electron paramagnetic resonance (EPR) spectra were employed to calculate the radical content in poly(MAA-co-TEMPO) copolymers and monitor further oxidation to poly(MAA-co-TEMPO$^+$). Figure 4.4 shows the EPR spectra of poly(MAA-co-TMPMA), poly(MAA-co-TEMPO) and poly(MAA-co-TEMPO$^+$). As can be observed, the signal intensity for poly(MAA-co-TMPMA) is absent due to the absence of radical species. On the other hand, poly(MAA-co-TEMPO) exhibited a hyperfine split spectrum at g = 2.005, which correlates with the oxygen-centred unpaired electron [42]. A sharp intense signal with mass normalized EPR intensity of 14.24 was obtained, which correlates with the high radical content of poly(MAA-co-TEMPO). Moreover, the mass normalized EPR intensity reduced to 0.26 after the chemical oxidation of poly(MAA-co-TEMPO) to poly(MAA-co-TEMPO$^+$), suggesting that almost 98% of the nitroxide radicals have been oxidized to the corresponding oxoammonium cations.

The redox behaviour of poly(MAA-co-TEMPO) was also assessed by cyclic voltammetry (CV) in 0.1 M NaOH solution. As displayed in Figure 4.5, both anodic and cathodic peaks were observed at about 0.6V and 0.39V, respectively, versus Ag/AgCl screen-printed electrode. The CV results confirm the reversible redox-switching of poly(MAA-co-TEMPO) to poly(MAA-co-TEMPO$^+$).

The suitability of the prepared copolymers as potential superplasticizers has been assessed by studying the flow characteristics with Portland cement (CEM I 52.5 N). The flow curves of the copolymers before and

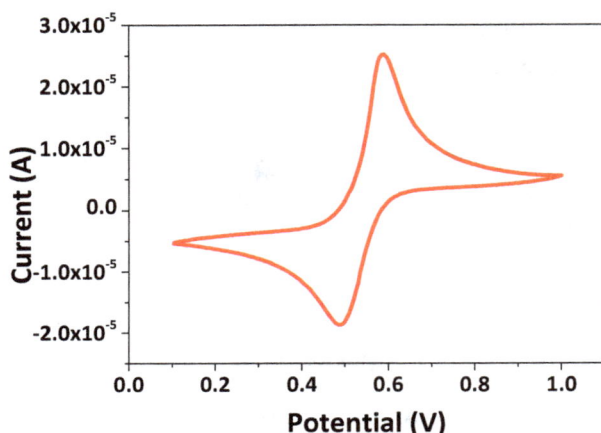

Figure 4.5 CV data of poly(MAA-co-TEMPO) in 0.1 M NaOH solution (10 mg/ml) at a scan rate of 50 mV/s.

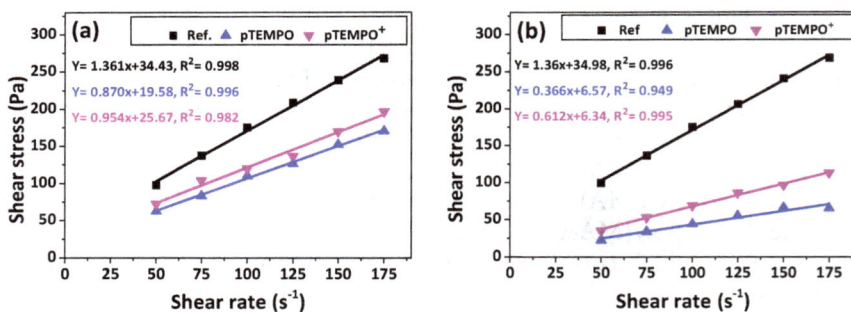

Figure 4.6 Flow curves of the copolymers before and after chemical oxidation (a) polymer conc.=0.1% (bwoc), and (b) polymer conc.=0.3% (bwoc) at W/C=0.35; pTEMPO = poly(MAA-co-TEMPO) and pTEMPO⁺=poly(MAA-co-TEMPO⁺).

after oxidation were investigated to assess the suitability of these redox copolymers as smart superplasticizers for active concrete rheology control. The Bingham model was used to determine the yield stress and plastic viscosity (Figure 4.6). Reference cement paste, without added polymer, showed yield stress and plastic viscosity of about 35 Pa and 1.25 Pa.s, respectively. The addition of 0.1% and 0.3% poly(MAA-*co*-TEMPO) by weight of cement led to a lower yield stress and plastic viscosity that depended on the polymer concentration. The cement pastes with 0.1% poly(MAA-*co*-TEMPO) showed yield stress of 20 Pa and plastic viscosity of 0.87 Pa.s, indicating better fluidity of the cement pastes. Increasing the poly(MAA-co-TEMPO) content in the cement pastes to 0.3% bwoc sharply decreased both yield stress and plastic viscosity to 6.57 Pa and 0.37 Pa.s, respectively. The addition of poly(MAA-co-TEMPO⁺) resulted in higher yield stress and plastic viscosity than poly(MAA-co-TEMPO) at a low dosage. However, when the poly(MAA-co-TEMPO⁺) dosage was 0.3% bwoc, a similar yield stress value as observed for poly(MAA-co-TEMPO) was obtained, but the plastic viscosity of poly(MAA-co-TEMPO⁺) containing pastes was almost twice the value for poly(MAA-co-TEMPO). The flow curves data suggest that poly(MAA-co-TEMPO) polymers could be used as potential superplasticizers, and their action could be controlled by poly(MAA-co-TEMPO)↔ poly(MAA-co-TEMPO⁺) switchability.

Zeta potential measurements were also performed for cement suspensions with different concentrations of poly(MAA-co-TEMPO) and poly(MAA-co-TEMPO⁺) to assess the stability of the cement suspensions containing these polymers and the effect of TEMPO oxidation on the charge characteristics. Cement particles usually show a very low zeta potential (i.e. close to 0 mV) as they tend to agglomerate instantly due to their electrostatic potentials [43]. Figure 4.7 shows that the addition of poly(MAA-co-TEMPO) increases the absolute zeta potential value of cement suspensions up to -18 mV and -23 mV depending on the concentration of poly(MAA-co-TEMPO). On the other hand, the addition of poly(MAA-co-TEMPO⁺) also increases the negative zeta potential in the range of –10 mV to –22 mV. It is clear that both polymers get adsorbed on the cement particles; however, the adsorption capacity of poly(MAA-co-TEMPO) is higher than

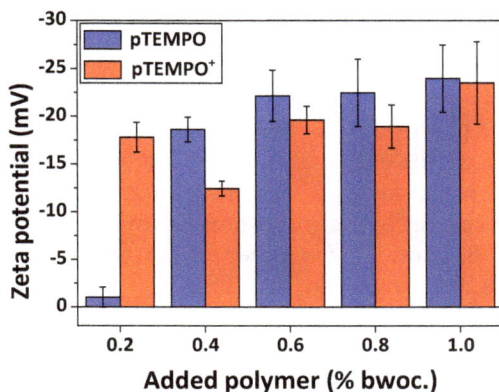

Figure 4.7 Zeta potential measurements of cement particles in the presence of poly(MAA-co-TEMPO) copolymers at different polymer concentrations (% bwoc) at W/C= 50; the average of five measurements was represented; pTEMPO = poly(MAA-co-TEMPO) and pTEMPO⁺=poly(MAA-co-TEMPO⁺).

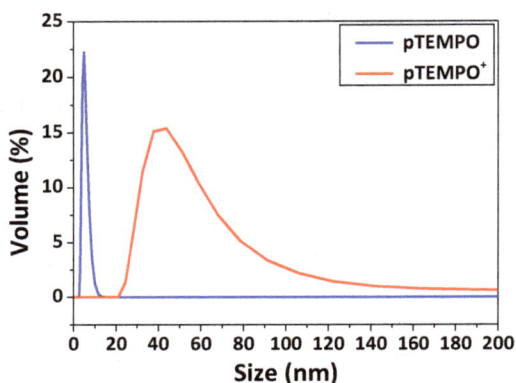

Figure 4.8 Dynamic light scattering measurements of pTEMPO; poly(MAA-co-TEMPO) and pTEMPO⁺; poly(MAA-co-TEMPO⁺) at a polymer concentration of 5 mg/ml in cement pore solution.

that for poly(MAA-*co*-TEMPO⁺), indicating that the dispersion ability of poly(MAA-*co*-TEMPO) is higher than that of poly(MAA-*co*-TEMPO⁺).

Next, dynamic light scattering (DLS) was used to assess the solubility of both poly(MAA-*co*-TEMPO) and poly(MAA-*co*-TEMPO⁺) in cement pore solution (Figure 4.8). Poly (MAA-*co*-TEMPO) showed a polymer size of 5 nm while poly(MAA-*co*-TEMPO⁺) had a size of 55 nm, indicating that poly(MAA-*co*-TEMPO⁺) tends to aggregate in cement pore solution, confirming that the oxidation of poly(MAA-*co*-TEMPO) could result in different physical properties of the copolymers. This finding agrees with the results of the flow curves where cement with poly(MAA-*co*-TEMPO) showed a higher fluidity than cement with the less soluble poly(MAA-*co*-TEMPO⁺).

The addition of poly(MAA-*co*-TEMPO) to cement paste induced a slight retardation effect on the cement hydration (Figure 4.9), while poly(MAA-*co*-TEMPO⁺) showed a slightly lower retardation effect than poly(MAA-*co*-TEMPO) on the cement hydration process. Nonetheless, these retardation effects are both minor and are not expected to strongly influence the cement hydration process.

4.3.1.2   Redox switchability behaviour

Time-dependent rheometry was investigated to study the structural evolution of cement pastes containing poly(MAA-*co*-TEMPO) when triggered by an external electric field of 1V, using the setup shown in Figure 4.10.

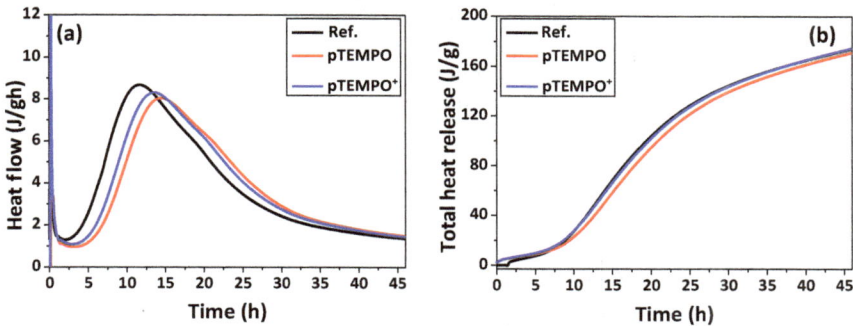

Figure 4.9 Time-dependent heat flow (a) and total heat (b) liberated by cement pastes (W/C = 0.4) at a temperature of 20 °C in the presence of 0.1% bwoc of pTEMPO; poly(MAA-co-TEMPO) and pTEMPO⁺; poly(MAA-co-TEMPO⁺) copolymers and compared to reference cement paste, as determined by isothermal calorimetry.

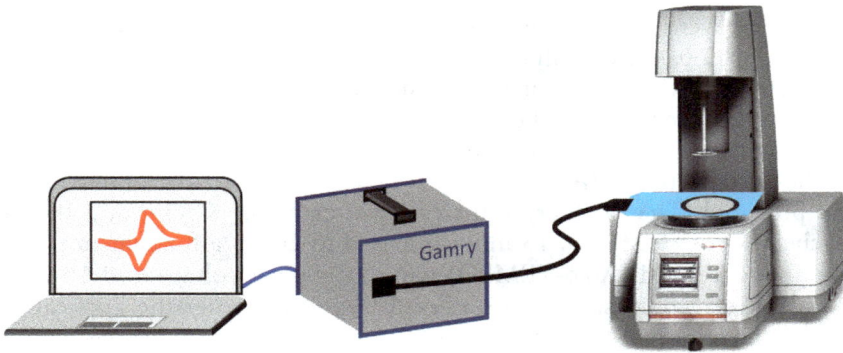

Figure 4.10 Experimental setup for the small-amplitude oscillatory shear (SAOS) test for cement pastes under the effect of an external electric field of 1 V.

Figure 4.11 displays the storage modulus (G') curves related to the elastic portion of the viscoelastic cement pastes. The G' value of reference cement paste, without added polymer, increased gradually over time and showed similar values under 0 V and 1 V triggers. This indicates that a low potential value (1 V) does not significantly influence the cement paste structure evolution. The addition of poly(MAA-co-TEMPO) (1% bwoc) resulted in lower G' values in the first 150 s, indicating superplasticizing behaviour, after which a pronounced increase in the G' was observed, leading to G'

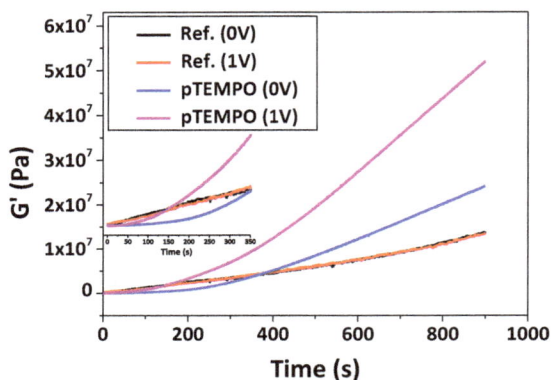

Figure 4.11 Storage modulus of cement pastes containing 1% pTEMPO; poly(MAA-co-TEMPO) (bwoc) before and after applying an electric potential of 1 V as compared to blank cement paste at W/C = 0.35. The percent standard deviation of three measurements lies between 10.5% and 16%.

values that were higher than those for blank cement pastes. On the other hand, cement pastes containing poly(MAA-*co*-TEMPO) in the presence of a 1V stimulus behaved similarly at the start of the test. However, after 50 s, a sharp increase in the G' was observed, which can be attributed to the cement pastes rearrangements, i.e. desorption and/or aggregation of the formed poly(MAA-*co*-TEMPO⁺) in response to the applied electric potential. The G' sharply increased after about 6 min, and the G' reached about 50 MPa in 15 min compared to 20 MPa for cement pastes containing poly(MAA-*co*-TEMPO) without applied voltage and 10 MPa for blank cement pastes. This observed large effect after the application of 1 V confirms that the redox-responsive superplasticizers can be used to induce ASC.

Figure 4.12 displays the adsorption mechanism of poly(MAA-*co*-TEMPO) and poly(MAA-*co*-TEMPO⁺) on cement surfaces. In Figure 4.12a, the polymer chains are extended and could be easily adsorbed to the cement particles via electrostatic attractions due to the presence of the COO⁻ groups. The steric hindrance effects of TEMPO pendant groups stabilize and disperse the cement particles. On the other hand, the oxidation of the nitroxide polyradical leads to the formation of copolymers with both negative and positive charges on the polymer chain. These polymers have both anionic and cationic groups that could interact with each other through electrostatic attraction forces inducing polymer dehydration and aggregation [44] that lead to high structural build-up due to the fast dispersion loss of cement pastes (Figure 4.12b).

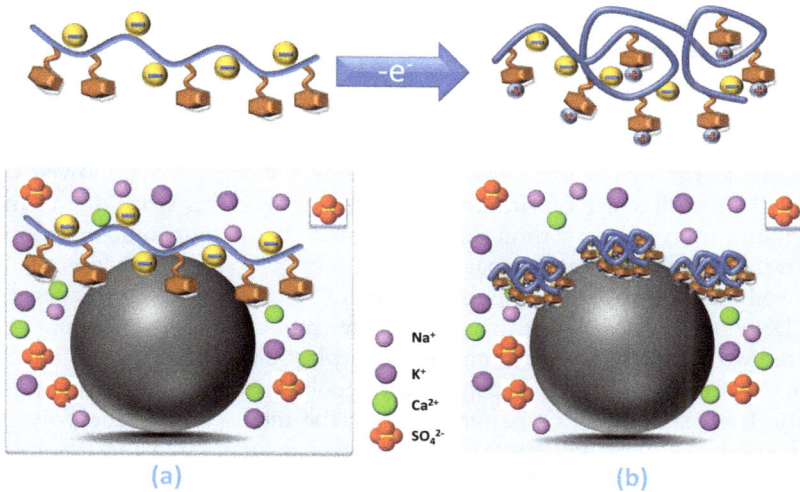

Figure 4.12 Schematic representation of the mode of action for (a) poly(MAA-co-TEMPO) shows an extended polymer chain adsorbed into cement particles, and (b) poly(MAA-co-TEMPO⁺) forms coils due to the intramolecular interactions between the positive and negative charges in the polymer chains causing dehydration and aggregation. (For reasons of clarity, cement particles and polymers are not drawn at the same scale).

4.3.2 Magneto-responsive polymer composites for ARC

4.3.2.1 Introduction

Magnetic responsive materials have potential breakthrough applications in many fields [34]. Magnetic materials have widely been exploited in a diverse range of applications, such as magneto-rheological fluids (MRFs) [45], catalysis [46] and biomedical applications [47, 48]. Magnetic materials exhibit an immediate response to an applied magnetic field [49]. Employing MNPs in the concrete industry as a magnetic responsive material has been extensively investigated by Jiao et al. [50] (see also Chapter 3). The combination of magnetic and polymeric materials could produce composites with outstanding magneto-responsive features [34]. Magnetic responsive polymer composites are composed of MNPs dispersed in or functionalized with polymeric materials and can be triggered by external magnetic fields [51]. One example of materials whose properties, such as the mechanical yield stress, can be tuned in response to external magnetic fields is MRFs [52]. The MRF instantly transforms from a liquid-like to a solid-like state upon applying external magnetic fields, providing active control over their rheological properties, such as increasing the shear yield stress [53]. However, magneto-responsive polymer composites have not yet been established as potential materials to actively control the rheological properties of cementitious materials.

In this work, Fe_3O_4 NPs have been functionalized with terpolymers containing dopamine side chains. Fe_3O_4 NPs of different sizes have been considered for this study. However, initial screening results showed that Fe_3O_4 NPs with a size of 100 nm are well-suited for this work due to their high saturation magnetization (Ms) and clear response that could be reflected from the response of cement pastes containing Fe_3O_4 NPs. Previous reports showed that the magnetic response of small Fe_3O_4 NPs is much weaker than the large ones [54, 55]. For this purpose, we prepared Fe_3O_4 NPs that were coated with dopamine-functionalized superplasticizing terpolymers as magnetic responsive materials to control the rheology of cementitious materials in the fresh state [56]. Dopamine ensures the interaction of the polymers with the Fe_3O_4 NPs while methacrylic acid is incorporated for adsorption onto cement particles, and $PEGMA_{500}$ is utilized as a steric comonomer to induce steric stabilization and superplasticizing of the cement paste.

4.3.2.2 Synthesis of poly(MAA-co-PEGMA$_{500}$-co-DopMA) terpolymers

To have an efficient and stable assembly of nanoparticles in a polymer matrix, the polymer needs to have functional groups that bind strongly to the nanoparticle surface. In this regard, we introduced the bidentate ligand dopamine to bind iron oxide NPs into superplasticizing copolymers containing both MAA and $PEGMA_{500}$ comonomers (Figure 4.13). These terpolymers are based on different hydrophilic monomers, employing anionic MAA, non-ionic $PEGMA_{500}$ and ionizable dopamine HCl. The theoretical overall polymerization degree of the terpolymers was 100 but with varying comonomer ratios. The different synthesis conditions, compositions and the main features of different polymers are summarized in Table 4.1.

It has been reported that catechol groups show high activities in free radical scavenging, so they are often protected when monomers containing catechol groups (e.g. dopamine HCl) are used in radical polymerization. To limit the ability of catechol groups to donate hydrogen to scavenge free radicals during polymerization, the polymerization of unprotected DopaMA is usually performed in high polar aprotic solvents such as DMF, DMSO and acetonitrile, where the catechol groups are proposed to form non-covalent bonds (i.e. hydrogen bonds) with the solvents and thus limit the interaction of catechol groups with the propagating radicals [57, 58]. Therefore, free radical polymerization (FRP) was employed in DMF to prepare pDopaMA-functionalized terpolymers with MAA and $PEGMA_{500}$, employing Azobisisobutyronitrile (AIBN) as the radical initiator, enabling

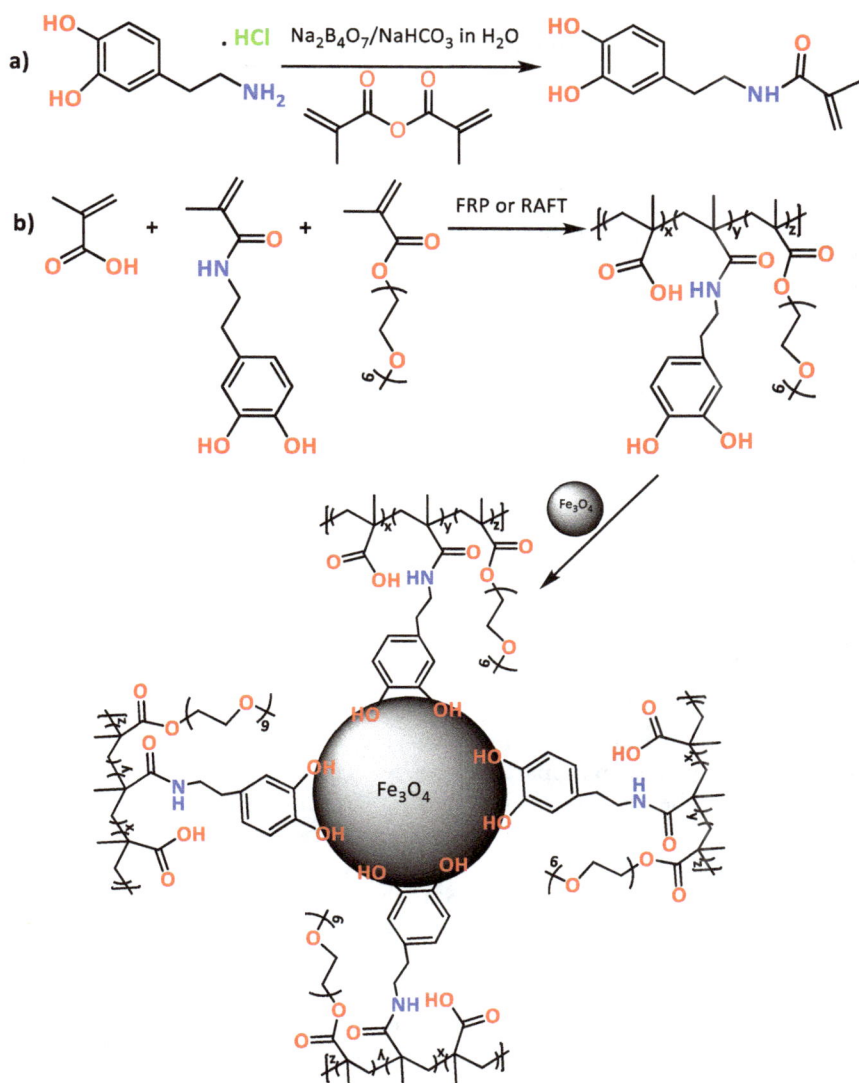

Figure 4.13 Schematic representation for the synthesis of MNPs-functionalized poly(MAA-co-PEGMA$_{500}$-co-DopaMA);pDopaMA.

Table 4.1 Conditions for pDopaMA synthesis and main structural details of the obtained poly(MAA-*co*-PEGMA$_{500}$-*co*-DopaMA) terpolymers

							Size^c (nm)	
Entry	Technique	MAA (mol.%)	PEGMA$_{500}$ (mol.%)	DopaMA (mol.%)	Mn (SEC)^a (KDa)	Đ^b	In H$_2$O	In pore solution
1	RAFT	45	45	10	29.2*	1.27	5.20	8.30
2	FRP	45	45	10	33.8	2.25	4.97	11.1
3	RAFT	40	40	20	35.7*	1.21	4.90	6.58
4	FRP	40	40	20	46.9	2.56	5.00	10.9

Notes:
ᵃ   Measured by size-exclusion chromatography (SEC) using PMMA standards in DMA.
ᵇ   Measured by SEC (M$_w$/M$_n$).
ᶜ   Measured by dynamic light scattering in water and cement pore solution (5 mg/ml).
*Values are shown after post-amidation.

the successful preparation of terpolymers with different compositions, having a polymer dispersity (Đ) of 2.25–2.56.

4.3.2.3   *Preparation and magnetic properties of polymer-coated Fe$_3$O$_4$ NPs*

Fe$_3$O$_4$ MNPs are utilized in this work to induce magnetic field responsiveness. The average nanoparticle size, as claimed by the provider, is 100 nm; However, the transmission electron microscopy (TEM) measurements revealed an average size of about 50 and broad size distribution (20 nm–110 nm), as shown in Figure 4.14. It is obvious that the nanoparticles tend to agglomerate, as observed from DLS analysis, where the z-average for the nanoparticles in both water and ethanol was over 700 nm. The preparation of Fe$_3$O$_4$ NPs functionalized with different dopamine-based terpolymers was achieved via a simple ligand exchange process in acetonitrile, in which the dopamine-based copolymers are grafted to the NPs (Figure 4.13). It has been reported that dopamine forms an initial structure coordinated to the surface of Fe$_3$O$_4$ NPs due to the improved orbital overlap of the five-membered ring and the reduced steric environment of the iron complex [59]. A stable chelating complex at the surface of the nanoparticle is obtained due to the existence of undercoordinated sites at the nanoparticle surface [60]. Rajh and coworkers confirmed that the surface of unmodified metal oxides particles is composed of undercoordinated metal-binding sites that are removed when binding monocyclic enediol ligands such as dopamine to the particle surface, thereby restoring the undercoordinated metal binding

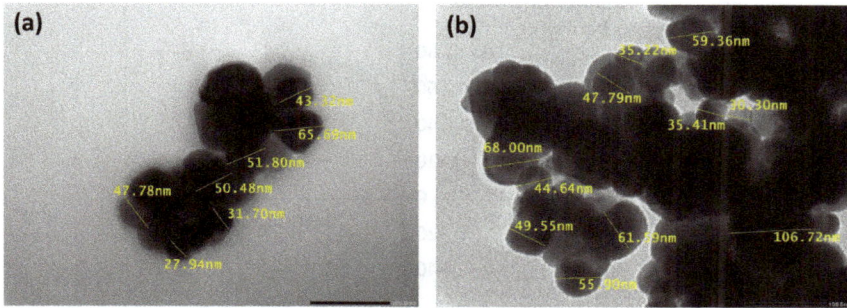

Figure 4.14 TEM images of (a) bare Fe3O4 NPs and (b) polyDopmaMA-coated Fe3O4 NPs.

sites to the bulk-like lattice structure and causing a tight binding of dopamine to iron oxides [60, 61].

The magnetic properties of the polymer-NPs composites were studied by the vibrating sample magnetometry (VSM) measurement of field-dependent hysteresis loops (M-H) at a temperature of 300 K and applied field ranging from 0 to 10,000 Oe. The saturation magnetization (Ms), coercivity (Ce) and remanence (Mr) values calculated from the M–H curves for both bare and coated MNPs are presented in Table 4.2. In general, Fe_3O_4 NPs show either superparamagnetic (SPM) or ferrimagnetic (FM) behaviour [62]. For the material to exhibit SPM, the coercivity (Ce) and remanent magnetization (Mr) values should be negligible. The ferrimagnetic materials exhibit a hysteresis loop when measuring the magnetization value (emu/g) versus the applied magnetic field (Oe). Figure 4.15 presents the FM hysteresis loops of bare and pDopaMA-coated Fe_3O_4 NPs. It is worth noting that these NPs are not exhibiting an SPM behaviour due to the large size of the NPs (i.e. 100 nm). As can be seen from Figure 4.15 and Table 4.2, the saturation magnetization (Ms) for bare Fe_3O_4 NPs is 77.05 emu/g, which showed a slight increase after different polymer coatings on Fe_3O_4 NPs. On the other hand, the Ce and Mr values of bare Fe_3O_4 NPs found to be about 158.34 Oe and 14.71 emu/g, respectively. After the coating of Fe_3O_4 NPs with pDopaMA, there was only a slight difference in both Ce and Mr, indicating that the coating did not affect the magnetic properties of Fe_3O_4 NPs.

4.3.2.4 Effect of poly(MAA-co-PEGMA$_{500}$-co-DopaMA) on cement hydration

Isothermal calorimetry (ITC) can be used to study cement hydration kinetics at a constant temperature, whereby the heat flow release can be monitored over time [63]. The effect of the different poly(MAA-*co*-PEGMA$_{500}$-*co*-DopaMA)

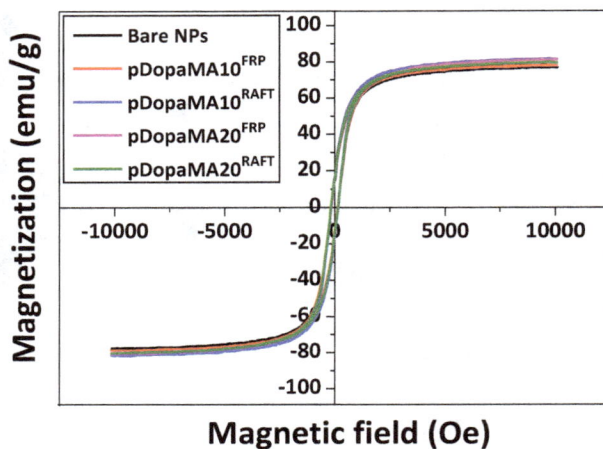

Figure 4.15 Hysteresis loops of bare Fe_3O_4 NPs and pDopaMA-coated Fe_3O_4 NPs.

Table 4.2 The saturation magnetization (Ms), coercivity (Ce), and remanence (Mr) values as calculated from the M-H curves for both bare Fe_3O_4 NPs and pDopaMA-coated Fe_3O_4 NPs

Entry	Ms (emu/g)	Mr (emu/g)	Ce (Oe)	Mr/Ms
Bare Fe_3O_4 NPs	77.05	14.72	158.34	0.191
pDopaMA10^{FRP}-NPs	78	13.90	158.09	0.178
pDopaMA10^{RAFT}-NPs	81.57	16.15	159.01	0.198
pDopaMA20^{FRP}-NPs	81.16	14.67	162.24	0.181
pDopaMA20^{RAFT}-NPs	79.68	15.07	161.43	0.189

terpolymers on cement hydration is illustrated in Figure 4.16. As can be seen, the terpolymers induce significant retardation of the cement hydration process, compared to blank cement paste, with a dormant period of about 3.5 to 4 h, as commonly observed for superplasticizers. More importantly, the effect of the different compositions on the cement hydration is similar, indicating that all prepared compositions have a similar effect on the cement hydration process.

4.3.2.5   Dispersing performance of poly(MAA-co-PEGMA$_{500}$-co-DopaMA)

The fluidity of cement pastes containing different poly(MAA-co-PEGMA$_{500}$-co- DopaMA) terpolymers was assessed by mini-slump tests at a W/C=0.35,

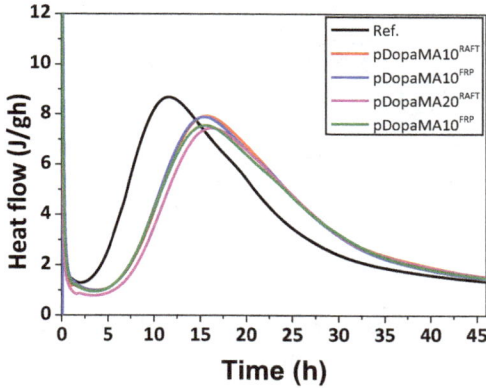

Figure 4.16 Heat flow calorimetry of cement pastes containing poly(MAA-co-PEGMA$_{500}$-co-DopaMA) copolymers compared to blank cement paste; W/C=0.4 and polymer conc.=0.1% (bwoc).

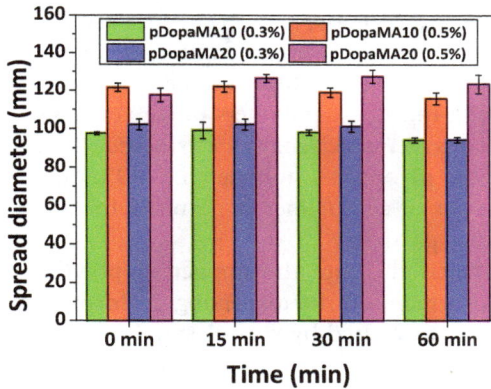

Figure 4.17 Effect of pDopaMA terpolymers on the dispersing performance of cement paste.

as displayed in Figure 4.17. The blank cement paste showed a 65–70 mm spread diameter. Initial screening measurements showed that the terpolymers did not show a significant dispersing performance at low polymer concentration (i.e. below 0.2 % by weight of cement). When increasing the polymer concentrations (i.e. 0.3% by weight of cement), the cement pastes showed high plasticity with spread diameters between 97 mm and 121 mm. Furthermore, when increasing the dopamine content in the polymers (i.e.

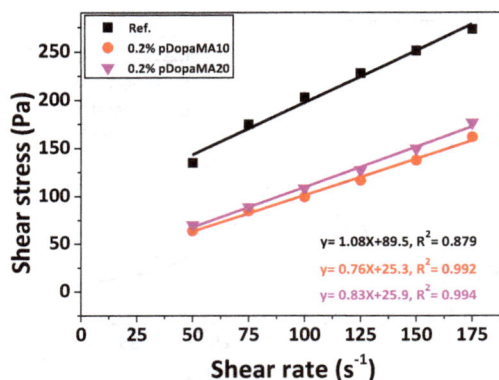

Figure 4.18 Flow curves of cement paste with pDopaMA terpolymers at W/C=0.35.

20 mol.%), a slight increase in the spread diameter could be observed at the same dosage.

On the other hand, when further increasing the polymer concentrations from 0.3% (bwoc) to 0.5% (bwoc), the spread diameters of cement pastes increased significantly. Moreover, the cement pastes containing different pDopaMA polymers showed high slump retention over 60 minutes, confirming that the polymers could be used as concrete superplasticizers. It is worth mentioning that these polymers only exert good dispersing power at higher polymer dosage, which may be attributed to the lower COO^- ionic group content in the pDopaMA polymer (i.e. 40 mol.%– 45 mol.%), and thus a larger amount of terpolymer is required for adsorption and hence dispersion of cement particles.

In order to further illustrate the impact of poly(MAA-*co*-PEGMA$_{500}$-*co*-DopaMA) terpolymers on the dispersing performance in cement paste, the flow curves were measured by rheometry. As presented in Figure 4.18, the reference cement paste without added polymer showed yield stress and plastic viscosity of 89.5 Pa and 1.08 Pa.s, respectively. When adding 0.2% (bwoc) pDopaMa terpolymers, the yield stress and plastic viscosity decreased sharply, which indicates that these terpolymers can significantly plasticize cement paste. Further increasing the polymer concentration to 0.3% (bwoc) led to a further decrease in the yield stress and plastic viscosity, which is in agreement with the mini-slump results (not shown).

4.3.2.6 *Evaluation of magneto-responsive rheological behaviour of cement paste*

Different rheological parameters such as elastic or storage modulus (G'), viscous or loss modulus (G''), and phase shift angle (δ) can be estimated by

the amplitude sweep test, and these parameters provide relevant information on the structural build-up during the cement hydration process [7, 64].
The small-amplitude oscillatory shear (SAOS) technique is a non-destructive
method that could effectively monitor the changes in the structure of fresh
cement pastes [65]. In this regard, the paste has to be deformed at low strain
amplitude (i.e. within the linear viscoelastic region [LVR]) below the critical
strain. This critical strain is defined as the cut-off point between the linear
and non-linear domain of the strain-sweep [66]. In this work, the amplitude
sweep tests have been used to monitor the rheological changes of cement
pastes upon the addition of terpolymers, Fe_3O_4 NPs and terpolymer coated
Fe_3O_4 NPs in response to the application of external magnetic fields. The
ASC features of cement pastes could be assessed in this way.

4.3.2.7   Rheological properties of reference cement paste

The storage modulus (G') and loss modulus (G") curves of reference cement
paste without added terpolymer were determined as a function of strain
without any triggers (Figure 4.19) at a water to cement ratio (w/c) of 0.45.
It could be noticed that the cement paste showed a linear viscoelastic region
(LVR), where G' is independent of the strain amplitude when the strain is
below 0.01%, indicating that the material is highly structured. However, at
higher strains, both G' and G" decrease with increasing strain amplitude,
indicating that increasing the strain above the critical strain value disrupts
the network structure of the cement paste [67]. It could also be observed that
G" was smaller than G' at the LVR, which indicates a solid-like response.
However, when the strain goes beyond 0.01%, G" increases until it reaches

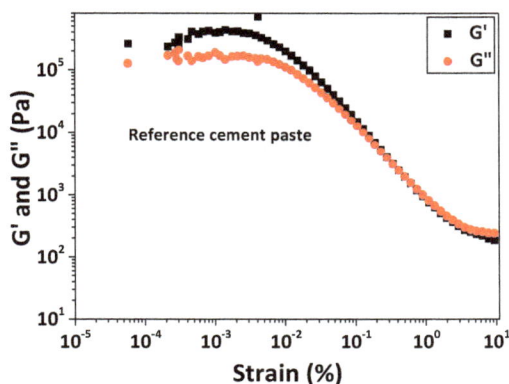

Figure 4.19 Storage and loss moduli as a function of strain for reference cement paste
without added terpolymer at 2 Hz and W/C= 0.45.

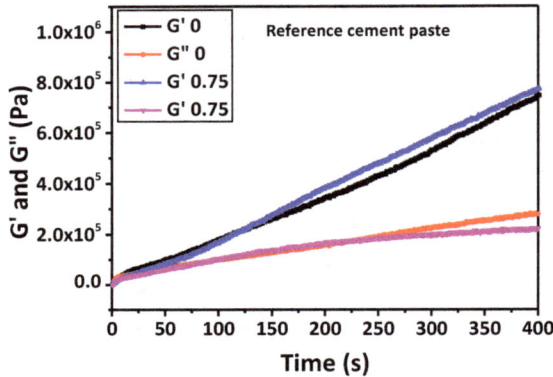

Figure 4.20 Results of time sweep measurements of blank cement pastes at W/C =0.45; magnetic field strengths (B) of 0 T and 0.75 T have been applied.

a higher value than G', indicating a change of the cement paste behaviour to a fluid-like response [64].

4.3.2.8   Magneto-rheological properties of reference cement paste

SAOS test results for reference cement paste under constant magnetic field strengths of 0 T and 0.75 T are depicted in Figure 4.20. The storage modulus of the reference cement pastes at 0 T exhibited a linear increase and reached 500 kPa after 400 s. The increase in G' value could be attributed to the structural build-up due to the flocculation of cement particles, cement thixotropy and early cement hydration [68]. When a magnetic field of 0.75 T was applied, the G' of the cement paste was similar to that without an applied magnetic field, indicating that the cement paste is not magneto-responsive. This behaviour was expected due to the absence of magnetic materials that could be triggered in response to the applied magnetic fields.

4.3.2.9   Magneto-rheological behaviour of cement paste with Fe_3O_4 NPs

The structural build-up evolution of cement pastes containing 3% Fe_3O_4 NPs is presented in Figure 4.21. This cement paste showed a higher G' than G", confirming the domination of the solid-like behaviour throughout the entire test. The storage modulus follows a linear increase in time, while the loss modulus shows a gradual increase until 200 s, followed by a plateau. This behaviour indicates the formation of solid-like structures during the

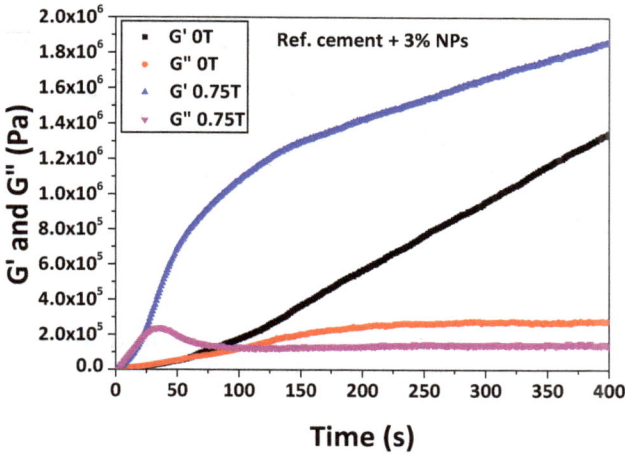

Figure 4.21 Results of time sweep measurements of cement pastes containing Fe_3O_4 NPs at W/C =0.45; magnetic field strengths (B) of 0 T and 0.75 T have been applied.

SAOS test at a 0 T magnetic field. Upon applying an external magnetic field of 0.75 T, the storage modulus sharply increased from very low values to about 1200 kPa within the first two minutes compared to 500 kPa for cement pastes without a magnetic field. The presence of magnetic NPs in the cement paste induces the formation of NP chains in the presence of magnetic fields, increasing the viscosity of cement pastes, and thus a faster structural build-up can occur [68]. Furthermore, the loss modulus of the cement paste under a magnetic field exhibits an initial bump, and then the loss modulus decreases until it reaches a steady state. This bump in the loss modulus could be attributed to the rearrangement of nanoparticles under the effect of external magnetic fields [69, 70].

The phase shift angle (δ) is another parameter that can be used to describe the viscoelastic behaviour of the material. It describes the ratio between the elastic (G") and the viscous (G') behaviour as in the following equation:

$$\tan(\delta) = \frac{G''}{G'} \tag{4.2}$$

Figure 4.22 shows the progress of the phase shift angle (δ) over time for cement pastes containing 3% Fe_3O_4 NPs in the absence (a) and presence (b) of the magnetic field. For cement pastes containing 3% Fe_3O_4 NPs, the phase shift angle equals 55° (i.e. $\tan(\delta)=1.43$), which means that the cement paste shows more viscous than elastic properties. It is also evident that the δ decreases gradually over time while increasing the storage modulus, and

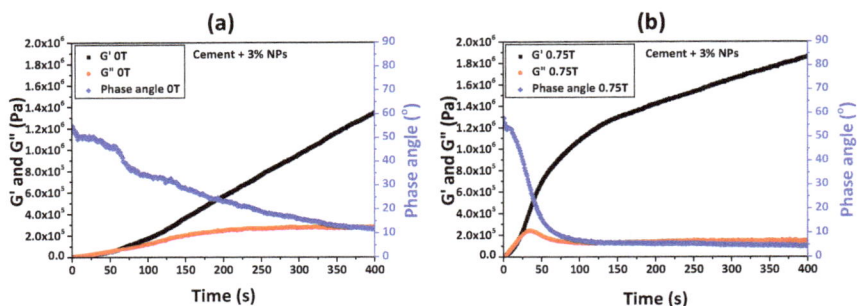

Figure 4.22 Results of time sweep measurements of cement pastes containing Fe$_3$O$_4$ NPs
at W/C =0.45; magnetic field strengths (B) of 0 T and 0.75 T have been applied.

thus the paste changes its state from more viscous to more solid-like behaviour due to the structural build-up. On the other hand, under the influence of an external magnetic field (i.e. 0.75 T), the phase shift angle sharply decreases from 55° to below 5° in just 100 s, indicating a faster transition from viscous to elastic behaviour for the cement paste. This sharp transition is a result of rapid structural build-up due to the alignment of Fe$_3$O$_4$ NPs to the magnetic field applied.

4.3.2.10   *Rheological properties of cement paste with poly(MAA-co-PEGMA$_{500}$-co-DopaMA)*

It is essential to know the effect of free polymer addition on the structural build-up of the cement paste to be able to assess the contribution from other parameters such as the applied magnetic field and the NPs addition. Figure 4.23 shows the effect of terpolymer addition to the cement paste in the absence of an external magnetic field at two different polymer concentrations (i.e. 0.1% and 0.5% by weight of cement). The terpolymers prepared by RAFT polymerization show a similar SAOS behaviour regardless of the dopamine-HCl content in the terpolymers (Figure 4.23a). The polymers at low concentrations did not affect the viscoelastic properties of the cement paste compared to the reference cement paste. However, the addition of a larger amount of terpolymer (i.e. 0.5% bwoc) highly affected the viscoelastic properties of the cement paste. As shown in Figure 4.23b, the storage modulus (G') has a sharp increase that starts directly at the start of the test and approaches 3.3 MPa in 400 s, compared to 0.5 MPa for the reference cement paste. On the other hand, the loss modulus (G") increased gradually at the beginning and then decreased to a very low value until it reached a steady state. This phenomenon could be clearly observed by the change of the phase shift angle (δ) from 20° to almost 0°, which indicates

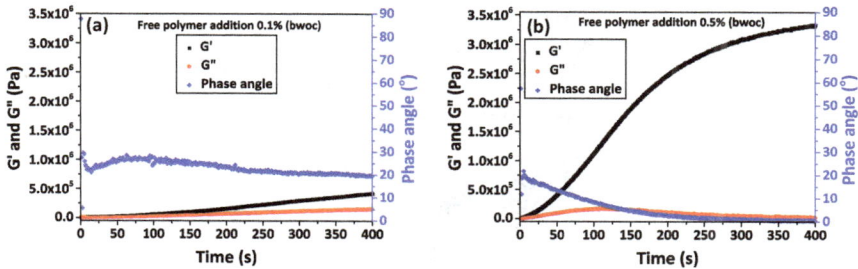

Figure 4.23 Results of time sweep measurements of cement pastes containing free poly(MAA-co-PEGMA$_{500}$-co-DopaMA) terpolymer at a concentration of (a) 0.1% bwoc and (b) 0.5% bwoc at w/c= 0.45; all DopaMA-based terpolymers show a similar trend.

the domination of elastic behaviour after the addition of high polymer concentration that contributes to fast structural build-up. This might be related to the fact that the terpolymers have many functional groups such as -OH, -COOH and -NH that can adhere to the cement particles through various active groups such as Al_2O_3, CaO, MgO and Fe_2O_3 on the cement surface [71, 72]. More importantly, the presence of dopamine structure in the terpolymers results in high adhesion capacity with cement particles so that the movement of polymers is constrained and the structural build-up is quickly achieved, perhaps even by bridging multiple cement particles. Similar general trends have been found for terpolymers prepared by FRP (data not shown here).

4.3.2.11 Magneto-rheological behaviour of cement paste containing pDopaMA coated Fe$_3$O$_4$ NPs

The prepared terpolymers can bind to the Fe_3O_4 NPs surface via the catechol groups on the one hand and can bind to cement particles via the adsorption of the COO^- groups onto the positively charged cement particles on the other hand (Figure 4.24). This interconnection could form a sort of interconnected network structure, and thus the whole system could induce a response to the applied external magnetic fields, changing its rheological behaviour and build-up evolution.

As a final step, the magneto-rheological behaviour of cement paste in the presence of the terpolymer-coated NPs was studied. Figure 4.25 shows the structural build-up evolution of cement pastes containing terpolymers coated Fe_3O_4 NPs in response to external magnetic fields. Figure 4.25a and 4.25b show the evolution of the storage modulus (G'), loss modulus (G") and phase shift angle (δ) for cement pastes containing Fe_3O_4 NPs coated

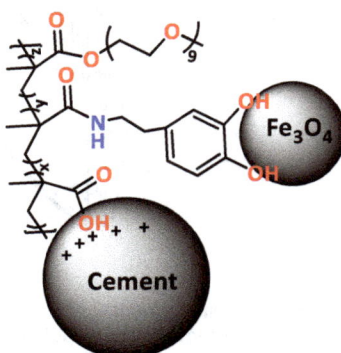

Figure 4.24 A schematic representation of the adsorption of poly(MAA-co-PEGMA$_{500}$-co-DopaMA) to both Fe$_3$O$_4$ NPs and cement particles. The chemical structures and particles are not drawn to a scale.

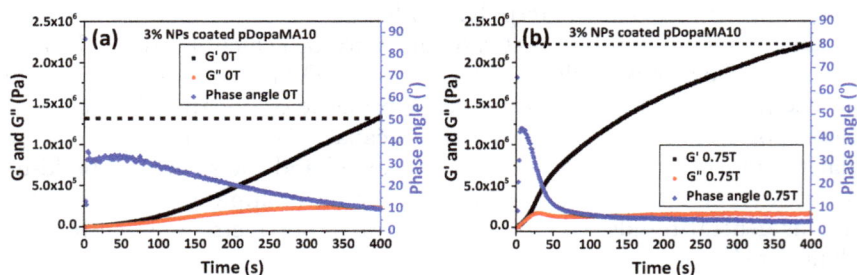

Figure 4.25 Results of time sweep measurements of cement pastes containing Fe$_3$O$_4$ NPs coated with terpolymers containing 10 mol% DopaMA at W/C =0.45; magnetic field strengths (B) of 0 T and 0.75 T have been applied.

with terpolymers containing 10 mol% DopaMA in the absence and presence of external magnetic fields, respectively. As can be observed, in the absence of a magnetic field (Figure 4.25a), the G' value increases linearly with time and reaches a value of about 1.2–1.3 MPa after 400 s. In the meantime, the G" curve follows a negligible increase with time, demonstrating the domination of solid-like behaviour that could also be observed from the low phase angle value throughout the experiment. The evolution of G', G" and δ for cement paste containing Fe$_3$O$_4$ NPs coated with terpolymers containing 10 mol% DopaMA in response to an applied external magnetic field of 0.75 T is exhibited in Figure 4.25b. At the beginning of the test, both G' and G" increased rapidly in response to the external magnetic field and exhibited a sharp increase in G', which correlates with the immediate response of the

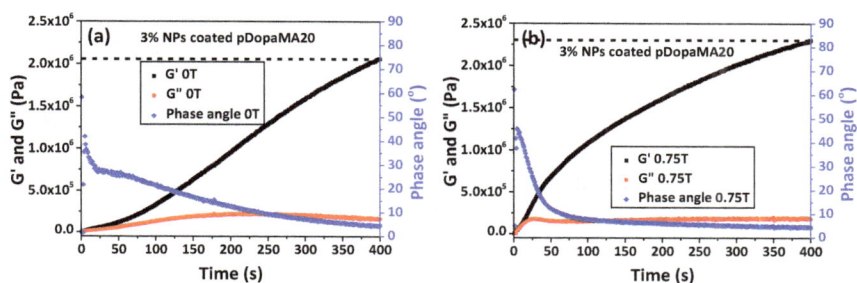

Figure 4.26 Results of time sweep measurements of cement pastes containing Fe₃O₄ NPs-coated pDopaMA20 polymers at W/C =0.45; magnetic field strengths (B) of 0 T and 0.75 T have been applied.

polymer composites to the external magnetic field. This increase in G' and the bump in G" show that the particles rearrange themselves to be aligned with the applied magnetic field. G' value reaches about 2.2 MPa. As such, the cement pastes show an obvious response when compared to the results without a magnetic field.

It is important to understand the effect of dopamine composition in the terpolymer on the structural build-up of cement pastes containing thereof. In this regard, another composition of the terpolymers containing 20 mol.% dopamine has been utilized to coat the Fe_3O_4 NPs. Figure 4.26 presents the evolution of G' and G" of cement pastes containing 3% Fe_3O_4 NPs coated with terpolymers having 20 mol% DopaMA. The terpolymers with 20 mol% DopaMA obtained by RAFT polymerization showed an immediate sharp increase in the G' curves in response to the applied magnetic field. Moreover, very high G' values of about 2 MPa and 2.3 MPa before and after applying the magnetic field have been obtained, respectively. Fe_3O_4 NPs-coated pDopaMA20 showed a sharp increase in the G' curves even without applying external magnetic fields. This phenomenon may result from the high dopamine content (i.e. 20 mol.%), where the terpolymers seem to show higher adhesion capacity, and thus higher G' values could be obtained (higher structural build-up evolution).

4.4 SUMMARY

In this chapter, the concept of smart superplasticizers has been introduced. These novel polymeric materials can induce rheological changes on-demand in response to applied external stimuli. Two different stimuli were introduced as potential stimuli for active rheology control (ARS). The first example represents a redox-responsive superplasticizer that can be used as cement components or added to dry cement during the mixing process. These copolymers showed excellent performance as superplasticizers with high initial dispersion, workability and workability retention. CV measurements have confirmed the redox responsive behaviour. Furthermore, SAOS measurements showed that these copolymers exhibit a distinct and clear response to the applied electric field (1V), indicated by the sharp increase in the G' curves (structural build-up evolution) when compared to both reference cement paste and cement pastes containing copolymers without applying such electric field. This confirms the active control of rheological properties of cement pastes, especially ASC, in the presence of such copolymers in response to the applied external electric fields. Despite the interesting behaviour of these redox-responsive polymers on a small scale, the applicability of redox-responsive materials on a large scale may face challenges regarding the feasibility of the response behaviour, the implementation of screen-printed electrodes on a big scale, the associated cost and whether the cost–benefit ratio is worthy or not.

The concept of ASC was extended by employing magneto-responsive polymer composites that could immediately respond to applied external magnetic fields. Functionalization of Fe_3O_4 NPs with the prepared terpolymers allowed for rheology control of cement pastes containing this composite due to their immediate response to the applied external magnetic fields. When applying an external magnetic field, the polymer composites align in response to the applied magnetic field and chains, thus accelerating the structural build-up evolution and facilitating the ASC. The developed magnetic responsive polymers could be of great interest and offer an immediate response to control the rheological behaviour of cementitious materials.

REFERENCES

[1]   K.W. Day, *Concrete mix design, quality control and specification*. 2006. CRC Press.

[2]   P. Bartos, Workability of special fresh concretes. *Special concretes-workability and mixing*. 1993, CRC Press, p. 14–18.

[3]   D. Marchon, S. Kawashima, H. Bessaies-Bey, S. Mantellato, and S. Ng, Hydration and rheology control of concrete for digital fabrication: Potential admixtures and cement chemistry, *Cement and Concrete Research* 112 (2018): 96–110.

[4]   Q. Yuan, D. Zhou, K.H. Khayat, D. Feys, and C. Shi, On the measurement of evolution of structural build-up of cement paste with time by static yield stress test vs. small amplitude oscillatory shear test, *Cement and Concrete Research* 99 (2017): 183–189.

[5]   C.F. Ferraris, Measurement of the rheological properties of cement paste: a new approach, *International RILEM Conference on the Role of Admixtures in High Performance Concrete, RILEM Publications SARL*, 1999, p. 333–342.

[6] H. Elaqra, and R. Rustom, Effect of using glass powder as cement replacement on rheological and mechanical properties of cement paste, *Construction and Building Materials* 179 (2018): 326–335.

[7] M. Nehdi, and S. Al Martini, Estimating time and temperature dependent yield stress of cement paste using oscillatory rheology and genetic algorithms, *Cement and Concrete Research* 39(11) (2009): 1007–1016.

[8] N. Roussel, H. Bessaies-Bey, S. Kawashima, D. Marchon, K. Vasilic, and R. Wolfs, Recent advances on yield stress and elasticity of fresh cement-based materials, *Cement and Concrete Research* 2019, 124 (2019): 105798.

[9] D. Jiao, R. De Schryver, C. Shi, and G. De Schutter, Thixotropic structural build-up of cement-based materials: A state-of-the-art review, *Cement and Concrete Composites* (2021): 122: 104152.

[10] A.M. Mostafa, and A. Yahia, New approach to assess build-up of cement-based suspensions, *Cement and Concrete Research* 85 (2016): 174–182.

[11] A.M. Mostafa, and A. Yahia, Physico-chemical kinetics of structural build-up of neat cement-based suspensions, *Cement and Concrete Research* 97 (2017): 11–27.

[12] Y. Liang, N. Hilal, P. Langston, and V. Starov, Interaction forces between colloidal particles in liquid: Theory and experiment, *Advances in Colloid and Interface Science* 134 (2007): 151–166.

[13] I. Papayianni, G. Tsohos, N. Oikonomou, and P. Mavria, Influence of superplasticizer type and mix design parameters on the performance of them in concrete mixtures, *Cement and Concrete Composites* 27(2) (2005): 217–222.

[14] M. Ezzat, X. Xu, K. El Cheikh, K. Lesage, R. Hoogenboom, and G. De Schutter, Structure-property relationships for polycarboxylate ether superplasticizers by means of RAFT polymerization, *Journal of Colloid and Interface Science* 553 (2019): 788–797.

[15] V. Morin, F.C. Tenoudji, A. Feylessoufi, and P. Richard, Superplasticizer effects on setting and structuration mechanisms of ultrahigh-performance concrete, *Cement and Concrete Research* 31(1) (2001): 63–71.

[16] S. Alsadey, Effect of superplasticizer on fresh and hardened properties of concrete, *Journal of Agricultural Science and Engineering* 1(2) (2015): 70–74.

[17] N. Roussel, Rheological requirements for printable concretes, *Cement and Concrete Research* 112 (2018): 76–85.

[18] L. Huang, P. Yan, and Y. Liu, Effect of alkali content in cement on the fluidity and structural build-up of plasticized cement pastes, *Construction and Building Materials* 253 (2020): 119180.

[19] G. Kocak, C. Tuncer, and V. Bütün, pH-Responsive polymers, *Polymer Chemistry* 8(1) (2017): 144–176.

[20] M. Wei, Y. Gao, X. Li, and M.J. Serpe, Stimuli-responsive polymers and their applications, *Polymer Chemistry* 8(1) (2017): 127–143.

[21] D. Wang, M.D. Green, K. Chen, C. Daengngam, and Y. Kotsuchibashi, Stimuli-responsive polymers: Design, synthesis, characterization, and applications, *International Journal of Polymer Science* 2016 (2016): 6480259.

[22] E. Cabane, X. Zhang, K. Langowska, C.G. Palivan, and W. Meier, Stimuli-responsive polymers and their applications in nanomedicine, *Biointerphases* 7(1) (2012): 9.

[23] E. Galati, M. Tebbe, A. Querejeta-Fernández, H.L. Xin, O. Gang, E.B. Zhulina, and E. Kumacheva, Shape-specific patterning of polymer-functionalized nanoparticles, *ACS Nano* 11(5) (2017): 4995–5002.

[24] M. Czaun, L. Hevesi, M. Takafuji, and H. Ihara, A novel approach to magneto-responsive polymeric gels assisted by iron nanoparticles as nano cross-linkers, *Chemical Communications* 18 (2008): 2124–2126.

[25] T. Manouras, and M. Vamvakaki, Field responsive materials: Photo-, electro-, magnetic-and ultrasound-sensitive polymers, *Polymer Chemistry* 8(1) (2017): 74–96.

[26] H. Okuzaki, *Electroresponsive polymer* (2014). Berlin, Heidelberg: Springer. 1–6.

[27] L.T. Strover, J. Malmstroem, and J. Travas-Sejdic, Graft copolymers with conducting polymer backbones: a versatile route to functional materials, *The Chemical Record* 16(1) (2016): 393–418.

[28] N. Casado, G. Hernández, H. Sardon, and D. Mecerreyes, Current trends in redox polymers for energy and medicine, *Progress in Polymer Science* 52 (2016): 107–135.

[29] N. Casado, G. Hernandez, H. Sardon, and D. Mecerreyes, Current trends in redox polymers for energy and medicine, *Progress in Polymer Science* 52 (2016): 107–135.

[30] M. Khodeir, S. Antoun, E. Van Ruymbeke, and J.F. Gohy, Temperature and redox-responsive hydrogels based on nitroxide radicals and oligoethyleneglycol methacrylate, *Macromolecular Chemistry and Physics* 221(6) (2020): 1900550.

[31] K.-A. Hansen, and J.P. Blinco, Nitroxide radical polymers–a versatile material class for high-tech applications, *Polymer Chemistry* 9(13) (2018): 1479–1516.

[32] K. Koshika, N. Sano, K. Oyaizu, and H. Nishide, An ultrafast charge-able polymer electrode based on the combination of nitroxide radical and aqueous electrolyte, *Chemical Communications* (7) (2009): 836–838.

[33] N. Casado, and D. Mecerreyes, Redox polymers for energy and nanomedicine, *Royal Society of Chemistry* 2021. Croydon, UK: Polymer Chemistry series No. 34.

[34] J. Thévenot, H. Oliveira, O. Sandre, and S. Lecommandoux, Magnetic responsive polymer composite materials, *Chemical Society Reviews* 42(17) (2013): 7099–7116.

[35] J. Zhang, R. Srivastava, and R. Misra, Core– shell magnetite nanoparticles surface encapsulated with smart stimuli-responsive polymer: synthesis, characterization, and LCST of viable drug-targeting delivery system, *Langmuir* 23(11) (2007): 6342–6351.

[36] R. Li, Q. Xiao, P.-a. Yang, H. Wang, and L. Liu, Magnet-induced deformation enhanced adhesion based on magneto-responsive polymer: Theoretical analysis and experimental verification, *Materials & Design* 194 (2020): 108905.

[37] T. Kang, F. Li, S. Baik, W. Shao, D. Ling, and T. Hyeon, Surface design of magnetic nanoparticles for stimuli-responsive cancer imaging and therapy, *Biomaterials* 136 (2017): 98–114.

[38] N. Casado, and D. Mecerreyes, Introduction to redox polymers: Classification, characterization methods and main applications, In: *Redox polymers for energy and nanomedicine*, (2021). Croydon, UK: Royal Society of Chemistry. 1–26.

[39] J. Winsberg, S. Muench, T. Hagemann, S. Morgenstern, T. Janoschka, M. Billing, F.H. Schacher, G. Hauffman, J.-F. Gohy, and S. Hoeppener, Polymer/zinc hybrid-flow battery using block copolymer micelles featuring a TEMPO corona as catholyte, *Polymer Chemistry* 7(9) (2016): 1711–1718.

[40] T. Janoschka, A. Teichler, A. Krieg, M.D. Hager, and U.S. Schubert, Polymerization of free secondary amine bearing monomers by RAFT polymerization and other controlled radical techniques, *Journal of Polymer Science Part A: Polymer Chemistry* 50(7) (2012): 1394–1407.

[41] K. Zhang, Y. Hu, L. Wang, J. Fan, M.J. Monteiro, and Z. Jia, The impact of the molecular weight on the electrochemical properties of poly (TEMPO methacrylate), *Polymer Chemistry* 8(11) (2017): 1815–1823.

[42] H. Nishide, S. Iwasa, Y.-J. Pu, T. Suga, K. Nakahara, and M. Satoh, Organic radical battery: Nitroxide polymers as a cathode-active material, *Electrochimica Acta* 50(2–3) (2004): 827–831.

[43] K. Yoshioka, E.-i. Tazawa, K. Kawai, and T. Enohata, Adsorption characteristics of superplasticizers on cement component minerals, *Cement and Concrete Research* 32(10) (2002): 1507–1513.

[44] M. Burgess, K. Hernández-Burgos, J.K. Schuh, J. Davila, E.C. Montoto, R.H. Ewoldt, and J.n. Rodríguez-López, Modulation of the electrochemical reactivity of solubilized redox active polymers via polyelectrolyte dynamics, *Journal of the American Chemical Society* 140(6) (2018): 2093–2104.

[45] M.A. Portillo, and G. Iglesias, Magnetic nanoparticles as a redispersing additive in magnetorheological fluid, *Journal of Nanomaterials* 2017 (2017): 9026219.

[46] A.H. Lu, E.e.L. Salabas, and F. Schüth, Magnetic nanoparticles: Synthesis, protection, functionalization, and application, *Angewandte Chemie International Edition* 46(8) (2007): 1222–1244.

[47] X. Li, J. Wei, K.E. Aifantis, Y. Fan, Q. Feng, F.Z. Cui, and F. Watari, Current investigations into magnetic nanoparticles for biomedical applications, *Journal of Biomedical Materials Research Part A* 104(5) (2016): 1285–1296.

[48] V.F. Cardoso, A. Francesko, C. Ribeiro, M. Bañobre-López, P. Martins, and S. Lanceros-Mendez, Advances in magnetic nanoparticles for biomedical applications, *Advanced Healthcare Materials* 7(5) (2018): 1700845.

[49] B. Issa, I.M. Obaidat, B.A. Albiss, and Y. Haik, Magnetic nanoparticles: Surface effects and properties related to biomedicine applications, *International Journal of Molecular Sciences* 14(11) (2013): 21266–21305.

[50] D. Jiao, *Active rheology control of cementitious materials using magnetic field*, Ghent University, 2021.

[51]   E.C. Frachini, and D.F. Petri, Magneto-responsive hydrogels: preparation, characterization, biotechnological and environmental applications, *Journal of the Brazilian Chemical Society* 30(10) (2019): 2010–2028.

[52]   P. Ilg, Stimuli-responsive hydrogels cross-linked by magnetic nanoparticles, *Soft Matter* 9(13) (2013): 3465–3468.

[53]   G. Zhang, Y. Li, H. Wang, and J. Wang, Rheological properties of polyurethane-based magnetorheological gels, *Frontiers in Materials* 6 (2019): 56.

[54]   Q. Li, C.W. Kartikowati, S. Horie, T. Ogi, T. Iwaki, and K. Okuyama, Correlation between particle size/domain structure and magnetic properties of highly crystalline Fe3O4 nanoparticles, *Scientific Reports* 7(1) (2017): 1–7.

[55]   R. Ge, X. Li, M. Lin, D. Wang, S. Li, S. Liu, Q. Tang, Y. Liu, J. Jiang, and L. Liu, Fe3O4@ polydopamine composite theranostic superparticles employing preassembled Fe3O4 nanoparticles as the core, *ACS Applied Materials & Interfaces* 8(35) (2016): 22942–22952.

[56]   F. Liaqat, M. Tahir, H. Huesmann, P. Daniel, M. Kappl, G.K. Auernhammer, D. Schneider, I. Lieberwirth, K. Char, and G. Fytas, Ultrastrong composites from dopamine modified-polymer-infiltrated colloidal crystals, *Materials Horizons* 2(4) (2015): 434–441.

[57]   J. Yang, J. Keijsers, M. van Heek, A. Stuiver, M.A.C. Stuart, and M. Kamperman, The effect of molecular composition and crosslinking on adhesion of a bio-inspired adhesive, *Polymer Chemistry* 6(16) (2015): 3121–3130.

[58]   Q. Zhang, G. Nurumbetov, A. Simula, C. Zhu, M. Li, P. Wilson, K. Kempe, B. Yang, L. Tao, and D.M. Haddleton, Synthesis of well-defined catechol polymers for surface functionalization of magnetic nanoparticles, *Polymer Chemistry* 7(45) (2016): 7002–7010.

[59]   M.D. Shultz, J.U. Reveles, S.N. Khanna, and E.E. Carpenter, Reactive nature of dopamine as a surface functionalization agent in iron oxide nanoparticles, *Journal of the American Chemical Society* 129(9) (2007): 2482–2487.

[60]   T. Rajh, L. Chen, K. Lukas, T. Liu, M. Thurnauer, and D. Tiede, Surface restructuring of nanoparticles: An efficient route for ligand–metal oxide crosstalk, *The Journal of Physical Chemistry B* 106(41) (2002): 10543–10552.

[61]   M. Mazur, A. Barras, V. Kuncser, A. Galatanu, V. Zaitzev, K.V. Turcheniuk, P. Woisel, J. Lyskawa, W. Laure, and A. Siriwardena, Iron oxide magnetic nanoparticles with versatile surface functions based on dopamine anchors, *Nanoscale* 5(7) (2013): 2692–2702.

[62]   M.D. Nguyen, H.-V. Tran, S. Xu, and T.R. Lee, Fe3O4 Nanoparticles: Structures, synthesis, magnetic properties, surface functionalization, and emerging applications, *Applied Sciences* 11(23) (2021): 11301.

[63]   J. Hu, Z. Ge, and K. Wang, Influence of cement fineness and water-to-cement ratio on mortar early-age heat of hydration and set times, *Construction and Building Materials* 50 (2014): 657–663.

[64] M. Choi, K. Park, and T. Oh, Viscoelastic properties of fresh cement paste to study the flow behaviour, *International Journal of Concrete Structures and Materials* 10(3) (2016): 65–74.

[65] W. Mbasha, I. Masalova, R. Haldenwang, and A. Malkin, The yield stress of cement pastes as obtained by different rheological approaches, *Applied Rheology* 25(5) (2015): 9–19.

[66] Q. Yuan, X. Lu, K.H. Khayat, D. Feys, and C. Shi, Small amplitude oscillatory shear technique to evaluate structural build-up of cement paste, *Materials and Structures* 50(2) (2017): 1–12.

[67] A. Franck, Understanding rheology of structured fluids, Book of TA instruments (2004): 1–11. www.tainstruments.com/pdf/literature/AAN01 6_V1_U_StructFluids.pdf

[68] S.D. Nair, and R.D. Ferron, Real time control of fresh cement paste stiffening: Smart cement-based materials via a magnetorheological approach, *Rheologica Acta* 55(7) (2016): 571–579.

[69] L. Mohan, C. Pellet, M. Cloitre, and R. Bonnecaze, Local mobility and microstructure in periodically sheared soft particle glasses and their connection to macroscopic rheology, *Journal of Rheology* 57(3) (2013): 1023–1046.

[70] D. Jiao, K. Lesage, M.Y. Yardimci, K. El Cheikh, C. Shi, and G. De Schutter, Rheological properties of cement paste with nano-Fe3O4 under magnetic field: Flow curve and nanoparticle agglomeration, *Materials* 13(22) (2020): 5164.

[71] S. Lv, High-performance superplasticizer based on chitosan. *Biopolymers and biotech admixtures for eco-efficient construction materials*, Elsevier 2016, p. 131–150.

[72] C. Zhang, Y. Zhou, H. Han, H. Zheng, W. Xu, and Z. Wang, Dopamine-triggered hydrogels with high transparency, self-adhesion, and thermoresponse as skinlike sensors, *ACS Nano* 15(1) (2021): 1785–1794.

Chapter 5

Application of active rheology control to pumping of cementitious materials

Robin De Schryver, Mert Yucel Yardimci, Karel Lesage, and Geert De Schutter

CONTENTS

5.1   Theoretical pumping concepts   171
    5.1.1   Poiseuille flow   172
    5.1.2   Lubrication flow   175
    5.1.3   Dry friction flow   177
5.2   Experimental ARC pumping setup   177
    5.2.1   Monitoring equipment   178
    5.2.2   Active control unit   181
5.3   Magneto-rheological ARC pumping experiment   181
5.4   Summary   184
References   186

5.1   THEORETICAL PUMPING CONCEPTS

As briefly introduced in Chapter 2, pumping of concrete or cementitious suspensions in general is an indispensable processing technique. In spite of numerous investigations on concrete pumping, the flow behaviour of concrete during pumping is not fully understood. Especially the character and formation of the so-called lubrication layer is not well understood, whether it concerns a distinct fluid layer or rather a gradual transition zone from pipe wall towards the central bulk. Nevertheless, depending on the type of cementitious suspension, three different pipe flow types can occur. For cement pastes a homogeneous flow, known as Poiseuille flow, occurs. For flowable or saturate concretes, lubrication or slippage flow manifests. Thirdly, for unsaturated 'rather dry' concretes, dry friction flow may occur. Irrespective of the flow type, the shear stress τ in the pipe during pumping is governed by the universal shear stress law as a function of radius r given in equation 5.1, which is derived from the longitudinal force equilibrium as illustrated in Figure 5.1 for a given pipe section with length L and radius R and which is driven by a pressure difference Δp.

DOI: 10.1201/9781003289463-5

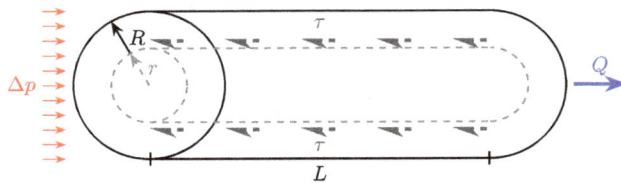

Figure 5.1   Poiseuille pipe flow configuration expresses a longitudinal force equilibrium between pressure loss Δp and internal shear stress τ as a function of pipe radius r, during a discharge Q. Source: after [6].

$$\tau(r) = \frac{r}{2}\frac{\Delta p}{L} \tag{5.1}$$

Before elaborating on the application of active rheology control (ARC) to pumping of smart cementitious suspensions, this section gives a brief overview of the theoretical background of the three different pipe flow types, i.e. Poiseuille flow, lubrication or slippage flow, and dry friction flow. Since a magneto-rheological cementitious paste suspension was used for the ARC pumping experiments, for which no significant lubrication or slippage occurred, the flow behaviour of the pumping experiment is governed by the first flow type, i.e. Poiseuille flow.

5.1.1   Poiseuille flow

The pipe flow problem was named after Poiseuille who conducted extensive research on the laminar flow of Newtonian fluids in cylinder tubes [1–5]. Even though Hagen also independently derived the analytical solution for Newtonian pipe flows, Poiseuille's research was more extensive, including experiments on water, ether, alcohol and blood. Even though the analytical solution was named after both Hagen and Poiseuille, the flow problem for non-Newtonian fluids is mostly referred to as Poiseuille flow. One of the main assumptions for Poiseuille flow is that the material is homogeneous. Based on the assumption that the flow is steady-state and one-dimensional, the velocity profile in the pipe can be derived after radial integration of shear rate $\dot{\gamma}$. This is at least the case if the universal shear stress can be solved for the shear rate as a function of its rheological properties. Once a velocity profile can be obtained, a cylindrical surface integral should be solved to find the overall volumetric discharge or flow rate Q. Last but not least, the total required power during pumping can be found by the surface integration of the velocity and applied driving pressure, which is equal to the product of the discharge Q and the pressure difference itself Δp (i.e. $\partial W / \partial t = Q\Delta p$). Of course one still needs to take into account special losses, gravitational losses and pump efficiency for the totally required pumping power. As a matter of fact, based on the required power (either total or per unit length), one can determine the heat that is introduced in the system.

Poiseuille flow solutions can be used to quantify the pumping behaviour of cementitious paste suspensions. Even though paste suspensions can mostly be described by Bingham flow behaviour, they may exhibit more complicated flow behaviour, e.g. shear-thickening, thixotropy, etc. Nevertheless, a few Poiseuille flow solutions exist for non-Newtonian fluids, which are summarized in Table 5.1. The most important one is the solution for a Bingham fluid, which is known as the Buckingham–Reiner equation and is named after its developers [6–9]. As a matter of fact, insights can be formulated from the non-dimensional expression of these Poiseuille flow solutions. For yield stress fluids, it is particularly useful to make use of the pressure number Pn given in equation 5.2 to do so.

$$Pn = \frac{\tau_w}{\tau_0} = \frac{\Delta p}{\tau_0} \frac{R}{2L} \tag{5.2}$$

The pressure number is in fact the non-dimensional form of the shear stress at the pipe wall τ_w, cf. equation 5.1. As such, no flow can occur when the pressure number Pn is less than 1, since pipe flow can only onset when the shear stress at the pipe wall boundary exceeds the yield stress. Alternatively, one could make use of the Bingham number Bn, which is equal to twice the inverse pressure number. Expressing the Poiseuille flows as such reveals important dimensioned proportionalities which govern the pipe flow. A Poiseuille flow solution therefore typically consists of a geometrical factor containing π, a dimensioned proportionality (written between brackets in Table 5.1) and a remaining part describing the flow behaviour as a function of the pressure number. As such, the most important flow proportionality for Bingham flows is the Bingham shear rate proportion τ_0 / μ. This also holds true for a thixotropic Bingham fluid in accordance with a model proposed by Roussel [10]. Moreover, it contains a critical shear rate below which no stable flow or solution can be attained. From the non-dimensional analysis of the expressions [6], one can derive that the critical shear rate $\dot{\gamma}_c$ is equal to the geometrical average of the Bingham shear proportion and the thixotropic shear proportion ($\dot{\gamma}_c = \sqrt{\frac{\tau_0}{\mu} \frac{1}{T\alpha}}$).

Based on the non-dimensional expressions, one can interpret the significance of a certain flow behaviour depending on the flow regime, as illustrated in Figure 5.2. For instance, when the pressure number is equal to 2, it means that the yield stress equally contributes to the pressure loss as the viscous shearing effects. Another example is that self-compacting concrete

Table 5.1 Overview of solutions for Poiseuille flow and non-Newtonian extensions. Extensions for yield stress (τ_0) fluids are denoted as a function of the non-dimensional pressure number Pn

Newtonian $\tau = \mu\dot{\gamma}$	$Q = \dfrac{\pi R^4}{8\mu}\dfrac{\Delta p}{L}$
Ostwald-de Waele $\tau = K\dot{\gamma}^n$	$Q = \pi\left(\dfrac{n}{3n+1}\right)\left(\dfrac{1}{K}\right)^{\frac{1}{n}}\left(\dfrac{\Delta p}{2L}\right)^{\frac{1}{n}}R^{\frac{3n+1}{n}}$
Bingham $\tau = \tau_0 + \mu\dot{\gamma}$	$Q = \dfrac{\pi}{12}\left[\dfrac{R^3\tau_0}{\mu}\right](3Pn - 4 + Pn^{-3})$
Herschel–Bulkley $\tau = \tau_0 + K\dot{\gamma}^n$	$Q = \dfrac{n\pi}{n+1}\left[R^3\left(\dfrac{\tau_0}{K}\right)^{\frac{1}{n}}\right]\dfrac{Pn^{-3}(Pn-1)^{\frac{n+1}{n}}}{(2n+1)(3n+1)}\cdot$ $\left((n+1)(2n+1)Pn^2 + 2n(n+1)Pn + 2n^2\right)$
Modified Bingham $\tau = \tau_0 + \mu\dot{\gamma} + c\dot{\gamma}^2$	$Q = \dfrac{2\pi}{105}\left[\dfrac{R^3\tau_0}{\mu}\right]MBn^{-4}Pn^{-3}\cdot$ $\begin{pmatrix} 16(\omega-1) \\ +4MBn(-\omega Pn+(7-6\omega)) \\ +2MBn^2\left(3Pn^2 + 8Pn + (24\omega - 35)\right) \\ +MBn^3\left(5Pn^3(6\omega-7) - 6\omega Pn^2 - 8\omega Pn + (35-16\omega)\right) \end{pmatrix}$
Simplified Roussel $\begin{cases}\tau = \tau_0(1+\lambda) + \mu\dot{\gamma} \\ \dfrac{\partial\tau}{\partial t} = \dfrac{1}{T} - \alpha\dot{\gamma}\lambda\end{cases}$	$Q = \dfrac{\pi}{48}\left[\dfrac{R^3\tau_0}{\mu}\right]Pn^{-3}\cdot$ $\begin{pmatrix} 6Pn^4(\cos(\phi)+1) \\ -8Pn^3\left(\cos(\phi)^3+1\right) \\ +3Pn^2\left(\cos(\phi)\left(8\cos(\phi)^2 - \rangle^2 + 2\rangle - 9\right)\right) \\ +6Pn\left(\cos(\phi)\left(-4\cos(\phi)^2 - 3\rangle^2 + 6\rangle + 1\right)\right) \\ +\cos(\phi)\left(8\cos(\phi)^2 + 2|\rangle^2 - 42\rangle + 15\right) + 2\rangle^3(-3\rangle + 4) \\ -3(\rangle - 1)^2(\rangle^2 - 2\rangle + 5)\ln\left(\dfrac{Pn-1}{\rangle - 1}(\cos(\phi)+1)\right) \end{pmatrix}$

Source: [6, 11].

Notes:

With pressure number $Pn = \dfrac{\Delta p}{\tau_0}\dfrac{R}{2L}$.

With modified Bingham parameters $MBn = \dfrac{4c}{\mu}\dfrac{\tau_0}{\mu}$ and $\omega = \sqrt{1 + MBn(Pn-1)}$.

With simplified Roussel parameters $\Lambda = \sqrt{\dfrac{4}{T\alpha}\dfrac{\mu}{\tau_0}}$ and $\sin(\phi) = \dfrac{\Lambda - 1}{Pn - 1}$.

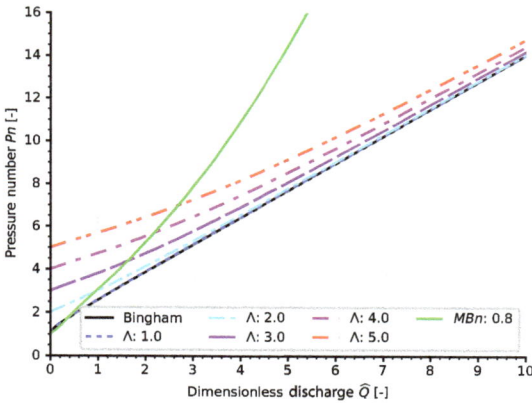

Figure 5.2 Non-dimensional discharge diagram for a spectrum of material parameters of cementitious suspensions, based on Poiseuille flow extensions (i.e. Bingham, modified Bingham and simplified Roussel). Source: after [6, 11].

(SCC) has a higher plastic viscosity than a traditional concrete. Thereby, during pumping at the same discharge Q, the non-dimensional discharge $\hat{Q}$ for the SCC will be higher than for the traditional concrete, despite its lower yield stress. Hence, the required pumping pressure will be higher for the SCC compared to the traditional concrete. Of course, this under the assumption that the lubrication action is the same for both and the bulk of both concretes is sheared. If not, it is obvious that a higher SCC viscosity would dominate the lubricating flow. Many other insights can be formulated from these non-dimensional expressions. For further reading one can consult the literature [6, 11].

5.1.2   Lubrication flow

The flow of concretes and mortars is, however, more complicated than Poiseuille flows. It comprises slippage or lubrication flow. Kaplan was the first pioneer to investigate the lubrication or slippage mechanisms of concrete pumping, and many others followed [12–24]. Even though the flow is more complicated, Poiseuille flow can easily be extended to slippage or lubrication flow as illustrated in Figure 5.3. Indeed, instead of starting from a zero velocity at the slippage interface R_i or lubrication boundary R_b, the Poiseuille flow velocity contribution of the bulk simply starts from the slippage or lubrication velocity. Hence, one can simply combine the flow contributions of slippage or lubrication and of additional bulk Poiseuille flow.

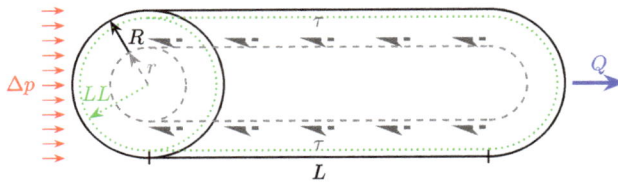

Figure 5.3 Pipe flow configuration in case lubrication flow is present, driven by pressure loss Δp and internal shear stress τ as a function of the pipe radius r. Two fluid phases are considered contributing to a discharge Q, i.e. a lubrication layer (LL) and a bulk fluid. Source: originating from [6].

However, one should bear in mind that lubrication or slippage flow comprises two different flow regimes. In the first flow regime only slippage occurs at the pipe wall interface or only the lubrication layer is sheared. In the second flow regime a flow contribution is added from shearing of the bulk concrete. Therefore, analytical Poiseuille flow solutions are still useful for lubrication or slippage flow types. However, one needs to adopt the expressions to account for slippage or lubrication action. In the literature, two expressions were derived in which the concrete bulk was considered as a Bingham fluid: the first one based on a slippage interface by Kaplan, the second one by consideration of a Bingham lubrication fluid of finite thickness [6, 25]. Respective slippage or lubrication flow solutions are summarized in Table 5.2.

Similar to Poiseuille flows of yield stress fluids, once again the pressure number Pn can be used to retrieve insights into the pumping behaviour depending on the occurring discharge Q. Furthermore, since the flow regime is twofold, either with or without bulk shearing, an additional pressure number can be defined. For slippage flow one can define a pipe wall interface pressure number ($Pn_{LL} = \dfrac{\Delta p}{\tau_{0LL}} \dfrac{R}{2L}$) with respect to the slippage interface yield stress τ_{0LL} and viscous constant η_{LL}. In addition, for lubrication flow one can define a bulk pressure number ($Pn_b = \dfrac{\Delta p}{\tau_0} \dfrac{R_b}{2L}$) with respective yield stress τ_{0LL}, plastic viscosity μ_{LL} and thickness e_{LL} which governs the flow until the bulk radius R_b. Nevertheless, the (bulk) pressure number Pn (Pn_b) is still important. When the pressure number is lower than 1, the bulk does not contribute to the flow, i.e. the first flow regime. Otherwise, the bulk contributes to the flow, i.e. the second flow regime. Further insights can be retrieved from these non-dimensional expressions for which literature can be consulted [6, 11].

Table 5.2 Overview of lubrication flow solutions

Kaplan slippage $\begin{cases} \tau = \tau_{0LL} + \eta_{LL}V_s & r = R_i \\ \tau = \tau_0 + \mu\dot{\gamma} & r < R_i \end{cases}$	$Q = \begin{cases} \pi\left[\dfrac{R^3\tau_{0LL}}{R\eta_{LL}}\right](Pn_{LL}-1) & Pn < 1 \\[2ex] +\dfrac{\pi}{12}\left[\dfrac{R^3\tau_{0LL}}{R\eta_{LL}}\right]\left[\dfrac{\tau_0}{\tau_{0LL}}\right]\left[\dfrac{R\eta_{LL}}{\mu}\right](3Pn-4+Pn^{-3}) & Pn \geq 1 \end{cases}$
Bingham lubrication $\begin{cases} \tau = \tau_{0LL} + \mu_{LL}\dot{\gamma} & r \geq R_b \\ \tau = \tau_0 + \mu\dot{\gamma} & r < R_b \end{cases}$	$Q = \begin{cases} \pi\left[\dfrac{R^3\tau_{0LL}}{\mu_{LL}}\right]\left(3Pn_{LL}\left(1-\beta^4\right)-4\left(1-\beta^3\right)\right) & Pn_b < 1 \\[2ex] +\dfrac{\pi}{12}\left[\dfrac{R^3\tau_{0LL}}{\mu_{LL}}\right]\left[\dfrac{\tau_0}{\tau_{0LL}}\right]\left[\dfrac{\mu_{LL}}{\mu}\right]\beta^3\left(3Pn_b-4+Pn_b^{-3}\right) & Pn_b \geq 1 \end{cases}$

Source: [6].

Notes:

With Kaplan slippage parameters $Pn_{LL} = Pn\dfrac{\tau_0}{\tau_{0LL}} = \dfrac{\Delta p}{2L}\dfrac{R}{\tau_{0LL}}$ and $R_i \approx R$.

With Bingham lubrication parameters $Pn_b = Pn\beta = \dfrac{\Delta p}{2L}\dfrac{R_b}{\tau_0}$ and $\beta = \dfrac{R_b}{R} = 1 - \dfrac{e_{LL}}{R}$.

5.1.3 Dry friction flow

A last flow type which may occur is dry friction flow. This typically occurs for unsaturated or 'rather dry' concretes. When insufficient cement paste is available in the mixture to lubricate suspended coarse and small aggregates, internal particle friction will govern the flow behaviour. The dominance of the inter-particle friction forces can be estimated by means of the Bagnold number [6, 26, 27]. A Bagnold number beyond 450 indicates rather dry granular flow behaviour [6, 26, 27]. As such, the flow behaviour will be of rather granular character and will no longer be independent of the applied pressure, e.g. by Coulomb friction behaviour. Since the third pipe flow behaviour also depends on the applied pressure, the longitudinal pressure loss evolves exponentially. Apart from thixotropically induced blocking, this is one of the main causes of blocking in pipelines. As a consequence, when pressure is applied to a granularly blocked concrete pipe water or cement paste will be squeezed out of the system. Hence, if granular blocking occurs, it is better to locally vibrate, rheologically activate or disassemble the pipe.

5.2 EXPERIMENTAL ARC PUMPING SETUP

To induce ARC during pumping, a specialized small-scale pumping circuit was designed, referred to as the small pumping system (SPS). The SPS was equipped with versatile monitoring devices for the purpose of gaining fundamental insights into pumping flow behaviour by acquisition of pumping data. Apart from monitoring, the SPS was also equipped with specially designed activation control units enabling active rheology control. The two following subsections therefore outline the used monitoring equipment and active control units.

5.2.1   Monitoring equipment

Several monitoring devices were installed and used to acquire pumping data, i.e. pressure sensors, a magnetic flow meter, a mass load cell, a temperature sensor and an ultrasonic velocity profiler (UVP). All of the data were recorded by a data acquisition system.

Parallel to the electromagnetic flow meter, as illustrated in Figure 5.4, the load cell was used to calibrate the electromagnetic flow meter for different cementitious suspensions. Unlike the conventional use of an electromagnetic flow meter of water, high differences in electric charges may exist for different cementitious mixture compositions. As such, an electromagnetic flow meter should be handled with care and be calibrated for each different mixture composition. Hence, a mass load cell was used to obtain the mass discharge $\dot{m}$, which was converted into the volumetric flow rate Q by means of measuring the density ρ of the suspension.

In addition, pressure sensors were inserted in special T-connectors (cf. Figure 5.4), each connecting different pipe sections. Philosophically, the sensing membrane of the pressure sensors should coincide with the virtually extended pipe wall, so that the pressure could be measured at the pipe wall itself. Practically, however, the sensing membrane was still slightly

Figure 5.4 Illustration of equipped monitoring devices in the small pumping system (SPS). (Left) Electromagnetic flow meter. (Centre) Load cell from which a container was supported to measure the mass flow rate. (Right) Pressure sensor screwed in a T-connection of the pipeline. Source: originating from [28].

embedded in the tubular column of the T-connector. Nevertheless, the static pressure can still be measured via a manometer tube principle in accordance with Bernouilli's principle. In that way, pressure losses during pumping can be monitored at desired strategic locations. However, although the height of the tubular column in the T-connector is very small (in order of magnitude of a few millimetres), its filling with a cementitious suspension may slightly reduce the measured pressure. If reduced pressure measurement potentially occurs, this is due to stiffness of the T-connector filling and thereby partial load transfer to the T-connector walls instead of the sensing membrane of the pressure sensor. Hence, the pressure sensors should be inserted as close as possible to the virtually extended pipe walls.

Lastly, a state-of-the-art UVP was equipped as illustrated in Figure 5.5. At a given location in the pipe, the flow velocity can be measured based on the Doppler principle of sound wave reflections in the direction of the flow. On the one hand, the Doppler device angle should be sufficient to allow for the Doppler effect to take place. On the other hand, the inclination angle should be kept to a minimum not to have distorted measurement along an inclined radial plane. As such ultrasound wave impulse responses in the form of frequency shifts and time delays can be used to reconstruct an entire velocity profile as a function of the pipe radius.

A schematic of the final SPS layout is depicted in Figure 5.6, whose practical outlook is illustrated in Figure 5.7. The numbers indicate respective locations of pressure sensors. Apart from the monitoring equipment, special active control units were designed and put in position between pressure sensors 2 and 3.

Figure 5.5 Illustration of equipped ultrasonic velocity profiler. (Left) Pipe section on which acoustic transducers are mounted under an inclined angle. (Right) Close-up of acoustic transducers by which ultrasound pulses travel from one sensor to another and back. Source: originating from [6].

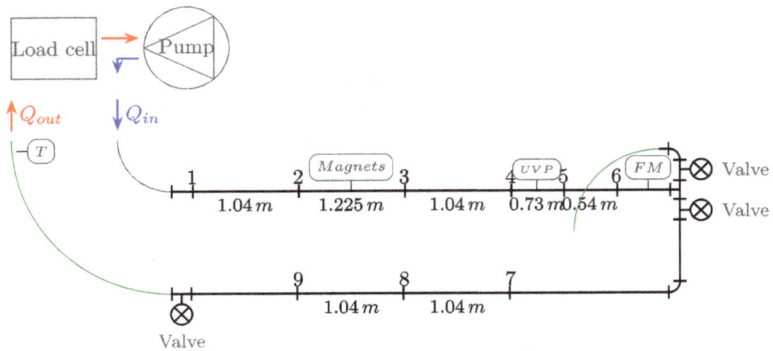

Figure 5.6 Schematic of the small pumping system (SPS) used for the magneto-rheology experiments. The pressure sensors are numbered at their corresponding location. The flow meter is indicated by FM, the ultrasonic velocity profiler by UVP and the temperature sensor by T. (Top) Layout during pumping phase 1, without magnetic field. (Bottom) Layout during pumping phase 2, with magnetic field. Plain pipe section 2–3 was replaced by a magnetic control unit consisting of static, permanent magnet pairs. Source: originating from [6].

Figure 5.7 Illustration of the SPS used for magneto-rheological pumping experiments. An auger pump is located at the bottom left, above which a container is supported from a load cell to measure the mass flow rate ṁ. Source: originating from [6].

5.2.2 Active control unit

To enable active rheology control applied to pumping, a special control unit was designed. A constant magnetic field perpendicular to the flow direction was established by use of permanent, static magnets. Alternatively, electromagnets could also be used to create a controllable magnetic field.

As illustrated in Figure 5.8, a special section was designed, which consisted of a transparent Plexiglas (polymethyl methacrylate) box in which permanent, static magnets could be fixed in pairs on opposite sides of the pipe section. Furthermore, it was designed in such a way that each sequence could individually be rotated around the axis of the pipe. In that way, the magnetic field could be alternated or rotated as cement paste longitudinally moves through the pipe. However, in current experiments the static magnets were identically positioned. Different configurations may be explored in future.

5.3 MAGNETO-RHEOLOGICAL ARC PUMPING EXPERIMENT

To investigate ARC applied to pumping, a pumping experiment was conducted in which a cementitious magneto-rheological paste was activated by the special magnetic control unit. The pumping experiment consisted of two phases. Firstly, the paste was pumped through the plain pumping circuit and, secondly, the same mixture was pumped through the same circuit after one section was replaced by a magnetic control unit. The magnetic control unit was a specially designed pipe section – of identical geometry and material (i.e. stainless steel with a low magnetic permeability) – on which permanent, static magnets were mounted. In that way, a rheological reference could be obtained from the first pumping phase, after which the pure magneto-rheological response could be obtained during the second pumping phase, i.e. the change in pressure loss of the replaced pipe section. Furthermore, to eliminate effects due to thixotropy, hydration and temperature, a pressure loss difference was computed between the section of interest (i.e. sections 2–3 in Figure 5.6) and a plain, non-magnetic reference section (i.e. sections 3–6 in Figure 5.6). Hence, the pure magneto-rheological response could be measured from the change in pressure loss section difference (i.e. 2-3_3-6) as depicted in Figure 5.9, independent of thixotropy, hydration and temperature effects. It is imminent from that a significant change in pressure loss occurs due to magnetic activation in the magnetic section, compared to another section where no magnetic control unit was inserted.

Although the confidence interval due to sensor accuracy did not allow to distinguish a different response for a different flow rate Q, arguments exist to speculate on the response as a function of the flow rate Q. For instance,

(a)   (b)

(c)   (d)

(e)

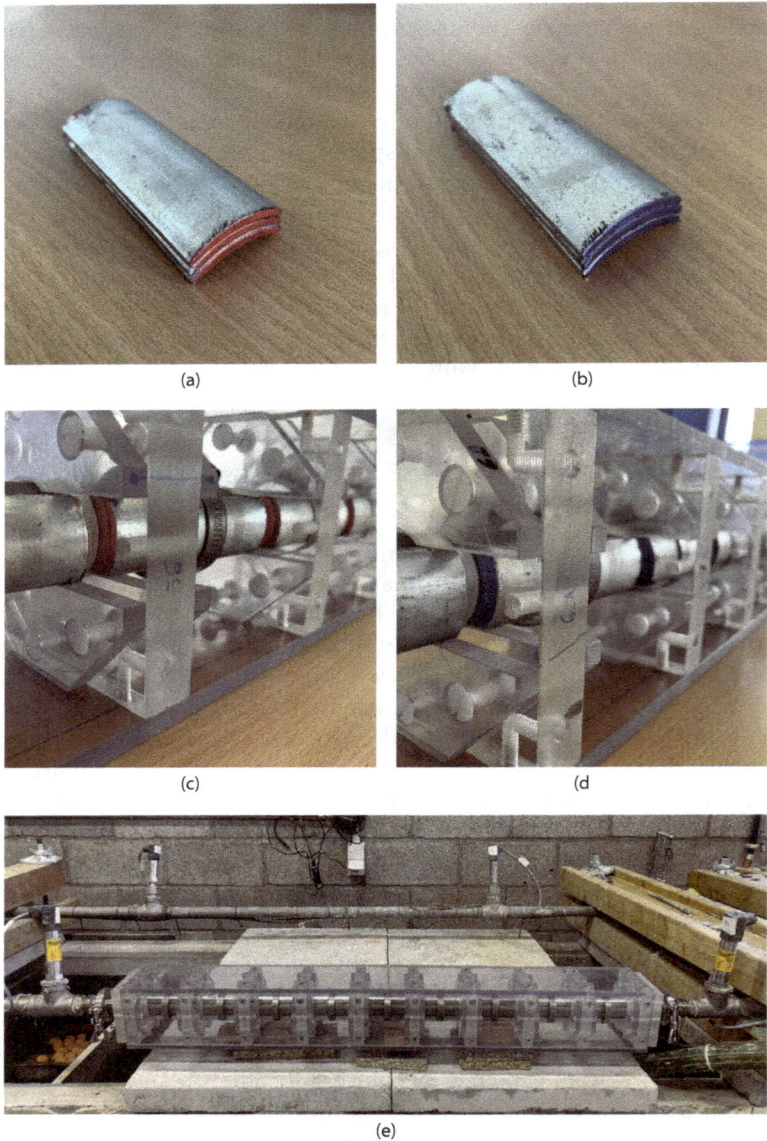

Figure 5.8 Illustration of the custom designed magnetic control unit. (a) Permanent arc magnets with an upward facing magnetic north pole. (b) Permanent arc magnets with an upward facing magnetic south pole. (c–d) Illustration of magnetic placement box. (e) Full magnetic section, with in total nine repeating units, each with three magnet pairs corresponding to 0.169 T. Source: originating from [6].

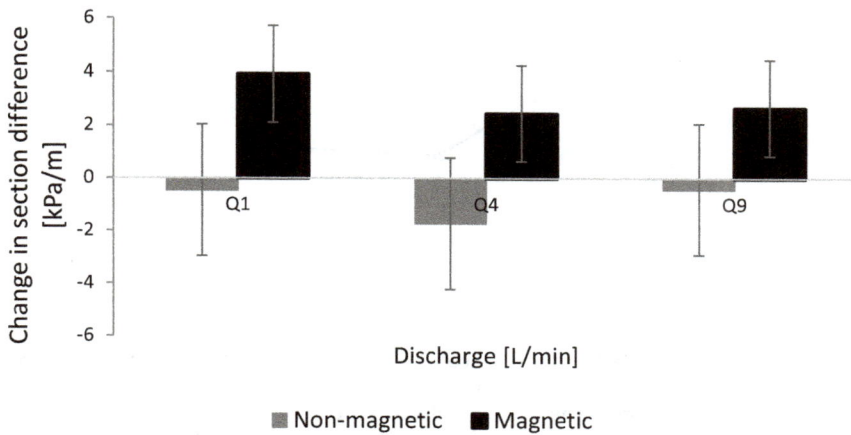

Figure 5.9 Change in pressure loss section difference before and after replacing section 2–3 with a magnetic control unit. The section difference between the section of interest (2–3) and another plain section (3–6) captures how the imposed magnetic field (B = 0.169 T) invoked an active response. Apart from the main section of interest (i.e. magnetic section 2–3), also a reference section 3–4 was considered for verification only. The reliability is indicated by confidence interval bars. Source: originating from [6].

the change in pressure loss was measured for the same section and hence the same pressure sensors and the measured pressure loss was in the same order of magnitude. If one assumes that a similar error is made for the measurement, one could speculate on the evolution as a function of the discharge Q. Doing so in Figure 5.10 it may be speculated that the magnetic response is higher for the lower discharge, after which it decreases for the intermediate discharge and slightly increases for the highest discharge. Nevertheless, further experimentation over a broader and more refined range of discharges should verify this speculation, since the currently available results show no statistically significant influence due to the high uncertainty on the obtained results [6].

A visual inspection was also conducted as illustrated in Figure 5.11. After the circuit was carefully rinsed with water, magnetic structures, known as spikes, formed inside the pipe parallel to the magnetic field (vertical). After the pumping circuit was cleaned with sponge balls, residual magnetic spikes were still attached to the pipe walls. This is due to magnetic attraction forces caused by strong magnetic gradients in the vicinity of the walls where the static magnets were mounted.

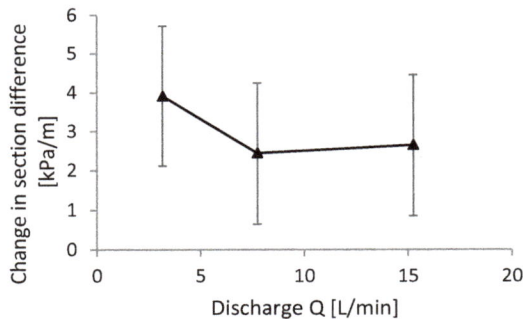

Figure 5.10 Statistically insignificant, speculated trend lines of change in pressure loss section difference before and after replacing section 2–3 with a magnetic control section. The section difference between a section of interest (2–3) and another plain section (3–6) captures how the imposed magnetic field (B = 0.169 T) invoked an active response. The reliability is indicated by confidence interval bars. Source: originating from [6].

In view of a numerical simulation of the magneto-rheological response through magnetic activation during pumping, a constitutive magneto-rheological ARC model has been developed. Further details are given in Chapter 8.

5.4   SUMMARY

In summary of applying ARC to pumping of cementitious suspensions, it can be concluded that active stiffening control could be successfully obtained for the pumping of a cementitious paste suspension. A significant response could be enabled and measured during pumping.

One should be aware, however, that the flow of cementitious suspensions can be more complicated than simple, homogeneous, Poiseuille flow. Indeed, mortars and concretes exhibit slippage or lubrication flow. Furthermore, if the cementitious suspension is too concentrated with aggregates, dry friction flow may occur with most likely granular blocking during pumping as a consequence.

Nevertheless, a custom pumping circuit was designed and used to investigate the fundamental flow behaviour of ARC applied during pumping. To do so, a small-scale pumping circuit, referred to as SPS, was equipped with several monitoring sensors and devices. The recorded data allowed to thoroughly analyse the obtained material response during pumping. More importantly, a special activating control unit was designed which allowed

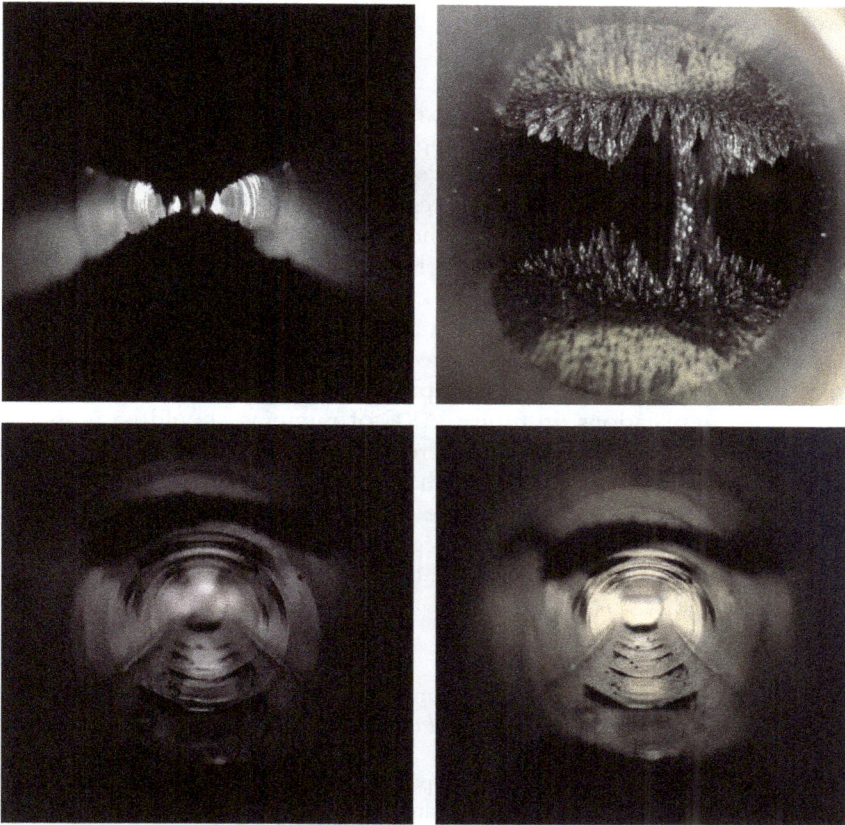

Figure 5.11 Visual inspections of the magnetic section 2–3 illustrating internal magnetic structuration. (a–b) After the circuit was gently rinsed with water. (c–d) After cleaning sponge balls were pumped through. Source: originating from [6].

to impose actively controlled rheological response. More specifically, a special pipe section was designed which consisted of a special box in which permanent, static magnets could be mounted. In that way, a static magnetic field could successfully be induced in a pipe section of interest, which allowed for active stiffening control.

Even though a single pumping experiment was conducted for a magnetorheological cementitious paste suspension by use of a magnetic fly ash, it was sufficient to observe a significant response during pumping. Although the accuracy of the employed pressure sensors did not allow to distinguish a different response as a function of the pumping flow rate, it could be speculated that a higher active stiffening response was observed for a lower discharge. Nonetheless, this should be verified by future, more versatile pumping experiments.

In future, active stiffening control during pumping has the potential to be applied to an activating pipe section in preparation to extrusion for 3D printing, shotcreting or other applications which require an increased stiffness in general. Contrary to active stiffening, active control of a reduction in flowability is another path of research that should be investigated. As such, it could be tested in future whether the use of an alternating or rotating magnet configuration or the use of electromagnets could induce magnetic micro-vibration. Future experiments could explore whether it could facilitate flow during pumping, or other applications where more flowability is desired. Moreover, a wider scope of materials could be explored where ARC could be applied during pumping. Firstly, although cement pastes have direct applications, in future, ARC of pumping should be up-scaled to more complicated mortar or concrete flows with associated challenges. Secondly, other magnetic additives than fly ash could be used in future to enable magneto-rheological response (cf. Chapter 3). Future research on active control could also be conducted for other forms of external stimuli, e.g. electrochemical control as illustrated in Chapter 4.

REFERENCES

[1] G. Hagen, Ueber die Bewegung des Wassers in engen cylindrischen Röhren, *Ann. Der Phys. Und Chemie.* 122 (1839): 423–442. doi:10.1002/andp.18391220304

[2] J.L.M. Poiseuille, Recherches expérimentales sur le mouvement des liquides dans les tubes de très-petits diamètres. I. Influence de la pression sur la quantité de liquide qui traverse les tubes de très-petits diamètres., *Comptes Rendus Académie Des Sci. Paris.* 11 (1840): 961–967. www.irphe.fr/~clanet/otherpaperfile/articles/Poiseuille/poiseuille1840a.pdf.

[3] J.L.M. Poiseuille, Recherches experimentales sur le mouvement des liquides dans les tubes de tres petits diametres; II. Influence de la longueur sur la quantite de liquide qui traverse les tubes de tres petits diametres; III. Influence du diametre sur la quantite de liquide, *Comptes Rendus Académie Des Sci. Paris.* 12 (1841): 112–115. www.irphe.fr/~clanet/otherpaperfile/articles/Poiseuille/poiseuille1841.pdf.

[4] S.P. Sutera, R. Skalak, the history of poiseuille's law, *Annu. Rev. Fluid Mech.* 25 (1993): 1–19. www.annualreviews.org/doi/pdf/10.1146/annurev.fl.25.010193.000245.

[5] P.-Y. Gires, Poiseuille et l'écoulement des liquides dans les capillaires, Bibnum, *Sci. l'ingénieur.* (2017): 0–11. http://journals.openedition.org/bibnum/1091.

[6] R. De Schryver, Active rheology control of cementitious materials: Numerical and experimental pumping investigation, PhD thesis, Ghent University, 2022.

[7]   E. Buckingham, On plastic flow through capillary tubes, *Proc. Am. Soc. Test. Mater.* (1921): 1154–1156. https://ci.nii.ac.jp/naid/10011031893/en/.

[8]   D. Feys, Interactions between rheological properties and pumping of self-compacting concrete, Ghent University, 2009. http://hdl.handle.net/1854/LU-948561.

[9]   J.F. Steffe, *Rheological methods in food process engineering*, Freeman Press, 1996. https://books.google.be/books?id=LrrdONuST9kC.

[10] N. Roussel, A thixotropy model for fresh fluid concretes: Theory, validation and applications, *Cem. Concr. Res.* 36 (2006): 1797–1806. doi:10.1016/j.cemconres.2006.05.025

[11] R. De Schryver, and G. De Schutter, Insights in thixotropic concrete pumping by a Poiseuille flow extension, *Appl. Rheol.* 30 (2020): 77–101. doi:10.1515/arh-2020-0103

[12] F. Chapdelaine, Étude Fondamentale et Pratique sur le Pompage du Béton, l'Université de Laval, 2007. https://corpus.ulaval.ca/jspui/bitstream/20.500.11794/18881/1/24177.pdf.

[13] T.-T. Ngo, E.-H. Kadri, F. Cussigh, R. Bennacer, and R. Duval, Practical tribometer to estimate pumpability of fresh concrete, *J. Asian Archit. Build. Eng.* 9 (2010): 229–236. doi:10.3130/jaabe.9.229

[14] E. Secrieru, D. Cotardo, V. Mechtcherine, L. Lohaus, C. Schröfl, and C. Begemann, Changes in concrete properties during pumping and formation of lubricating material under pressure, *Cem. Concr. Res.* 108 (2018): 129–139. doi:10.1016/j.cemconres.2018.03.018

[15] E. Secrieru, Pumpverhalten moderner Betone – Charakterisierung und Vorhersage Pumping behaviour of modern concretes – Characterisation and prediction (2018): 146.

[16] E. Secrieru, W. Mohamed, S. Fataei, and V. Mechtcherine, Assessment and prediction of concrete flow and pumping pressure in pipeline, *Cem. Concr. Compos.* 107 (2020): 103495. doi:10.1016/j.cemconcomp.2019.103495

[17] T.T. Ngo, E.H. Kadri, R. Bennacer, and F. Cussigh, Use of tribometer to estimate interface friction and concrete boundary layer composition during the fluid concrete pumping, *Constr. Build. Mater.* 24 (2010): 1253–1261. doi:10.1016/j.conbuildmat.2009.12.010

[18] T.T. Ngo, E.H. Kadri, F. Cussigh, and R. Bennacer, Measurement and modeling of fresh concrete viscous constant to predict pumping pressures, *Can. J. Civ. Eng.* 38 (2011): 944–956. doi:10.1139/l11-058

[19] T.-T. Ngo, E.-H. Kadri, F. Cussigh, and R. Bennacer, Relationships between concrete composition and boundary layer composition to optimise concrete pumpability, *Eur. J. Environ. Civ. Eng.* 16 (2012): 157–177. www.tandfonline.com/doi/citedby/10.1080/19648189.2012.666910?scroll=top&needAccess=true.

[20] H.D. Le, Etude de l'effet de la couche limite sur les profils de vitesses du béton pompé, PhD thesis, Université Cergy-Pontoise and Ghent University, 2014.

[21] H.D. Le, E.H. Kadri, S. Aggoun, J. Vierendeels, P. Troch, and G. De Schutter, Effect of lubrication layer on velocity profile of concrete in a pumping pipe, *Mater. Struct.* 48 (2015): 3991–4003. doi:10.1617/s11527-014-0458-5

[22] D. Feys, K.H. Khayat, A. Perez-Schell, and R. Khatib, Development of a tribometer to characterize lubrication layer properties of self-consolidating concrete, *Cem. Concr. Compos.* 54 (2014) 40–52. doi:10.1016/j.cemconcomp.2014.05.008

[23] D. Feys, K.H. Khayat, A. Perez-Schell, and R. Khatib, Prediction of pumping pressure by means of new tribometer for highly-workable concrete, *Cem. Concr. Compos.* 57 (2015): 102–115. doi:10.1016/j.cemconcomp.2014.12.007

[24] E. Secrieru, S. Fataei, C. Schröfl, and V. Mechtcherine, Study on concrete pumpability combining different laboratory tools and linkage to rheology, *Constr. Build. Mater.* 144 (2017): 451–461. doi:10.1016/j.conbuildmat.2017.03.199

[25] D. Kaplan, Pompage des bétons, École Nationale des Ponts et Chaussées, 2001. https://pastel.archives-ouvertes.fr/tel-01310219.

[26] A. Leonardi, F.K. Wittel, M. Mendoza, and H.J. Herrmann, Multiphase debris flow simulations with the discrete element method coupled with a Lattice-Boltzmann fluid, *III Int. Conf. Part. Methods – Fundam. Appl.* (2013): 276–287. doi:10.1.1.385.4415

[27] A. Leonardi, F.K. Wittel, M. Mendoza, and H.J. Herrmann, Coupled DEM-LBM method for the free-surface simulation of heterogeneous suspensions, *Comput. Part. Mech.* 1 (2014) 3–13. doi:10.1007/s40571-014-0001-z

[28] R. De Schryver, K. El Cheikh, K. Lesage, M.Y. Yardimci, and G. De Schutter, Numerical reliability study based on rheological input for Bingham paste pumping using a finite volume approach in OpenFOAM, *Materials (Basel).* 14 (2021): 1–23. doi:10.3390/ma14175011

Chapter 6

Application of active rheology control to concrete formwork leakage

Chizya Chibulu, Mert Yucel Yardimci, Karel Lesage, and Geert De Schutter

CONTENTS

6.1	ARC: A solution to formwork leakage problems	189
6.2	Theoretical concepts	191
	6.2.1 Factors affecting formwork leakage	191
	6.2.2 Principles of ARC for formwork leakage control	193
6.3	Development of set-up for experimental study of formwork leakage	195
6.4	Laboratory study on active stiffening control for formwork leakage	199
6.5	Reduction in flow rate due to magnetically induced blocking	201
6.6	Significance of flow area to particle size ratio (D_{gap}/d_p)	203
6.7	Correlation between MR response in rheometer and pressure-flow test	205
6.8	Reduction in filtration effects	208
6.9	Increasing the efficacy of ARC for formwork leakage control	209
	6.9.1 Optimization of magnetic particle concentration	209
	6.9.2 Adjustment of rheological properties of base fluid	210
	6.9.3 Maximization of magnetic field strength	211
	6.9.4 Challenging effects of formwork pressure	213
6.10	Further research and outlook	214
6.11	Summary	216
References		217

6.1 ARC: A SOLUTION TO FORMWORK LEAKAGE PROBLEMS

Formwork/forms are the moulds into which fresh concrete is placed until it hardens. Formwork is used for the containment and structural support of fresh concrete. The forms are used to position, align, size, shape and provide a good surface finish on the concrete element. Different structural elements such as beams, columns, walls, foundations and ceilings can be constructed using formwork. The use of self-compacting concrete (SCC) presents new possibilities for the creation of intricately shaped structures with good quality surfaces [1]. Formwork design and installation play an important role to accomplish this.

SCC in comparison to traditional concrete can be placed more easily due to its excellent flow properties. This opens the possibilities to increase the placement rate and speed up the construction process. Additionally, it eliminates the need for external vibration owing to its self-levelling nature, further increasing the casting rate [2, 3, 4]. On the downside, an increased rate of rise of the concrete within the forms generally translates into an increase in the pressure exerted on the forms [5, 2, 3, 4]. In fact, when SCC is pumped from the bottom of the formwork, this pressure can even exceed hydrostatic pressure. This is because the constant pump pressure keeps the concrete in motion, preventing it from exhibiting thixotropic behaviour [4, 6] which has widely been reported to be a dominant factor in reducing formwork pressure [7, 8, 9, 10, 11] . Due to this high pressure, the probability of excessive formwork deformation or failure also increases [6]. The high fluidity of SCC, coupled with increased formwork pressure and possible formwork deformation, provides the perfect recipe for excessive formwork leakage. As such close attention must be paid to the design of the formwork [2, 3]. The design and construction of the formwork must take into consideration not only the rigidity required to prevent formwork collapse, but also the tightness of the joints to avoid excessive leakages that could lead to honeycombing [5]. A study by University of Leuven and Belgian Building Research Institute [12] in collaboration with 18 industrial partners was undertaken to support and enhance on-site application of SCC. In addition to the material cost of SCC and excessive formwork pressure, a major concern raised by the contractors was thus unsurprisingly the formwork tightness requirements and the consequential need for supplementary sealing actions to prevent formwork leakage. Although several research studies exploring the relationship between SCC and formwork pressure have been conducted, there is currently only limited information on formwork tightness requirements [12].

The transportation, placement, consolidation, and resulting formwork pressure are significantly affected by the rheological properties of the fresh concrete [13]. An appropriate match between the rheological properties and casting technique is essential to obtain the required quality and performance for the specific structure [14]. A fundamental understanding of the influence of rheological properties on the tightness of formwork is thus necessary to be able to develop formwork tightness criteria. Intuitively, it can be said that from the material perspective, formwork tightness can be improved by simply lowering the concrete fluidity. However, the material requirements for lower formwork leakage, and those for pumping and good form filling are in contradiction. During pumping, the concrete must be flowable enough to avoid high pumping pressure and pipe blockages [15], as well as to facilitate good form filling. However, once the concrete is cast in the formwork, a higher flowability as already mentioned could result in increased formwork pressure and excessive leakages of formworks. Lowering the concrete fluidity would also necessitate the use of external compaction methods to consolidate the concrete which could also give rise to formwork leakages. This dilemma cannot be solved using the current practices as we are currently limited to passively rely on the behaviour of the concrete. There are currently no active means of control that could influence the flow behaviour of the concrete once it is already mixed and flowing in the pumping pipes or formworks. Although external parameter such as the pumping pressure or vibration methods can be used to invoke an effect, we are still essentially unable to actively change the constitutive behaviour of the concrete and are limited to a pre-defined rheological response [16, 17].

The study of active stiffening control to reduce formwork leakage could greatly enhance the concrete casting procedure by contributing towards the design of performant leak-proof formwork systems with electro-magnetic activation equipment. This could push the casting process towards a more automated and well controllable process. To spearhead this innovative concept, we embarked on a fundamental study of pressure-driven flow of cementitious materials through a confinement to simulate tightness of formwork joints under pressure. A small laboratory-scale test set-up was developed to include active control mechanisms to determine the viability of the concept of ARC for formwork leakage. The theoretical concepts, experimental methodology and findings are detailed in this chapter.

6.2 THEORETICAL CONCEPTS

6.2.1 Factors affecting formwork leakage

Three main factors influencing formwork leakage can be considered: the width of gaps between formwork joints, the physical properties of the material and the resulting pressure exerted onto the formwork.

Primarily, for leakage to occur there needs to be a pathway along which the cementitious material can flow through to exit the forms. This could be in the form of a crack, separation between formwork panels or insufficiently sealed tie hole. These pathways may be formed because of poor fabrication, damage during construction or insufficient tightening of the formwork. To prevent or minimize leakages, the contact surfaces of adjacent formwork panels must be as close as possible. However, in practice it is difficult to achieve formwork joints with perfect connections and thus different sealing methods are used in industry. As illustrated in Figure 6.1, the thickness of

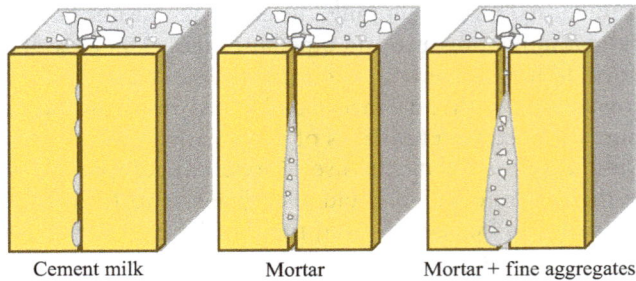

| Cement milk | Mortar | Mortar + fine aggregates |

Figure 6.1 Leakage of concrete through formwork joints. Source: after [18].

the gap between formwork panels affects the nature of flow through the gap, as well as the quantity and composition of the concrete leaking through the gaps [18].

Fluid flow through formwork joints can be worsened under the influence of high pressure. When formworks are improperly installed or are too flexible, displacement of the formwork panels can occur which would lead to leaks [19]. Provided concrete is in the plastic state, formwork pressure will exist. The vertical force generated by the self-weight of fresh concrete is transferred to the walls of the formwork and results in the compression of fresh concrete against the face of the forms [20]. To relieve the stress exerted, mortar may leak out through gaps which may be present at the concrete–formwork interface. For the pressure to begin to decrease, the material needs to build up an internal microstructure, be it due to chemical or physical phenomenon. During the early ages (plastic stage), the predominant influences are physical and depend on the degree of internal friction. Only after the dormant period does the chemical effect become more evident [8, 9, 11]. It therefore follows, for a faster rate of pressure decay, a faster rate of structural buildup is required [21]. A rapid build-up in structure of the bulk concrete however may not be appropriate in all instances. Take, for example, placement by injection or pumping techniques. If stoppage during casting were to occur, a material that builds up its internal structure too quickly would possibly lead to pipe blockages. This could result in the abuse of equipment ultimate capacity in order to resume casting. Another example would be in the case of multi-layer casting. Increased thixotropy heightens concrete sensitivity to delays between successive lifts. Weak interfaces may be created which could cause losses of mechanical strength [22]. ARC could provide a solution by using a localized approach, without negatively impacting the bulk concrete, by initiating structural build-up specifically at the formwork joints.

The ease with which the concrete flows through the formwork joints is dependent on rheological properties of the fluid. Mixtures with higher slump have been reported to exhibit more leakage, resulting in poor aesthetics [23]. Literature shows that fluidity of the concrete is influenced by the interstitial forces and distances between particles. Understanding the nature of the particle interactions is a step forward in determining what methods could be used to affect the flow between the formwork joints, especially if a localized approach is to be used to actively alter the rheological properties of the concrete in the vicinity of the joints. Fresh concrete, mortar or even cementitious pastes can be considered as a suspension, with rigid particles of various sizes suspended in the bulk fluid. Due to the heterogeneous nature of the material, different levels of fluidity can result in stable mixtures or the occurrence of segregation [24]. Studies on extrusion of cement-based materials show that such materials with high solid fractions behave as frictional plastic material and are thus highly sensitive to fluid filtration during slow flow and high-pressure gradients. During the extrusion process, a high-pressure gradient is created due to friction, which then induces liquid or cement paste filtration through the granular matrix. Different scales of filtration can be observed due to the different scales of heterogeneity (e.g. liquid phase filtering through cement grains or cement paste filtering through sand/aggregates) [25]. Rigid particles that are forced to flow from bulk region through a narrow constriction are known to exhibit different modes of flow depending on the material properties and boundary conditions. The particles may either flow steadily, intermittently or not at all. Blocking or clogging mechanisms of rigid particles in materials, such as suspensions and granular material, are primarily ruled by the ratio between the flow width (D_{gap}) and the particles size (d_p). In fact, there tends to be a threshold D_{gap}/d_p ratio at which full flow transitions into intermittent flow or blocking. At low D_{gap}/d_p, the particles are able to form stable arches across the constriction, thereby causing interruption of flow [26, 27, 28, 29]. This was also found to be the case in mortar extrusions where aggregate blocking was mainly affected by the D_{gap}/d_p ratio [30]. By effectively controlling this D_{gap}/d_p ratio using magnetically responsive additives, blocking could actively be induced to minimize formwork leakage.

6.2.2   Principles of ARC for formwork leakage control

Magnetorheological (MR) fluids form part of a set of fluids that are categorized as smart or actively controllable fluids. In the presence of an externally applied magnetic field, MR fluids can be transformed from a liquid into a semisolid [31, 32, 33, 34]. MR fluids are conventionally used in three principal modes of operation (Figure 6.2): valve mode (pressure-driven flow), direct shear mode and squeeze mode. The configuration in the valve mode best mimics that of formwork leakage, that is, pressure-driven

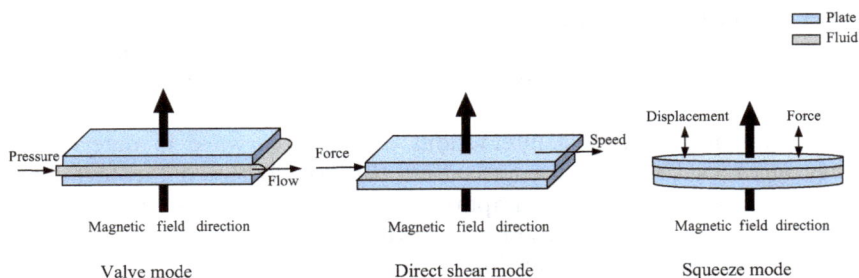

Figure 6.2 Basic operating modes for controllable fluid devices. Source: after [36].

flow between two parallel plates (formwork panels). In the valve mode, a magnetic field is applied perpendicular to the direction of fluid flow by placing two fixed magnetic poles on opposite sides of the flow channel. A controllable valve mechanism is created by varying the magnetic field, which in turn allows for the variation of the pressure threshold [35].

The externally applied magnetic field forces the suspended magnetic particles in the fluid to form chain-like structures or bunch together, thereby creating resistance to the free fluid motion [31, 36]. In the presence of the magnetic field, each of the particles becomes a dipole (north and south) and this tends to cause the formation of chains with neighbouring particles. The arrangement of the particle chains is dependent on the pattern of magnetic flux paths and, in response, the rheological behaviour depends on the strength of the magnetic field applied. The level of magnetic field required for rheological control is dependent on the formulation of the MR fluid, particularly the quality and quantity of the magnetizable particles. The yield stress capability of the fluid depends on the percentage and size of particles, as well as their magnetization characteristics. A higher percentage and larger sizes of these particles increase the yield stresses that can be achieved [32].

MR fluids are formulated using three basic components: a base fluid, magnetically polarizable particles suspended in the base fluid and stabilizing additives to overcome sedimentation of the particles [31, 32]. The resulting magneto-rheological behaviour is strongly dependent on the material selection. The selection of base fluid and magnetizable particles is in turn dependent on the specific application [31]. For the current application, the cementitious paste forms the base material. Magnetizable particles would then need to be added to the paste to make it magnetically responsive. Fly ash is composed of aluminosilicate, with ferric oxide and other metal oxides blended into its crystal lattice. This means that in the presence of a magnetic field, fly ash can be magnetized [37]. Fly ash can thus provide the

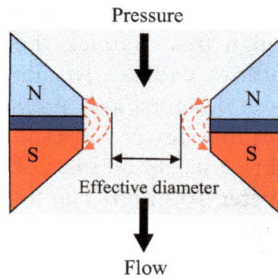

Figure 6.3 Illustration of magnetic gradient pinch (MGP) mode. Source: after [38].

magnetically polarizable particles to form a magnetically responsive fluid as previously described in Chapter 3. Other available sources of magnetizable particles include, but are not limited to, Fe_3O_4 iron particles, carbonyl iron particles, calcium aluminate cement or synthetic calcium aluminate aggregates.

Conventionally, MR fluids in the valve mode work by generating a field-dependent yield stress throughout the entire active volume of the fluid. The fluid behaves as a solid until the pressure supplied is sufficient to overcome the yield stress, after which the entire body of fluid flows as a plug. Goncalves and Carlson reported of a different type of valve called the magnetic gradient pinch (MGP) valve [38]. The difference between the two types of valves lies in the design of the magnetic circuit. In the MGP valve, the application of the magnetic field leads to a change in the effective diameter of the MGP orifice. Instead of solidifying the entire volume of fluid, only the fluid near the walls of the valve is solidified through the application of a non-uniform magnetic field as shown in Figure 6.3. The magnetic field strength then controls the inward distance from the wall in which solidification of the fluid occurs, thereby controlling the flow diameter [38]. At a sufficiently high magnetic field, the flow diameter could thus be blocked, and the flow of the fluid prevented. The working principles of MR fluids were thus implemented in the laboratory-scale test set-up to test the viability of ARC for formwork leakage control. This is detailed in the next section.

6.3 DEVELOPMENT OF SET-UP FOR EXPERIMENTAL STUDY OF FORMWORK LEAKAGE

The concept of confined flow through small openings coupled with MR technology was used to develop a device to study the flow of cementitious materials (pastes and mortars) through a confinement to imitate leakage though formwork joints. Since there is a direct relationship between the formwork gap width, the maximum size of particles in the fluid and the resulting flow regime through this channel, the test set-up was designed to include changeable outflow widths. To simulate different formwork pressures, the formwork was designed as a pressurized cylinder, similar to a previously developed portable device for evaluating formwork pressure [39]. The newly developed test set-up consists of a vertical steel column of circular cross section, diameter 100 mm. The top plate of the test cylinder is connected to an air compressor that supplies a regulated amount of air pressure into the device. The bottom plate of the test cylinder consists of interchangeable plates with various slit sizes so studies on the effect of different flow areas and geometries can be done. Figure 6.4 shows a schematic of the set-up, while Figure 6.5 shows a photograph. The test cylinder is filled with paste to a height of approximately 250 mm. A specified overhead pressure is then applied for a few seconds to obtain a stable pressure value. The slit at the bottom of the cylinder is then opened, allowing the material to flow out. The mass of the outflow is captured using an electronic balance equipped with data acquisition software. A calibrated pressure sensor is placed at the top plate of the test cylinder to measure the overhead pressure applied to correlate it with the resulting flow rate. The overhead pressure can be varied to simulate different casting rates or heights, while the effect of varying rheological properties can be investigated by varying the mix design of the test fluid, keeping all other test variables constant. If

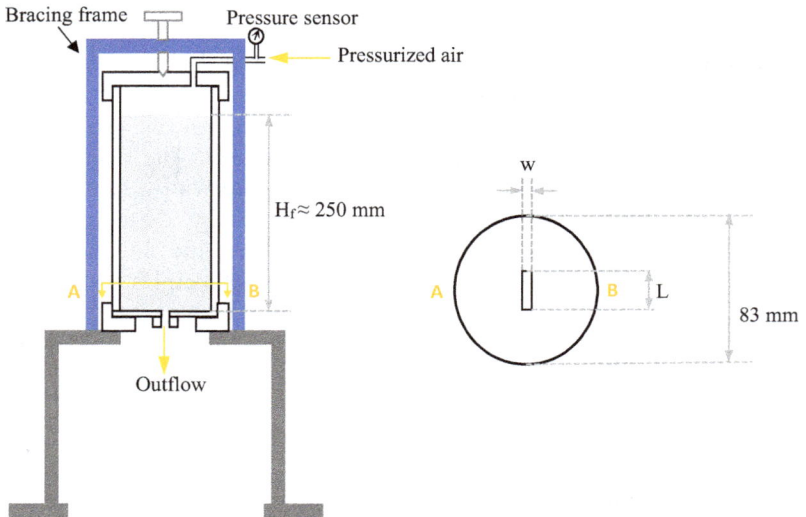

Figure 6.4 Experimental set-up for the study of leakage.

Figure 6.5 **A photograph showing the experiment set-up for the study of leakage (Courtesy C. Chibulu).**

desired, the rheological properties of the pastes can be modified by using workability boxes and a "vectorised rheograph" approach [40]. While this design concept may not be directly representative of the actual conditions in formwork, it was considered a sufficient base for a fundamental study to be conducted that could later be extrapolated to a more representative test operation.

To test the hypothesis of ARC for formwork leakage control, two different types of bottom plates were designed for each identical geometry: a reference plate in which no ARC was applied and a corresponding plate in which a magnetic field was generated across the flow channel to induce blocking. The magnetic field was applied by externally attaching two block magnets (dimensions 20 mm × 5 mm × 5 mm) on opposite ends of the outflow slit (see Figure 6.6). As observed from Figure 6.6, these static magnet blocks generate an external magnetic field approximately perpendicular to the flowing direction through the slit. To maintain the same geometry in the case of no magnetic field, two aluminium blocks (with the same dimensions as the magnets) were attached to the reference bottom plate of the same geometry. Prior to testing, the space between the magnets/aluminium blocks

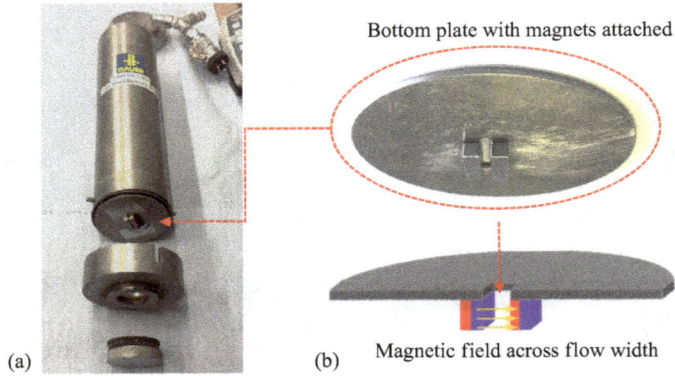

Figure 6.6 (a) Assembly of tests device and (b) schematic of bottom plate with magnets (Courtesy C. Chibulu).

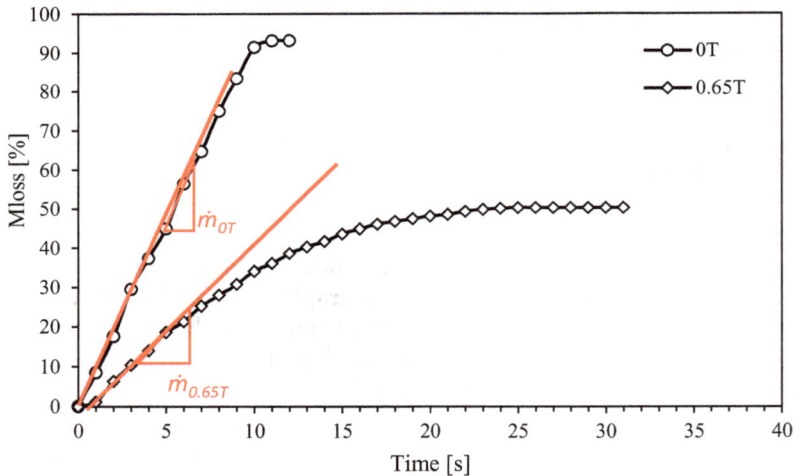

Figure 6.7 Example of extrapolation of initial mass flow rate.

was temporarily filled with a rubber filling to prevent the paste from prematurely forming chains before the bottom lid was opened. In doing so, the ability of the paste to form chains under pressure can be more accurately assessed and is in fact more representative to the actual case in formworks where the material would not be allowed to sit between the joints before flow is initiated.

The mass of the outflowing material was recorded as a function of time. Relative to the total mass initially present in the pressurized cylinder, the mass of accumulated out-flown material can be expressed as a percentage of mass loss (M_{loss}). The evolution of the percentage mass loss under various pressure levels could be plotted. The mass flow rate $\dot{m}$ was determined during the initial stage, when the slope of the graph was approximately purely linear (see Figure 6.7).

6.4 LABORATORY STUDY ON ACTIVE STIFFENING CONTROL FOR FORMWORK LEAKAGE

The mixture proportions using fly ash as a responsive additive are shown in Table 6.1. The water content and the total volume of powder were kept constant. Two pastes were selected to study the effect of fly ash content on the confined flow of paste under magnetic field. The pastes contained 40% fly ash by total volume of binder. The superplasticizer content was adjusted to attain slump values of 200 mm and 300 mmfor CF40L and CF40H pastes, respectively. The "L" in CF40L thus represents a low slump value, and the "H" in CF40 represents a high slump value.

Furthermore, eight pastes were prepared using Fe_3O_4 particles as responsive additive. The mixture proportions are shown in Table 6.2 and Figure 6.8. The water content and the total volume of powder were kept constant. The superplasticizer content was then adjusted to attain slump flow diameters of approximately 130 mm, 240 mm and 300 mm for the reference pastes with low, medium and high slump flow, respectively. The medium slump pastes Ref-M, M1, M2 and M3 were used to study the effect of nanoparticle concentration on the magnetic response, while Ref-L, L3, Ref-M, M3, Ref-H and H3 pastes were used to study the effect of the paste rheology on the magnetic response for a constant nanoparticle concentration.

To study the effects of addition of ultra-fine aggregates to the magnetic response of the material and its consequent effects on the blockage of flow, five pastes containing 0%, 10%, 20%, 30% and 40% ultra-fine quartz aggregate with D50 of 170 mm were also tested. The paste mixture proportions were kept constant and are as shown in Table 6.3 and Figure 6.9.

Mini-slump tests were performed immediately after mixing to make an initial flow characterization of the material. A truncated steel cone (mini-slump

Table 6.1 Mixture proportions and rheological properties of fly ash pastes tested

Mix No.	W/P* [-]	CEM [vol. % binder]	FA [vol. % binder]	SP [wt. % cement]	Mini-slump [mm]	Yield stress [Pa]	Viscosity [Pa·s]
CF40L	0.34	60	40	0.39	200	33	3.3
CF40H	0.34	60	40	0.72	300	2	1.8

Note: * W/P = water to powder mass ratio.

Table 6.2 Mixture proportions and rheological properties of Fe_3O_4 pastes tested

Mix. No	W/C ratio	SP [wt.% cement]	MNP [% paste mass]	Mini slump [mm]	Yield stress [Pa]	Viscosity [Pa·s]
Ref-L	0.4	0.2	0	130	142	0.90
L3	0.4	0.2	3	120	199	1.21
Ref-M	0.4	0.4	0	238	81	0.43
M1	0.4	0.4	1	235	100	0.47
M2	0.4	0.4	2	235	132	0.60
M3	0.4	0.4	3	230	144	0.60
Ref-H	0.4	0.6	0	300	30	0.21
H3	0.4	0.6	3	285	33	0.45

Note: L, M and H means low, medium and high slump flow, respectively.

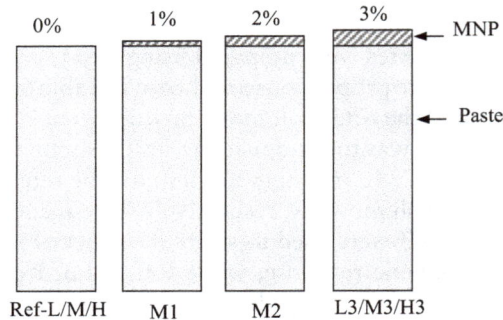

Figure 6.8 Mixture parameters for Fe_3O_4 pastes.

Table 6.3 Mixture proportions and rheological properties of Fe_3O_4 mortars

Mix. No	W/C	SP [wt. % cement]	VMA [wt. % cement]	Sand [% total volume]	NP [% total volume]	Mini slump [mm]
M0	0.55	-	0.025	0	1	255
M1	0.55	0.05	0	10	1	255
M2	0.55	0.15	0	20	1	255
M3	0.55	0.4	0	30	1	255
M4	0.55	1.0	0	40	1	255

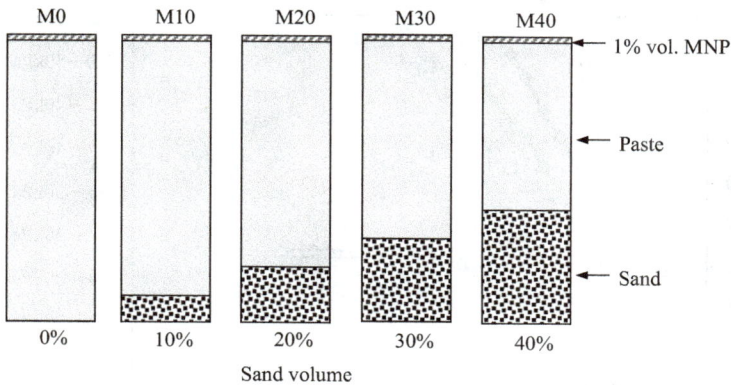

Figure 6.9 Mix parameters for Fe$_3$O$_4$ paste and mortar.

cone) with dimensions 70 mm (upper diameter), 100 mm (bottom diameter) and 60 mm (height) was positioned on a flat steel plate. The cone was filled without any compaction, and the spread diameter was recorded as the average of two measurements taken in perpendicular directions.

The developed pressurized cylinder was used to mimic formwork leakage under pressure. The outflow geometry selected was a rectangular cross-section, with dimensions 2 mm × 20 mm, and height of approximately 7 mm. The overhead pressure was varied between 10 and 50 kPa (0.1 and 0.5 bar), and the magnetic field strength was approximately 0.65 T and 0 T for the case with and without ARC, respectively.

A rotational parallel plate rheometer (Anton Paar – MCR 102) with 20 mm disc diameter was used to perform the rheological measurements. A fixed gap of 1 mm between the top and bottom plates was used and the temperature was kept constant throughout testing at 20±0.5 °C. Flow curve tests as well as oscillatory time sweep tests have been performed, following the principles given in Chapter 2. Resulting values for yield stress and viscosity are given in Table 6.2 and Table 6.3. A summary of the main findings of the laboratory study is given in the following sections.

6.5   REDUCTION IN FLOW RATE DUE TO MAGNETICALLY INDUCED BLOCKING

The percentage mass loss from the test cylinder is plotted as a function of time in Figure 6.10. Without magnetic field, the curves are generally linear with a plateau at the end, representing the end of the tests when no more material flows out. At the end of the test, the test cylinder was almost completely emptied, with only less than 10% of the material remaining in

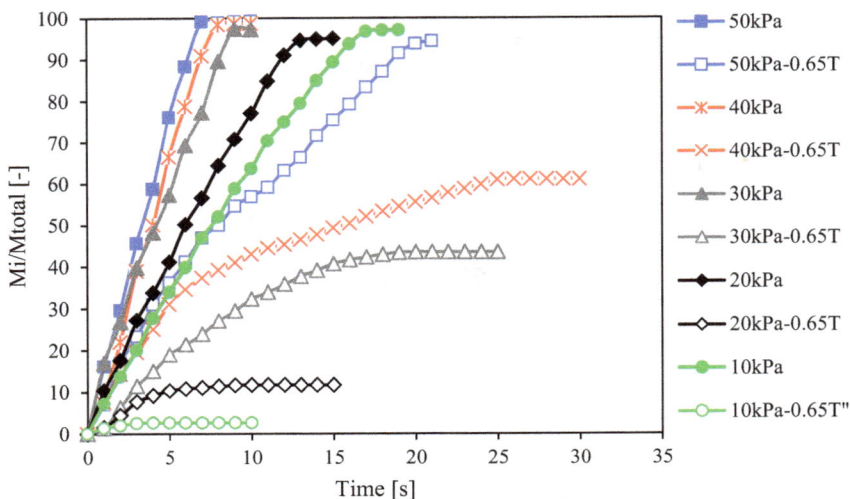

Figure 6.10 **Example of mass of outflow versus time graph with/without magnetic field application – H3 paste.**

the column. The remaining material corresponds to the dead zones at the bottom of the column where the material remains stationary [25, 41]. The linear evolution signifies a constant flow rate, irrespective of the height of fluid in the column. In contrast, when a magnetic field is applied, a considerable reduction in both the flow rate and the percentage mass loss is observed, most significantly in the low-pressure range (10 kPa and 20 kPa). Under magnetic field, the curves show an initial linear region, followed by a continuing decrease in slope indicating a gradual reduction in flow rate until the test cylinder was emptied or complete blockage occurred.

This difference in flow behaviour with and without ARC can be attributed to the formation of chains or agglomerates of the magnetic particles. Like in an MR fluid, when a static magnetic field is applied across the flow width, some of the magnetic particles align along the magnetic field lines, thereby forming chains and agglomerates. These chains and agglomerates give the fluid an additional yield stress component that is dependent on the magnetic flux density [35]. In this case, however, there is an even larger effect because of the dynamic nature of the problem in which the number of magnetic particles reaching the constriction increases with time. As the fluid flows through the slit, the magnetizable particles are attracted by the magnetic field and can deposit on the walls of the slit. The deposition of particles creates a dense network, pinching the flow path closer and closer together (as shown in Figure 6.11) until in some case complete blockage occurs.

Figure 6.11 Deposition of magnetizable particles as visible after gently rinsing off the paste from the bottom plate (Courtesy C. Chibulu).

Images of the flow patterns were captured and are shown in Figure 6.12 for the CF40H paste. The images show a relatively uniform flow pattern throughout the duration of the test in the case of no ARC. Under magnetic field however, there is an obvious decrease in the stream width, indicating a gradual reduction in the volume of paste exiting the test cylinder. This corresponds to the formation and deposition of magnetic cluster, as was further validated by additional experiments reported in [40].

6.6   SIGNIFICANCE OF FLOW AREA TO PARTICLE SIZE RATIO (D_{GAP}/D_P)

The D_{gap}/d_p ratio is one of the most important factors that dictates whether the resulting flow will be uninterrupted, partially, or completely blocked. This ratio can be adjusted by either changing the particle sizes suspended in the fluid or by altering the flow area. In this study, the ratio was adjusted by increasing the flow width from 1 mm to 2 mm. The results for the CF40H pastes are shown in Figure 6.13.

When the flow area is significantly smaller such that the D_{gap}/d_p ratio is below a critical threshold, blocking occurs even without magnetic field

B = 0 T; P = 10 kPa; 2 mm gap width

B = 0.65 T; P = 10 kPa; 2 mm gap width

Figure 6.12 Flow patterns with/without ARC – 2 mm gap, B= 0T, 0.65T, P = 10 kPa (Courtesy C. Chibulu).

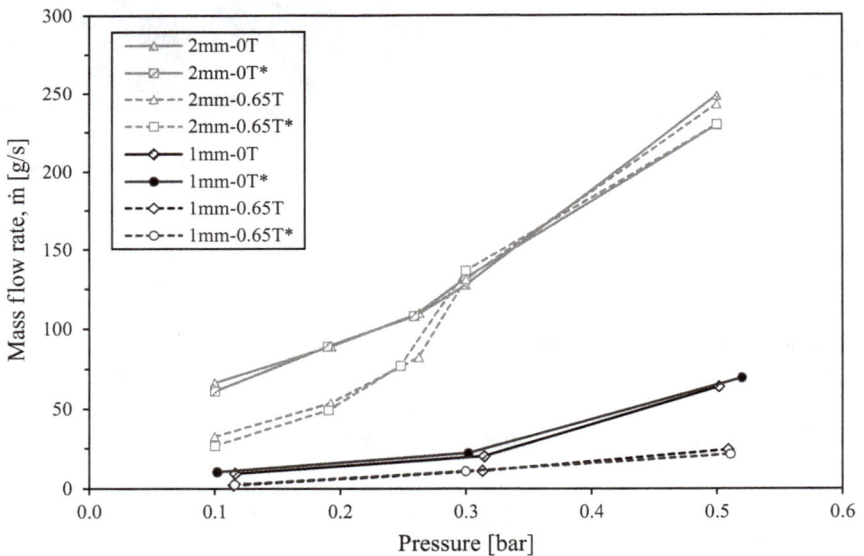

Figure 6.13 Effect of flow area on initial mass flow rate – CF40H paste. Source: [39].

application. When rigid particles flow through a constrained path, there is a convergence in flow towards the gap and many particles arrive at the outflow area simultaneously. Conversely, because of this constriction, the rate of particles exiting through the outflow can be lower than the rate of particles arriving at the outflow. This could then lead to an accumulation of

Figure 6.14 Image of the bottom plate after the test with variation in flow area and without magnetic field (left) A_1 – agglomeration of particles leading to blocking; (right) A_2 – no agglomeration of particles at the slit resulting in full flow. Source: [39].

particles and the formation and growth of agglomerates that in time block the flow. The larger the particles in relation to the flow area, the easier it is for these agglomerates of particles to form and consequently block the flow [42]. This aggregate build-up is said to be rather complex to analytic-ally or numerically describe due to changing conditions at the constriction with time. There is a continuous modification of the local geometry of the wall, the flow field and the nature of the interactions due to the presence of previously deposited particles [43]. Applying a magnetic field would in this case further boost formation of this dense agglomerate, driving the flow to block quicker.

When the D_{gap}/d_p ratio was significantly higher than the threshold, the pastes flowed continuously at all pressure values when no magnetic field was present. This resulted in almost complete emptying of the test cylinder. There is no agglomeration of particles at the outflow such as the case in the low D_{gap}/d_p case (Figure 6.14). When the ARC method was used, however, a reduction in flow rate and complete blockage of flow were observed at lower pressure values. The ARC method is hence most beneficial for cases in which the D_{gap}/d_p is above the critical threshold. The onus is on determining the range of this threshold value.

6.7 CORRELATION BETWEEN MR RESPONSE IN RHEOMETER AND PRESSURE-FLOW TEST

The relative change in viscosity (k_μ) was calculated as the ratio of the viscosity obtained under 0.65 T magnetic field ($\mu_{0.65T}$) over the viscosity without magnetic field (μ_{0T}). The relative change in storage modulus ($k_{G'}$)

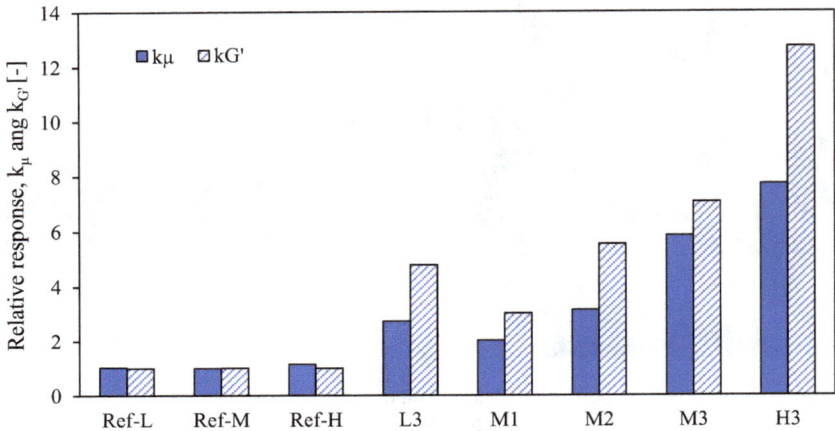

Figure 6.15 Relative change in viscosity and storage modulus due to applied magnetic field.

was also determined as the ratio of the storage modulus at 300 s obtained under magnetic field $G'_{0.65T}(300s)$, over the storage modulus without magnetic field $G'_{0T}(300s)$. The resulting values are graphically summarized in Figure 6.15. For the reference pastes, k_μ and $k_{G'}$ are approximately 1, indicating that the viscosity and storage modulus remain constant in the presence of the magnetic field. However, with the incorporation of MNPs, k_μ and $k_{G'}$ increase significantly under magnetic field. For the same MNP addition (3%), the paste rheology shows a clear effect on the relative change in viscosity and storage modulus. The k_μ and $k_{G'}$ value become more significant with decreasing stiffness. The highest relative change observed in H3 is due to the very low stiffness of the paste without magnetic field. This clearly illustrates that pastes with lower stiffness have a higher capacity of magnetic response under static conditions. Moreover, an increase in viscosity due to magnetic field is certainly beneficial for flow reduction.

The ARC method illustrated shows that a reduction in flow rate can be obtained under magnetic field in the pressure range of 10–50 kPa. However, the most significant reductions in percentage mass loss were only observed at 10 kPa and 20 kPa. As such, only the 10–20 kPa flow results were used to determine the relationship between the MR response and the pressure-driven flow. The k_μ and $k_{G'}$ value was used to relate the flow rate at 0 T ($\dot{m}_{0T}$) with that at 0.65 T ($\dot{m}_{0.65T}$). The experimental $\dot{m}_{0.65T}$ was correlated to an estimated value derived by dividing the experimental $\dot{m}_{0T}$ value by the magnetic response k_μ or $k_{G'}$. The results are plotted in Figure 6.16(a) and (b) for k_μ and $k_{G'}$ factors, respectively. The results show

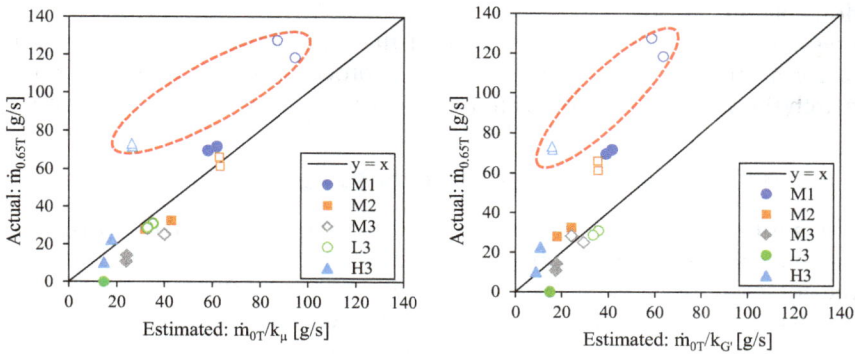

Figure 6.16 Relationship between MR response and pressure-driven flow at 10 and 20 kPa overhead pressure – MNP pastes.

that at low pressure values, the flow rate under magnetic field can be reasonably estimated using the magnetic response factors. At 20 kPa however, the M1 and H3 pastes which have the lowest concentration of MNPs and highest flowability respectively, significantly deviate from the line of equality. On the one hand, the low amount of MNPs in the M1 paste could give an overestimation of the magnetic response in the rheometer in comparison to the actual conditions in the pressure-driven flow tests. This may suggest that the method of estimation may be less accurate for pastes with very low concentrations of MNPs. On the other hand, the high fluidity of the H3 pastes makes it more sensitive to increases in flow rate under increased pressure. The magnetic response in the rheometer could thus give an overestimation of the potential flow reduction since the negative effect of the low viscoelastic properties under pressure is not evident. The reverse argument could also explain why the flow rates in the L3 pastes are slightly overestimated at 10 kPa.

It is unsurprising that at higher pressure (20 kPa), and therefore higher fluid velocity, the estimated flow rate deviates from the line of equality since the $k_{G'}$ factor is obtained in near-static conditions. In this case the $k_{G'}$ factor overestimates the effectiveness of the magnetically induced flow reduction. It is thus hypothesized that the use of the $k_{G'}$ factor in estimating the flow rate under magnetic field is limited to low pressure ranges, especially for highly flowable mixtures which are more sensitive to increases in pressure. Conversely, the k_{μ} factor is obtained under shear and would thus perhaps explain the less pronounced deviation in estimated flow rate values in comparison to those from the $k_{G'}$ factor. Nevertheless, as the pressure is increased, the shear rate in the pressure-driven tests presumably exceeds the shear rates used in the flow curve protocol, specifically in the highly flowable mixtures which are more sensitive to increases in flow rate under pressure. It can therefore be theorized that to obtain a more accurate k_μ factor at higher pressures, the shearing protocol in the flow curves should match the shearing conditions to avoid overestimation in flow reduction.

6.8   REDUCTION IN FILTRATION EFFECTS

The addition of ultrafine sand to the cementitious pastes showed that the volume concentration of fine aggregates influences the flow type as well as the potential for blocking (Figure 6.17). From 0% to 30% volume addition, the material behaves similar to paste and flows uniformly through the slit when no magnetic field is applied. However, at 40% volume concentration of fine sand, the flow behaviour changes, and filtration effects are observed. The fluid does not flow uniformly but instead only the liquid fraction is lost from the test cylinder. With increasing volume fraction of sand, the distance between sand particles decreases and their interactions increase. Due to the confinement at the outflow, as the particle interactions increase, the probability of blocking also increases which explains the filtration effect.

Without ARC, the 0% to 30% volume fraction had similar flow rates and flow behaviour. The material flows uniformly until the test cylinder is emptied. However, with 40% addition of fine sand, filtration effects were observed and only the liquid fractions were lost from the test cylinder. When the ARC method was applied to the 40% sand sample, the filtration effect was drastically reduced.

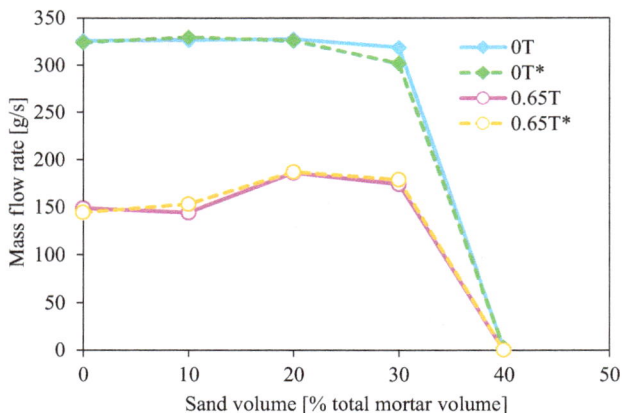

Figure 6.17 **Effect of volume fraction of aggregates on flow rate and magnetically induced blocking.**

6.9 INCREASING THE EFFICACY OF ARC FOR FORMWORK LEAKAGE CONTROL

6.9.1 Optimization of magnetic particle concentration

Firstly, the type and size of magnetic particles used influence the resulting magnetic response. Particles with a higher saturation magnetization would give a larger response to the applied magnetic field. Fe_3O_4 magnetic particles have a higher saturation magnetization than the fly ash powder since the fly ash does not contain 100% magnetizable particles. Therefore, the quantity of magnetizable particles to be used for a given set of conditions depends on the source of the particles.

Intuitively, increasing the number of magnetizable particles increases the magneto-rheological effect. This has already been illustrated by the relative change in viscosity and storage modulus detailed in section 6.7. in which the magnetic response was seen to increase with increasing amount of Fe_3O_4 nanoparticles in the paste volume. Similarly, increasing the concentration of magnetizable particles increases the probability or degree of blocking in the pressure-flow tests. As the concentration of MNPs was increased from 0% to 3%, the magnetic response of the cement pastes was visibly intensified (Figure 6.18). The reduction in flow rate was found to be proportional to the concentration of MNPs. The higher the concentration of MNPs, the greater the reduction in flow rate. Increasing the magnetizable particles

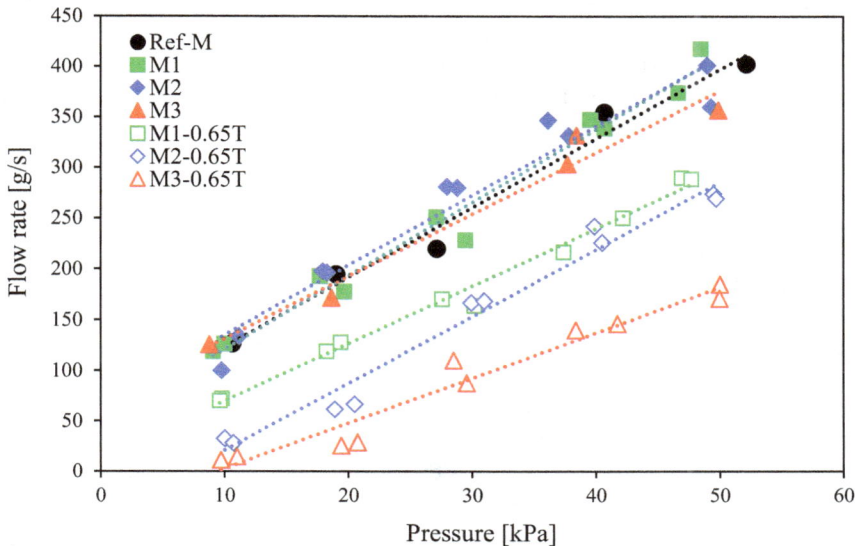

Figure 6.18 Effect of MNP concentration on initial mass flow rate – medium slump paste.

concentration increases the number of particles contributing towards the magnetic clustering or chain formation, resulting in the creation of larger sized clusters. In addition, the interaction between magnetic particles also increases due to the reduction in distance between particles, thereby enhancing their magnetic effect [44]. However, using a large quantity of magnetizable particles may not always be necessary, as the quantity of required particles depends also on the expected magnitude of formwork pressure. In Figure 6.18, at 10 kPa, the flow rate of the M2 and M3 pastes was quite similar. The use of 2% MNPs provided sufficient magnetic response to overcome the pressure-driven flow, meaning that increasing the amount of magnetizable particles to 3% may not be an efficient use of resources. However, as the pressure was increased, the M2 graph eventually diverged from the M3 graph and converged with the M1 graph signalling that the MNP concentration was insufficient to provide the same magnetically induced resistance as the M3 paste. This illustrates that the pressure conditions will need to be considered when selecting the required MNP concentration to increase the efficacy of the ARC method for formwork leakage.

6.9.2 Adjustment of rheological properties of base fluid

In addition to the D_{gap}/d_p ratio, the rheological properties of the fluid also play an important role for successful ARC for formwork leakage. In general, the stress exerted on the pastes must first exceed its yield stress for the material to flow. Once the yield stress is exceeded, the viscosity provides an additional resistance to flow [45]. In the situation of bottom-up pumping ofSCC, the concrete is constantly being sheared as the material fills the forms. The yield stress of the fluid is thus already exceeded, and the continuous shearing does not allow the material to experience the benefits of its thixotropic nature [46]. If gaps are present at the formwork interface during this form-filling phase, as far as rheological properties go, the ease with which the material will flow out will thus largely be influenced by the viscosity. For the same flow area, it is thus expected that increased viscosity and yield stress result in increased resistance to flow. This is clearly shown in Figure 6.19 which shows the differences in flow rate and the net mass loss due to decreased fluidity for the fly ash pastes and MNP pastes. The flow rate versus pressure curves of the low-slump flow pastes are shifted downwards from those of the higher-slump pastes, i.e. towards lower flow rates at all pressure levels. This is because the low-slump pastes require more stress to initiate flow than the higher-slump pastes which generally have a lower yield stress. Once that stress is exceeded, the low-slump pastes require higher stresses to maintain flow due to the increased resistance provided by their higher viscosity values.

Under static conditions, the resultant movement of the magnetizable particles due to the magnetic force is dependent on the fluidity of the

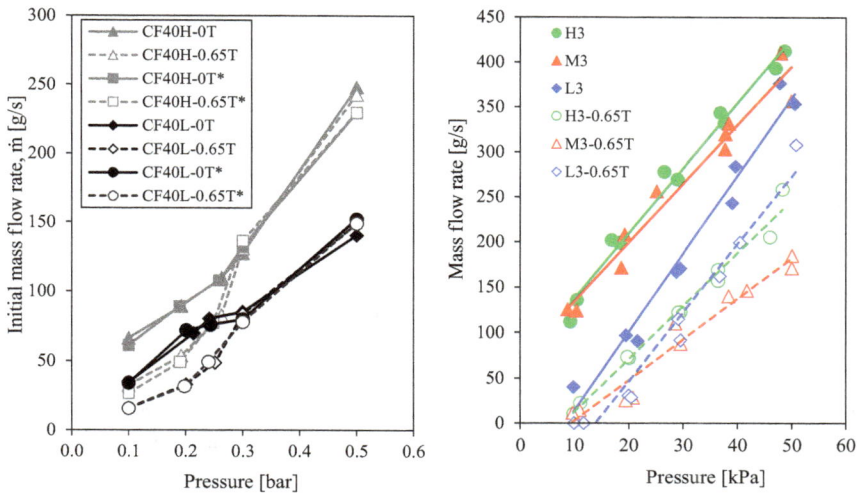

Figure 6.19 Effect of paste rheology on (a) fly ash pastes and (b) Fe_3O_4 pastes. Source: [39].

paste – a stiffer paste provides a greater resistance to movement. Under the dynamic conditions of the pressure-driven flow test, the downward velocity of the magnetizable particles is also affected by the fluidity of the paste – keeping the pressure constant, a stiffer paste would flow at a lower velocity. Furthermore, the residual pressure after pressure losses due to viscous effects is higher for the high-slump pastes compared to the low-slump pastes. The velocity of the high-slump pastes thus tends to be higher. Although a more fluid paste would have a higher magnetic response, it would also have a higher fluid velocity which could prevent the magnetizable particles from forming magnetic clusters. The rheological properties of the fluid evidently have a conflicting effect on the outcome of the ARC method. The dominant force would thus dictate the magnitude of flow reduction observed. It is therefore imperative that the rheological properties of the fluid be carefully selected. In the practice, this could maybe allow for higher slump pastes to be used for the purpose of easier pumping, without worrying about the increased risk of formwork leakage. However, more studies need to be carried out to reach this conclusion.

6.9.3 Maximization of magnetic field strength

Another method to increase the effectiveness of the ARC method would simply be to increase the magnetic field strength across the flow width. This

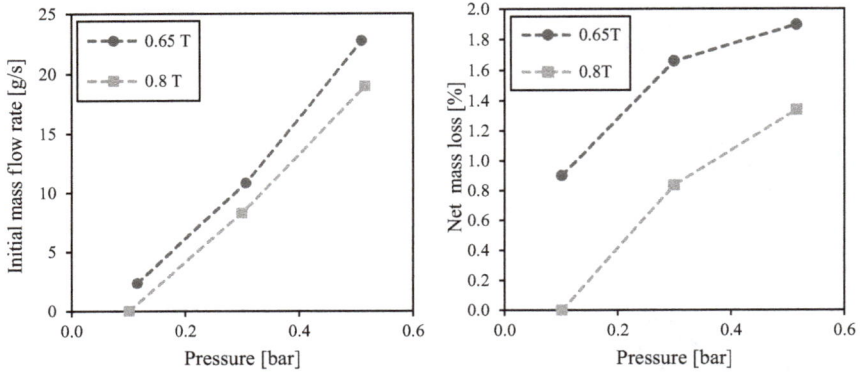

Figure 6.20 Influence of magnetic field strength on: (left) initial mass flow rate with/ without magnetic field; (right) percentage loss of total mass in the column. Source: [39].

was observed for flow tests which were carried out for the CF40H paste under magnetic fields of 0 T, 0.65 T and 0.8 T. The flow width was kept constant at 1 mm, and tests were carried at approximately 0.1 bar, 0.3 bar and 0.5 bar. Increasing the magnetic field strength increased the degree of blocking (Figure 6.20). By increasing the magnetic field strength from 0.65 T to 0.8 T (23% increase), the initial flow rate was decreased by 100%, 24%, 16% at 0.1, 0.3 and 0.5 bar, respectively. It should be noted however that due to the D_{gap}/d_p being sufficiently low to induce intrinsic blocking, and thus the net mass loss was already low even without magnetic field application.

A higher magnetic field strength implies a higher attraction force over a larger area around the magnet, as well as stronger interaction between magnetic particles [47, 48], thereby increasing the density of the agglomerates formed. Although higher magnetic field strength in principle increases the magnetic response, there should exist a threshold level at which no further benefits of increasing the magnetic field are observable. This is because the magnetic particles possess a saturation level of magnetization beyond which an external magnetic field cannot increase the magnetization of the particles any further [49]. In magnetic separation applications, a high magnetic field does not automatically imply better results and it is advised to determine an optimum value. The filtration of magnetic particles was found to increase with increasing magnetic field strength up to a threshold point beyond which the filtration efficiency remained constant [48]. In the context of ARC for formwork leakage control, further studies need to be conducted to inform on how to select the optimum magnetic field strength for a given set of conditions.

6.9.4   Challenging effects of formwork pressure

The amount of pressure applied has a predominant effect on the likelihood of the ARC method, leading to complete blockage of flow. This is because the pressure exerted on the fluid opposes the formation and deposition of the magnetic clusters at the slit. Clogging of rigid particles flowing in a constricted channel is said to be dependent on the balance between structuration and fragmentation mechanisms which in turn are caused by the particle-fluid and inter-particle interactions. These two mechanisms occur simultaneously but a dominance of structuration could lead to clogging [50]. Application of the magnetic field works in favour of structuration. Clustering of magnetizable particles acts as an initiator for the formation of larger agglomerates which eventually block the flow. A higher pressure thus of course relates to a greater negative influence on the formation of these clusters. As such, the ARC method was found to be the most effective at lower pressure values (10–20 kPa), with decreasing success as the pressure was increased to 50 kPa. At low pressures, complete blockage of flow was observed, with minimal amounts of material loss. However, with increasing pressure, the ARC method resulted in the reduction in flow rate but was insufficient to prevent material loss.In the case of the fly ash pastes (CF40H), the effect of flow width was studied using two different bottom plates with flow widths of 1 mm and 2 mm, and magnetic field strength of 0 T or 0.65 T. In the case of the 2 mm gap width, increasing the pressure led to the convergence of the 0.65 T and 0 T flow rate curves. This indicates that beyond the flow converging point X_{cp}, the ARC method has no effect on the flow rate. On the other hand, for the pastes containing MNPs, although the reduction in flow rate also decreased with increasing pressure, there was no flow converging point X_{cp} under the same flow conditions as the CF40H paste (Figure 6.21).

The fly ash had a lower saturation magnetization (1.74 emu/g) in comparison to the MNPs (77.56 emu/g), which explains the lower ARC potential. It can thus be hypothesized that the flow convergence point, which arises from the negative effects of pressure, is characteristic to the set of variables selected. The current results suggest that the converging point is dependent on the initial geometric conditions, i.e. the flow width (or generally D_{gap}/d_p), the concentration of magnetizable particles, as well as the magnetic field strength applied. Lower concentration of magnetizable particles as well as lower magnetic field strength could shift the converging point to lower pressure values.

During the form-filling process, formwork pressure gradually increases from 0 kPa to the maximum value as the height of the fluid increases. If blocking is magnetically induced during this initial form-filling stage when the pressure is still sufficiently low (i.e. when the pressure is still below the X_{cp} value), and the blockage is able to withstand the increasing pressure,

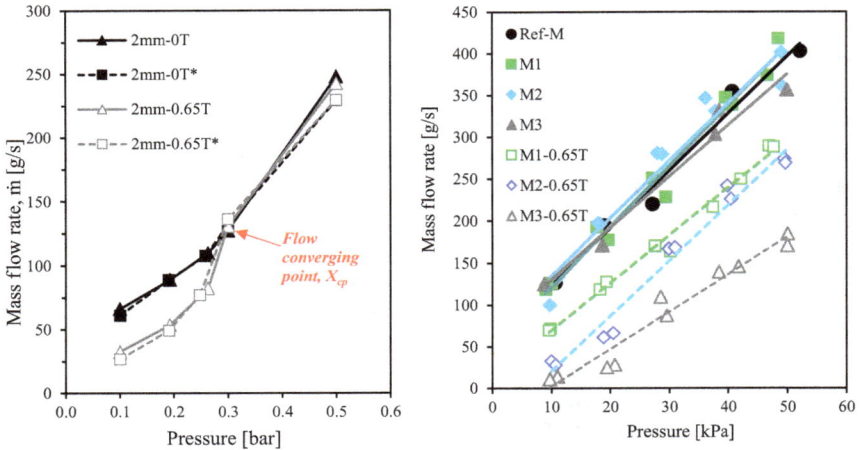

Figure 6.21 **Effect of pressure on flow rate reduction: (left) fly ash pastes; (right) Fe$_3$O$_4$ pastes. Source: [39].**

this could be beneficial for formwork leakage. Despite the existence of this threshold point for the CF40H paste, once the magnetic cluster was formed at 10 kPa, a stepwise increase in the pressure showed that the cluster was able to withstand pressures of up to 100 kPa, as illustrated in Figure 6.22. This is a promising result, but further studies are required to test the resistance of the magnetic cluster to shear.

6.10   FURTHER RESEARCH AND OUTLOOK

In this study, the leakage of cementitious material was only investigated through a rectangular cross-section. In reality, the gaps at the formwork interface could be of irregular geometry. This simplification, however, does not hinder the interpretation of the results that could be obtained. Assuming a regular flow path simplifies the characterization of the flow area and resulting flow behaviour. Furthermore, due to the restriction of the current design of the bottom plate (space constraints plus the use of static magnets), the number of flow widths that could be used without changing the magnetic field strength resulted in only two magnetic field strengths being assessed. To determine the optimal magnetic field strength required for a given set of parameters, further studies considering different magnetic field strength and flow widths are required.

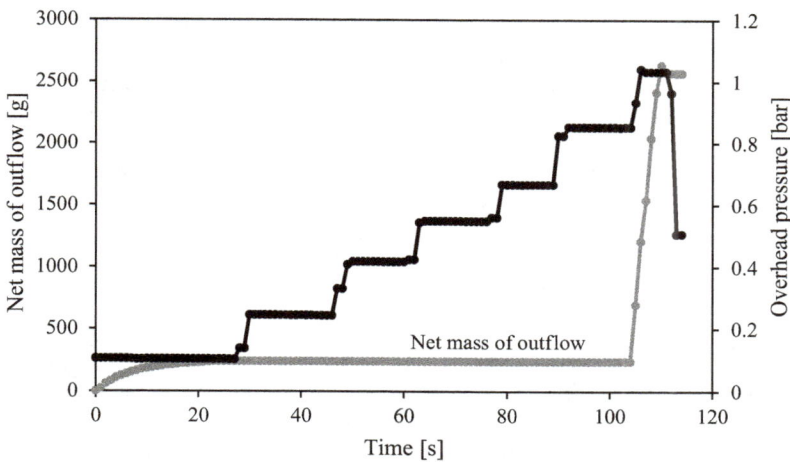

Figure 6.22 Test on the resistance of the magnetic clusters formed at 10 kPa to increased pressure – CF40H paste, 2 mm gap. Source: [39].

During the pressure-flow test, only the overhead pressure applied was measured; however, for future adaptations of the test cylinder, it would be advantageous to include a pressure measurement close to the outflow. This would provide useful information for estimating the pressure drop across the height of the cylinder, as well as give more accurate input for the estimation of shear rates at the outflow. A good estimation of the shear rates could inform the range of parameters for the rheometer experiments.

The study also only involved a limited number of selected mixtures and flow geometries. As is the nature of experimental investigations, it is not always practical or possible to encompass an exhaustive range of parameters. To increase the number of possible configurations and outcomes, future studies could thus benefit from the inclusion of numerical methods.

The outflow is located at the bottom of the test cylinder, implying predominantly normal stress being applied to the magnetic clusters formed. To investigate the effects of shearing on the formation and stability of the magnetic clusters, a vertical slit should be considered in future studies.

Preliminary studies were conducted on mortar samples in which the effect of aggregate inclusion on the MR response was investigated. The initial results indicate a reduction in filtration and consequently, improvement in the aesthetics when the ARC method was used. However, further studies on mortar and concrete are still necessary. This should include durability tests to check the effects of the ARC method on the pore structure and susceptibility of the finished surface to the ingress of harmful substances.

For practical or large-scale applications, several challenges can already be envisioned, such as the cost and logistics of using the said method. The potential added cost of the magnetic additives as well as design, installation and training of personnel to execute the method are all potential challenges for the implementation of ARC for formwork leakage control. Nevertheless, there already exist formwork systems based on magnetic clamping systems, giving promise to the practicality of introducing ARC for this application.

6.11   SUMMARY

Tightness of formwork joints under high formwork pressure is one of the concerning issues raised by contractors with regard to casting SCC. This cannot easily be addressed by simply designing less fluid concrete mixtures as this contradicts with the rheological requirements for concrete pumping and good form filling. There are currently no active methods of controlling the material properties during casting which can be used to combat formwork leakage of SCC under pressure. There is thus a need for research to bridge these gaps in knowledge to transform and optimize the current practices.

This chapter introduces a magnetic field-based active stiffening control methodology that could be beneficial for reducing formwork leakage during casting. As a demonstration of formwork leakage, pressure-driven flow through a narrow slit was investigated on magnetizable pastes containing fly ash, Fe_3O_4 nanoparticles and aluminate aggregate as responsive additives. The active stiffening control was incorporated by applying an external magnetic field at the outflow. The mass of the outflowing material was continuously recorded, and the flow rate and final net mass loss evaluated. Experimental results showed a reduction in flow rate after applying the external magnetic field due to the agglomeration of magnetizable particles at the outflow region. Some key parameters have been illustrated, including the ratio D_{gap}/d_p, type and concentration of magnetizable particles, viscosity, and yield stress of the fluid, pressure level and magnetic field strength.

In view of ARC based on magnetic field application, it is concluded that tailoring cementitious pastes to include magnetizable particles is a means to minimize or actively control confined flow of cementitious pastes. The presented methodology shows promising results for the larger initiative which is ultimately to actively control formwork leakage. The novel concept however requires further verification on mortar- or concrete-scale tests using a more representative test set-up.

REFERENCES

[1]   J. R. Desmyter, "Barriers to the application of cast-in-place self-compacting concrete," in *Fifth International RILEM Symposium*, Ghent, 2007.

[2]   A. Kashani and T. Ngo, "Production and placement of self-compacting concrete," in *Self-compacting concrete: Materials, properties, and applications*, Sawston, UK: Woodhead Publishing, 2020, pp. 65–81.

[3]   R. Gaimster and N. Dixon, "Self-compacting concrete," in *Advanced concrete technology*, Amsterdam, the Netherlands: Elsevier, 2003, pp. 9/1–9/23

[4]   A. Leemann, C. Hoffmann and F. Winnefeld, "Pressure of self-consolidating concrete on formwork," *Concrete International,* vol. 28, no. 2, pp. 28–31, February 2006.

[5]   RILEM Technical Committee 188, "Final report of RILEM TC 188-CSC 'casting of self compacting concrete'," *Materials and Structures*, vol. 39, pp. 937–954, 2006.

[6]   N. Gowripalan, P. Shakor and P. Rocker, "Pressure exerted on formwork by self-compacting concrete at early ages: A review," *Case Studies in Construction Materials*, vol. 15, p. e00642, 2021.

[7]   G. Ovarlez and N. Roussel, "A physical model for the prediction of lateral stress exerted by self-compacting concrete on formwork," *Materials and Structures*, vol. 39, no. 2, pp. 269–279, 2006.

[8]   P. Billberg, "Understanding formwork pressure generated by fresh concrete," in *Understanding the rheology of concrete*, N. Roussel (Ed.), Cambridge, Woodhead Publishing Limited, 2012, pp. 296–330.

[9]   J. Assaad and K. H. Khayat, "Variations of lateral and pore water pressure of self-consolidating concrete at early age," *ACI Materials Journal*, vol. 101, no. 4, pp. 310–317, July/August 2004.

[10] N. Roussel, "Thixotropy: from measurement to casting of concrete," in *Understanding the rheology of concrete*, N. Roussel (Ed.), Cambridge, Woodhead Publishing Limited, 2012, pp. 286–295.

[11] K. H. Khayat and A. Omran, "SCC formwork pressure - state-of-the-art: capturing existing knowledge on formwork pressure exerted by SCC," 2007.

[12] T. Cools, A. Van Gysel and P. Van Itterbeeck, "Tightness requirements for SCC formwork," in *SCC 2016 conference proceedings*, Washington DC, 2016.

[13] G. R. Lomboy, X. Wang and K. Wang, "Rheological behavior and formwork pressure of SCC, SFSCC, and NC mixtures," *Cement & Concrete Composites*, vol. 54, pp. 110–116, 2014.

[14] N. L. Thrane, "Modelling the flow of self-compacting concrete," in *Understanding the rheology of concrete*, N. Roussel (Ed.), Cambridge, Woodhead Publishing Limited, 2012, pp. 259–285.

[15] T. Craipeau, F. Toussaint, A. Perrot and T. Lecompte, "Experimental approach on a moving formwork," *Construction and Building Materials*, vol. 270, p. 121472, 2021.

[16] G. De Schutter and K. Lesage, "Active control of properties of concrete: A (p)review," *Materials and Structures (2018) 51:123*, vol. 51, p. 123, 2018.

[17] G. De Schutter, K. Lesage, V. Mechtcherine, V. N. Nerella, G. Habert and I. Agusti-Juan, "Vision of 3D printing with concrete – Technical, economic and environmental potentials," *Cement and Concrete Research 112*, vol. 112, pp. 25–36, 2018.

[18] J. Apers, "Het Uiterlijk Van Beton," *Dossier Cement*, 22 June 2000.

[19] G. De Schutter, *Damage to concrete structures*, Ghent: CRC Press, 2012.

[20] C. Djelal, Y. Vanhove and A. Magnin, "Tribological behaviour of self compacting concrete," *Cement and Concrete Research*, vol. 34, no. 5, pp. 821–828, 2004.

[21] K. H. Khayat and A. Omran, "SCC formwork pressure," Quebec, 2009.

[22] J. J. Assaad, "Correlating thixotropy of self-consolidating concrete to stability, formwork pressure, and multilayer casting," *Journal of Materials in Civil Engineering*, vol. 28, no. 10, p. 04016107, 2016.

[23] T. Cools, A. Van Gysel and P. Van Itterbeeck, "Tightness requirements for SCC formwork," In Int. RILEM Symposium on Self-Compacting Concrete (SCC2016), Washington, USA, 2016, pp. 627–636.

[24] W. L. Sluyter and P. C. Kreijger, "Qualitative explanation of the fluidity of fresh concrete," Eindhoven University of Technology, Report M/78/01, Eindhoven, the Netherlands, 1978.

[25] A. Perrot, C. Lanos, P. Estellé and Y. Melinge, "Ram extrusion force for a frictional plastic material: Model prediction and application to cement paste," *Rheol Acta*, vol. 45, pp. 457–467, 2006.

[26] A. Marin, H. Lhuissier, M. Rossi and C. J. Kähler, "Clogging in constricted suspension flows," *Physical Review*, vol. E 97, p. 021102(R), 2018.

[27] S. Xu, H. Sun, Y. Cai and X. Geng, "Studying the orifice jamming of a polydispersed particle system via coupled CFD–DEM simulations," *Powder Technology 368*, vol. 368, pp. 308–322, 2020.

[28] A. Ashour, S. Wegner, T. Trittel, T. Börzsönyi and R. Stannarius, "Outflow and clogging of shape-anisotropic grains in hoppers with small apertures," *Soft Matter*, vol. 13, pp. 402–414, 2017.

[29] Y. Tang, X. Yao, Y. Chen, Y. Z. D. S. Zhou, Y. Zhang, T. Zhang and Y. Peng, "Experiment research on physical clogging mechanism in the porous media and its impact on permeability," *Granular Matter*, vol. 22, p. 37, 2020.

[30] K. El Cheikh, S. Rémond, S. Khalil and G. Aouad, "Numerical and experimental studies of aggregate blocking in mortar extrusion," *Construction and Building Materials*, vol. 145, pp. 452–463, 2017.

[31] A. Hajalilou, S. A. Mazlan, H. Lavvafi and K. Shameli, *Field responsive fluids as smart materials*, Springer Nature, Singapore, 2016.

[32] A. G. Olabi and A. Grunwald, "Design and application of magneto-rheological fluid," *Materials and Design*, vol. 28, pp. 2658–2664, 2007.

[33] D. J. Carlson, "MR fluids and devices in the real world," *International Journal of Modern Physics B*, vol. 19, no. 7, 8 & 9, pp. 1463–1470, 2005.

[34] T. Manourasa and M. Vamvakaki, "Field responsive materials: Photo-, electro-, magnetic- and ultrasound-sensitive polymers," *Polymer Chemistry*, vol. 8, pp. 74–96, 2017.

[35] E. Gedik, H. Kurt, Z. Recebli and C. Balan, "Two-dimensional CFD simulation of magnetorheological fluid between two fixed parallel plates applied external magnetic field," *Computers & Fluids*, vol. 63, pp. 128–134, 2012.

[36] A. Grunwald and A. G. Olabi, "Design of magneto-rheological (MR) valve," *Sensors and Actuators A*, vol. 148, pp. 211–223, 2008.

[37] Y.-W. Li, C.-S. Zhao, X. Wu, D.-F. Lu and S. Han, "Aggregation mechanism of fine fly ash particles in uniform magnetic field," *Korean Journal of Chemical Engineering*, vol. 24, no. 2, pp. 319–327, 2007.

[38] F. D. Goncalves and J. D. Carlson, "An alternate operation mode for MR fluids – magnetic gradient pinch," *Journal of Physics: Conference Series*, vol. 149, p. 012050, 2009.

[39] C. Chibulu, K. El Cheikh, M. Y. Yardimci and G. De Schutter, "Experimental study of formwork tightness as a function of rheological properties of SCC," in *Rheology and Processing of Construction Materials. RheoCon 2019, SCC 2019. RILEM Bookseries*, vol. 23, V. Mechtcherine, K. Khayat and E. Secrieru (Eds.), Springer, Cham, 2019, pp. 476–481.

[40] C. Chibulu, M. Y. Yardımcı, D. Jiao, R. De Schryver, K. Lesage and G. De Schutter, "Active stiffening control by magnetically induced blocking in confined flow of fly ash pastes," *Construction and Building Materials*, vol. 313, no. 2, p. 125485, 2021.

[41] M. Alfi, N. Benarjee, D. Feys and J. Park, "Simulation of formwork filling by cement fluid: The effect of the formwork structure on yield-stress fluid," in *COMSOL*, Boston, 2013.

[42] N. Delouche, A. B. Schofield and H. Tabuteau, "Dynamics of progressive pore clogging by colloidal aggregates," *Soft Matter*, vol. 16, pp. 9899–9907, 2020.

[43] E. Dressaire and A. Sauret, "Clogging of microfluidic systems," *Soft Matter*, vol. 13, pp. 37–48, 2017.

[44] D. Jiao, K. Lesage, M. Y. Yardimci, K. El Cheikh, C. Shi and G. De Schutter, "Structural evolution of cement paste with nano-Fe3O4 under magnetic field- Effect of concentration and particle size of nano-Fe3O4," *Cement and Concrete Composites*, vol. 120, p. 104036, 2021.

[45] S. E. Chidiac and F. Mahmoodzadeh, "Plastic viscosity of fresh concrete – A critical review of predictions methods," *Cement & Concrete Composites*, vol. 31, pp. 535–544, 2009.

[46] P. Billberg, "Understanding formwork pressure generated by fresh concrete," in *Understanding the rheology of concrete*, Cambridge, Woodhead Publishing Limited, 2012, pp. 296–330.

[47] D. Jiao, K. Lesage, M. Y. Yardimci, K. El Cheikh, C. Shi and G. De Schutter, "Quantitative assessment of the influence of external magnetic field on clustering of nano-Fe3O4 particles in cementitious paste," *Cement and Concrete Research*, vol. 142, p. 106345, 2021.

[48] J. Svoboda, "The effect of magnetic field strength on the efficiency of magnetic separation," *Minerals Engineering*, vol. 7, no. 5/6, pp. 747–757, 1994.

[49] Y.-q. Lin, J. Li, X.-d. Liu, T.-z. Zhang, B.-c. Wen, Q.-m. Zhang and H. Miao, "Saturation magnetization and law of approach to saturation for

self-formed ionic ferrofluids based on MnFe2O4 nanoparticles," *Chinese Journal of Chemical Physics*, vol. 23, no. 3, pp. 325–330, 2010.

[50] K. Shahzad, W. Van Aeken, M. Mottaghi, V. K. Kamyab and S. Kuhn, "Aggregation and clogging phenomena of rigid microparticles in microfluidics," *Microfluidics and Nanofluidics*, vol. 22, p. 104, 2018.

Chapter 7

Application of active rheology control to 3D printing of cementitious materials

Dengwu Jiao, Karel Lesage, and Geert De Schutter

CONTENTS

7.1 General introduction   221
7.2 Theoretical foundations   223
7.3 Experimental simulation with rheometer tests   224
7.4 Print head design and outlook   229
7.5 Summary   230
References   230

7.1 GENERAL INTRODUCTION

Three-dimensional (3D) concrete printing is an innovative construction process for fabricating concrete components employing additive manufacturing (AM) technique. The potential benefits of 3D concrete printing include abandoning traditional formwork construction, printing on demand, reducing material wastage, using multi-materials according to strength requirements, etc. The set-up of 3D concrete printing generally consists of robot controller, mixer, pump, connecting hose, printer head and nozzle [1]. 3D concrete printing usually includes two steps. The first step is to prepare a mixture with appropriate rheological behaviour for pumping and extruding. The concrete mixture is then conveyed using a pump towards a printer head. Through the nozzle at the end of the printer head, the concrete can leave the printer and then deposit on the print surface layer by layer.

3D printing of concrete needs to meet the contradicting requirements regarding pumpability, extrudability and buildability [2, 3]. On the one hand, the material needs to flow easily to be pumped towards the printing unit and to be extruded through the nozzle. On the other hand, after the printed material reaches its final position in the printed layers, the yield stress of the mixture should be high enough to maintain its shape under its own weight and the load from upper layers [4–6]. This indicates that 3D printable concrete should have fast evolution of yield strength with time to transform from a pumpable mix to a highly yield strength mix in a very short period. Generally, there are two approaches to achieve this target, i.e. the addition of additives during the initial mixing of concrete and buildability enhancement at the print head by an intervention. However, there are some limitations of the former approach. For example, despite the fact that the rapid increase in the yield strength is beneficial to the buildability, such increase will possibly restrict the continuous pumping of fresh concrete [7]. Some other limitations such as different cycle time and the impact of process parameters are also often reported [8, 9]. On the other hand, interventions at the print head to rapidly increase the buildability do not affect the pumpability of the concrete mix as the rheological properties remain unchanged before extrusion. It is an effective solution to overcome the challenge of the opposing flowability requirements during pumping, extruding and building.

Buildability enhancement by an intervention is also defined as active rheology control or control on-demand of 3D printable concrete. The intervention applied in a print head includes physical vibration [10], magnetic field [4, 11, 12], ultrasonic [13], chemical [14, 15] and microwave heating [16], as summarized by Muthukrishnan et al. [7]. A schematic diagram summarizing the methods to achieve the active rheology control of 3D printable concrete is presented in Figure 7.1. Magneto-rheology control, by means of pre-adding responsive additives (such as chemical admixture [17] or magnetic particles [18, 19]) and applying a magnetic

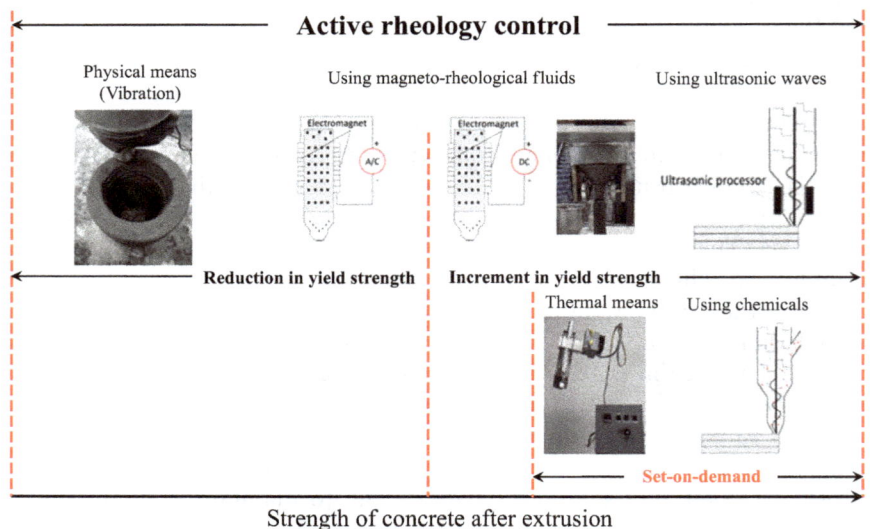

Figure 7.1 Summary of methods to achieve the active rheology control of 3D printable concrete. Source: after [7].

field signal, is a potential approach to meet the contradicting requirements of concrete properties in different casting processes. It is one of the promising methods of reaching active rheology control of 3D printable cementitious materials.

This chapter presents the application of magneto-rheology control in extrusion-based 3D concrete printing. The theoretical foundations relating to the magneto-rheology control of 3D printable concrete are illustrated. A rheological measurement attempt simulating the extrusion-based 3D printing processes is presented. Future work and perspectives of magneto-rheology control of 3D printable cementitious materials are also discussed.

7.2 THEORETICAL FOUNDATIONS

As stated in Chapter 3, for cementitious paste containing magnetic particles, applying an external magnetic field improves the early liquid-like behaviour due to the micro-agitation effect of the movement of magnetic particles [19], and enhances the solid-like properties after longer magnetization time because of the formation of magnetic clusters [20]. The rheological response depends on the viscoelasticity of paste medium, physical properties of magnetic particles and magnetic field type [21, 22]. After removing the external magnetic field, the magnetic clusters are disintegrated reversibly but some residual magnetic clusters possibly exist in the cementitious suspensions due to the viscoelastic properties of the suspension.

More specifically, a typical structural evolution of cementitious paste with the change of magnetic field from high value to zero, described by the development of storage modulus, is presented in Figure 7.2. It can be seen that the storage modulus exhibited a sudden decrease when the magnetic field changed from high to low strength value. This decrease indicates that the sudden decreasing magnetic field results in higher liquid-like properties. On the one side, the magnetic forces on the nano-Fe_3O_4 particles decrease with the decrease of magnetic field strength, resulting in the disintegration of parts of column-like structures. On the other side, the Brownian motion and electrostatic forces between solid particles gradually increase, and thus a reduction in storage modulus was observed. Comparing with the cementitious paste under constant magnetic fields, the slope of storage modulus with time was slightly higher compared to the cementitious paste after experiencing the decrease of magnetic field from high to lower value. For example, the storage modulus when the magnetic field decreased from 0.25 T to 0 T (i.e. t=600 s) was almost similar to that under constant magnetic field of 0 T. However, the difference in the storage modulus between these two modes (black and green lines) was about 150 kPa at 900 s. This phenomenon can be explained by the magnetic properties of nano-Fe_3O_4 particles. After the magnetic field is removed, there exist residual chain-like structures and clusters formed under external magnetic field. As a result, the

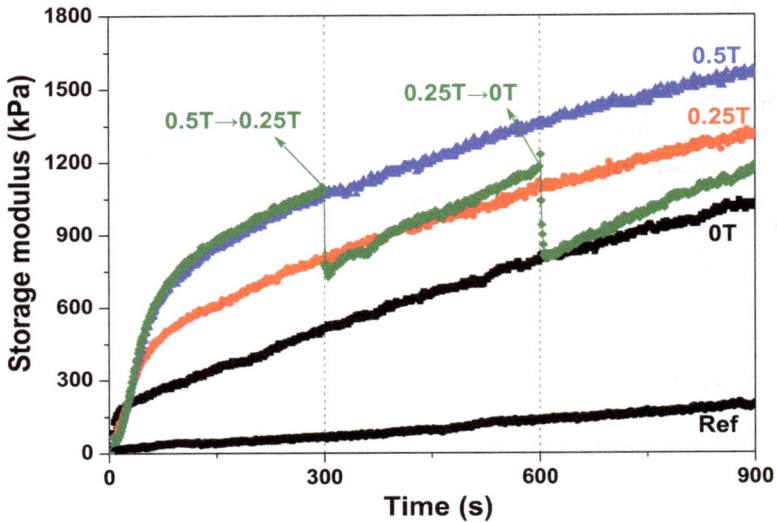

Figure 7.2 Structural evolution of cementitious paste (w/c=0.4, MNPs=5%) under step-decreased magnetic fields. Source: after [23].

interactions between solid particles are enhanced. Thus, a higher increase rate of storage modulus was obtained for the cementitious paste after experiencing the changing magnetic field from high to zero. This provides a potential approach of applying magneto-rheology control in 3D concrete printing.

It should be mentioned here that the residual magnetic clusters only exist in cementitious paste at resting state, while applying an additional shear stress might destroy the magnetic clusters completely. Indeed, a typical evolution of apparent viscosity of a limestone powder suspension under step-decreased magnetic field at constant shear of 10 s^{-1} is shown in Figure 7.3. It can be seen that the apparent viscosity exhibits a sudden decrease with the step-decrease of the magnetic field, followed by a gradual reduction behaviour to a new equilibrium. When the magnetic field decreases from 0.5 T to 0 T, the apparent viscosity reduces to the same level as that obtained without magnetic field.

7.3 EXPERIMENTAL SIMULATION WITH RHEOMETER TESTS

This section examines the potential application of magneto-rheology control in extrusion-based 3D concrete printing by rheological experiments on

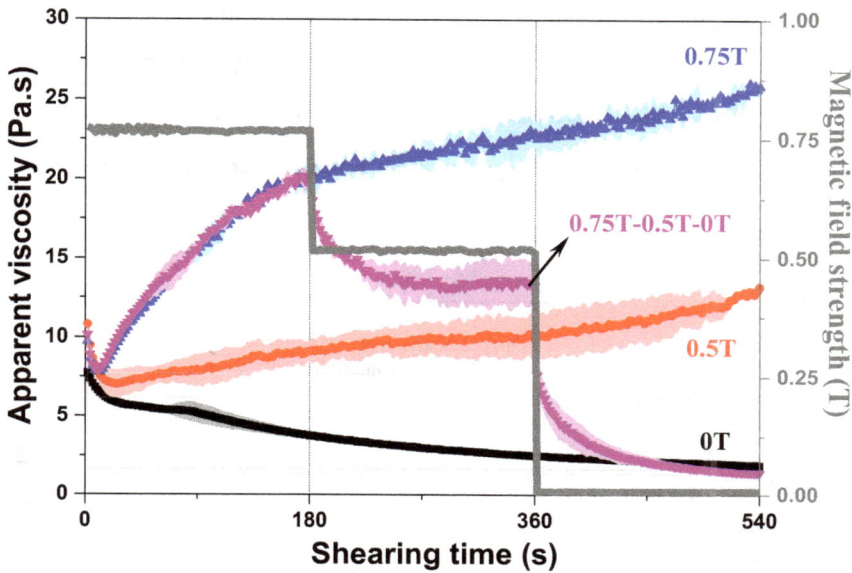

Figure 7.3 Evolution of apparent viscosity of limestone powder suspension (W/LP=0.4, MNPs=3%) under step-decreased magnetic field, at constant shear rate of 10 s^{-1}. Source: after [24].

cementitious paste with nano-Fe_3O_4 particles. The rheological test for examining the structural evolution of cementitious paste is performed by using a parallel plate rotational rheometer (MCR 102, Anton Paar) equipped with a magneto-rheological device (MRD). The diameter of the plate is 20 mm, and the gap between the parallel plates is fixed at 1 mm. The rheological protocol simulating 3D concrete printing is presented in Figure 7.4. Extrusion-based 3D concrete printing process includes mixing, pumping and extruding. From the viewpoint of rheological measurements, a pre-shearing with a shear rate of 100 s^{-1} for 30 s is used to simulate the mixing process. The pumping and extruding processes are simulated by a constant shearing with a relatively low shear rate of 20 s^{-1} for 60 s. The structuration of the paste after extrusion is described by the structural build-up obtained from oscillatory time sweep test, where the shear strain and frequency were fixed at 0.005% (within the linear viscoelastic region) and 2 Hz, respectively. For the normal 3D printing simulation, no magnetic field is applied. For the magneto-controlled 3D printing, a pulsed magnetic field (PMF) is applied towards the end of the shearing, lasting for 1 s or 5 s. The removal of the PMF occurs simultaneously with the termination of the shearing. The

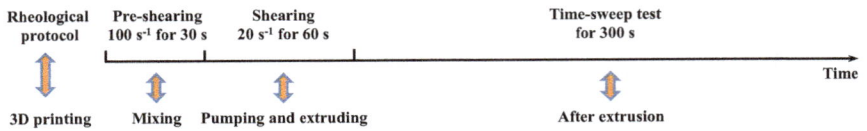

Figure 7.4 Rheological testing protocol simulating extrusion-based 3D concrete printing. Source: after [12].

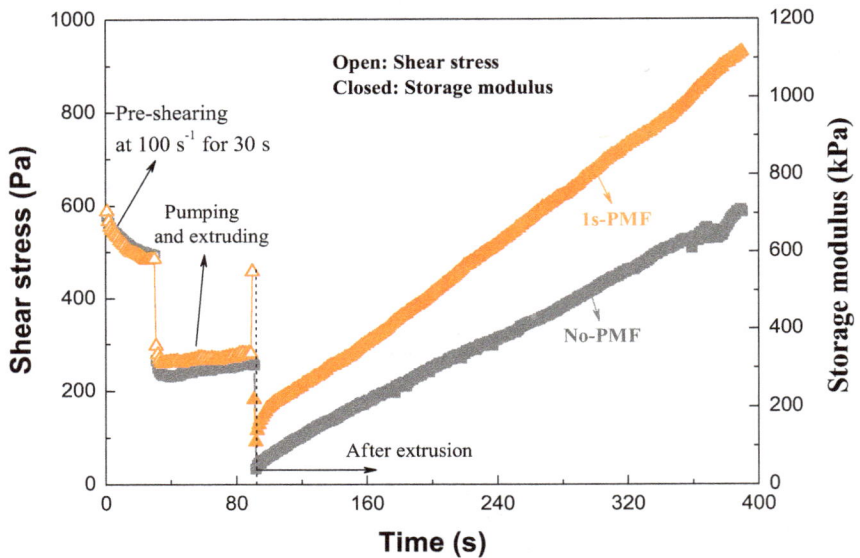

Figure 7.5 Typical evolution of shear stress and storage modulus of cementitious paste (w/c=0.35, MNPs (100 nm) = 3%). Source: after [12].

subsequent evolution of storage modulus, regarded as a post-effect of the applied magnetic field, is evaluated. During the rheological test, the temperature was maintained at 20 ± 0.5 °C, and the data was recorded every second.

Figure 7.5 presents a typical evolution of shear stress and storage modulus of cementitious paste with w/c of 0.4 and MNPs (100 nm) of 3%. For the situation without the PMF, the storage modulus increases linearly from ~50 kPa after terminating the shearing. This indicates the gradual increase of the internal structuration of the paste. If a PMF for 1 s (1s-PMF) is applied towards the end of shearing, the shear stress immediately jumps to a high value. When the shearing terminates and the PMF is removed, the microstructure starts to build up at a relatively higher initial strength. Indeed, the storage modulus shows a significant increase within the first ten seconds with an initial value higher than 100 kPa. With elapsing resting time, the storage modulus gradually increases with a higher growth rate than the case without applied PMF. The results indicate that the application of a PMF during extrusion in 3D printing will possibly lead to a higher evolution of structuration of cementitious paste with magnetic particles. The results in [12] also show that the faster structural build-up after extrusion is independent of the duration of the PMF, within the range considered (1 s or 5 s).

The results are explained by the distribution of magnetic nanoparticles. It is postulated that the nanoparticles are randomly distributed in the voids between cement particles during shearing, as shown in Figure 7.6 (a). When applying a magnetic field, the nanoparticles agglomerate to chains/clusters in a very short time [19, 20, 25], as depicted in Figure 7.6 (b). Furthermore, the flowing paste tends to facilitate the formation of magnetic chains/clusters [24]. This exerts a resistance to the shear flow, resulting in a jump increase in the shear stress. Note that the possible negative influence of the jumped shear stress to extrudability could be neglected due to the very short magnetic field. After removal of the magnetic field, the formed magnetic clusters are disassembled, while some residual chains/clusters remain in the suspension, as presented in Figure 7.6 (c), due to the intrinsic properties of the nanoparticles (i.e., high remanent magnetization (14.81 Am2/kg) of the nanoparticles with 100 nm [21]). The presence of the residual clusters, on the one side, results in an initial microstructural strength of the cementitious paste, and thus a relatively high initial storage modulus is observed. On the other side, the residual clusters enhance the interactions between solid particles, improving the structural evolution of the cementitious paste.

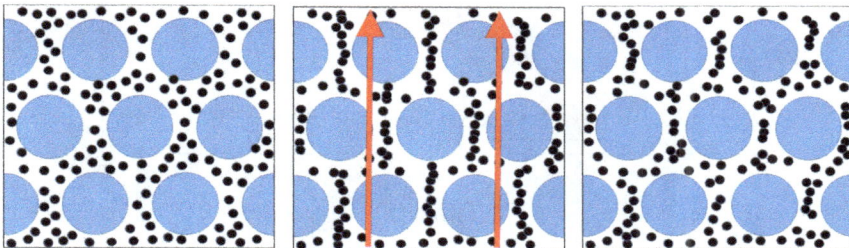

Figure 7.6 Schematic diagram of illustrating nanoparticles distribution (a) before; (b) apply and (c) remove the pulsed magnetic field. Blue cycles represent the cement particles, and dark dots indicate the nano-Fe_3O_4 particles.

We also attempted to validate the proposed hypothesis by examining the PMF-induced structural build-up of cementitious pastes containing two different types of nano-Fe_3O_4 particles with the same saturation magnetization but different remanent magnetization and particle size, as shown in Figure 7.7. We can see that the two mixtures have almost similar evolutions of storage modulus without the PMF. This is consistent with our previous findings [21], which show that changing the particle size of nano-Fe3O4 has negligible influence on the structural evolution without magnetic field. According to the proposed hypothesis, the increase in shear stress for the two mixtures should be similar in the magnitude while applying a PMF due to the same saturation magnetization. After removing the PMF, less residual magnetic clusters should exist in the MNP2-paste compared to the MNP1-paste because of the relatively lower remanent magnetization of MNP2 than MNP1. Therefore, MNP1-paste should exhibit faster structural build-up than that of MNP2-paste in the subsequent resting period. It can be observed from Figure 7.7 that all the experimental results are consistent with the expectations.

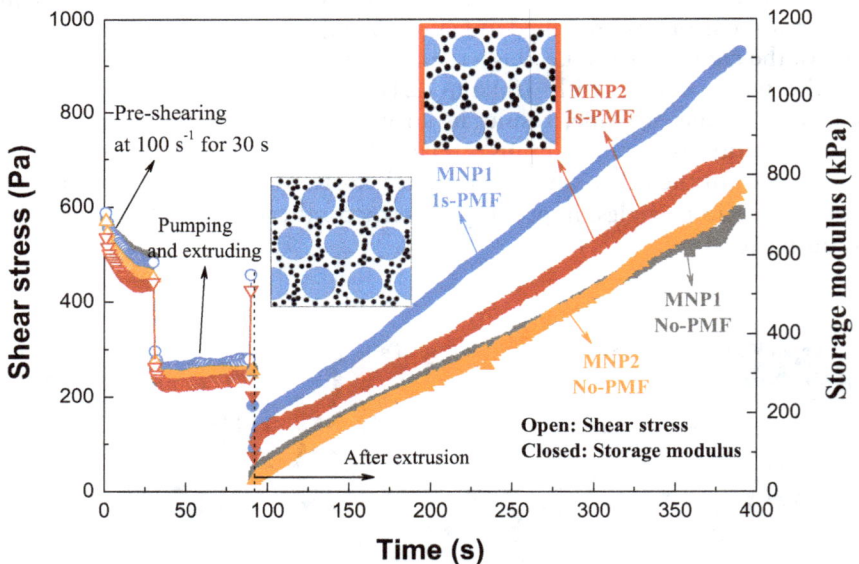

Figure 7.7 Effect of nano-Fe_3O_4 type on the magneto-induced structural build-up of cementitious paste (w/c=0.45, 3% nano-Fe_3O_4, MNP1 and MNP2 have particle size and remanent magnetization of 100 nm and 14.82 Am²/kg, and 200 nm and 10.23 Am²/kg, respectively). Source: after [12].

Overall, applying a PMF during extrusion process seems a potential method to actively improve the stiffness of cementitious materials after extrusion due to the intrinsic properties (i.e. high remanent magnetization) of magnetic particles.

7.4   PRINT HEAD DESIGN AND OUTLOOK

Compared to other intervention signals, improving the buildability by using magnetic particles and applying magnetic field is a relatively new area, and only few studies have been published in this area. The magnetic print-head design is still in infancy, and there are no published works reporting the development. Chen et al. [26] proposed a conceptual design of a print head consisting of an electromagnet, as presented in Figure 7.8. In this set-up, a safe direct current (DC) current with voltage of 12–36 V and frequency of 0–999 Hz can be generated from a 220V/50Hz current through a transformer and a pulse generator. The maximum strength of the electromagnetic field at voltage of 36 V can carry 300 kg of load against gravity. Theoretically, this set-up can be used to enhance the mobility of the 3D printing material by destroying any flos using vibration of magnetic particles, and increase the thixotropic structural build-up of the materials. To the author's best knowledge, however, no other publications were reported on this new technique. Besides, the electromagnetic heating effect is not taken into account in this device, and a cooling system is necessary to counteract the generated heat.

For the future perspectives, the magnetic print head used in real 3D concrete printing should be designed by using either an electric coil or a permanent magnet. For the electromagnet, the magnetic field strength can be controlled by the input of the current of the coil, and meanwhile, a cooling system should be installed to eliminate the additional heat generation. With

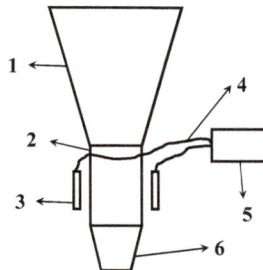

Figure 7.8 Conceptual magnetically print head design. 1. Storage hopper, 2. Pipe with non-magnetic field shielding material, 3. Magnetic field generator, 4. Conductor, 5. DC power supply with high-voltage pulses and adjustable amplitude, and 6. Nozzle. Source: after [26].

regard to the permanent magnet approach, the magnetic field strength can be adjusted by varying the distance between two magnets. In both cases, the field strength should be strong enough to penetrate uniformly throughout the extruder to obtain a homogeneous magnetic field. The magneto-rheology control of 3D printable concrete contributes to a smarter material in additive manufacturing, where the physical and/or chemical properties can be changed by the application of the stimulus. This adds one extra dimension to 3D concrete printing, leading to the development of four-dimensional (4D) printing [27].

7.5 SUMMARY

Applying interventions at the print head to rapidly increase the buildability does not affect the pumpability of the concrete mix before extrusion. Magneto-rheology control, by means of pre-adding responsive additives and applying a magnetic field signal, is a promising method of active rheology control of 3D printable cementitious materials. This chapter describes the theoretical foundations of magneto-rheology control in 3D concrete printing, and a conceptual examination is presented by rheological experiments. It is found that cementitious paste with nano-Fe_3O_4 particles shows higher structural build-up after removing the PMF which is applied before the time sweep test. This is correlated to possible residual magnetic clusters formed under magnetic field because of the presence of the remanent magnetization characteristic of the nanoparticles. These findings clear the path to actively improve the buildability of cementitious materials for 3D printing by introducing a short-pulsed magnetic field during extrusion. Further research regarding the magnetic print head or nozzle design for the magneto-rheology control of 3D concrete printing is prospected.

REFERENCES

[1] Gosselin, C., et al., Large-scale 3D printing of ultra-high performance concrete – a new processing route for architects and builders. *Materials & Design*, 2016. 100: p. 102–109.

[2] Reiter, L., et al., The role of early age structural build-up in digital fabrication with concrete. *Cement and Concrete Research*, 2018. 112: p. 86–95.

[3] Tao, Y., et al., Stiffening control of cement-based materials using accelerators in inline mixing processes: Possibilities and challenges. *Cement and Concrete Composites*, 2021. **119**: p. 103972.

[4] De Schutter, G., et al., Vision of 3D printing with concrete - Technical, economic and environmental potentials. *Cement and Concrete Research*, 2018. 112: p. 25–36.

[5] Yuan, Q., et al., A feasible method for measuring the buildability of fresh 3D printing mortar. *Construction and Building Materials*, 2019. 227 : p. 116600.

[6]   Jiao, D., et al., Thixotropic structural build-up of cement-based materials: A state-of-the-art review. *Cement and Concrete Composites*, 2021. 122 : p. 104152.

[7]   Muthukrishnan, S., S. Ramakrishnan, and J. Sanjayan, Technologies for improving buildability in 3D concrete printing. *Cement and Concrete Composites*, 2021. 122: p. 104144 .

[8]   Dressler, I., N. Freund, and D. Lowke, The effect of accelerator dosage on fresh concrete properties and on interlayer strength in shotcrete 3D printing. *Materials (Basel)*, 2020. 13: p.374.

[9]   Mechtcherine, V., et al., Extrusion-based additive manufacturing with cement-based materials – Production steps, processes, and their underlying physics: A review. *Cement and Concrete Research*, 2020. 132: p. 106037 .

[10] Sanjayan, J.G., R. Jayathilakage, and P. Rajeev, Vibration induced active rheology control for 3D concrete printing. *Cement and Concrete Research*, 2021. 140: p. 106293.

[11] Deshmukh, A.S., et al., Rheological response of magnetorheological cementitious inks tuned for active control in digital construction. *Aci Materials Journal*, 2021. 118(6): p. 263–274.

[12] Jiao, D., C. Shi, and G. De Schutter, Magneto-rheology control in 3D concrete printing: A rheological attempt. *Materials Letters*, 2021. 309: p. 131374.

[13] Vaitkevičius, V., E. Šerelis, and V. Kerševičius, Effect of ultra-sonic activation on early hydration process in 3D concrete printing technology. *Construction and Building Materials*, 2018. 169 : p. 354–363.

[14] Reiter, L., et al., Setting on demand for digital concrete – Principles, measurements, chemistry, validation. *Cement and Concrete Research*, 2020. 132: p. 106047.

[15] Tao, Y., et al., Mechanical and microstructural properties of 3D printable concrete in the context of the twin-pipe pumping strategy. *Cement and Concrete Composites*, 2022. 125: p. 104324.

[16] Muthukrishnan, S., S. Ramakrishnan, and J. Sanjayan. Buildability of geopolymer concrete for 3D printing with microwave heating. In *RILEM international conference on concrete and digital fabrication*. 2020. Springer.

[17] Lesage, K. and G. De Schutter, Admixture for a cementitious material to influence the rheology properties of the cementitious material. In *International Bureau, Universiteit Gent*. 2020. Belgium.

[18] Nair, S.D. and R.D. Ferron, Set-on-demand concrete. *Cement and Concrete Research*, 2014. 57: p. 13–27.

[19] Jiao, D., et al., Rheological behavior of cement paste with nano-Fe_3O_4 under magnetic field: Magneto-rheological responses and conceptual calculations. *Cement and Concrete Composites*, 2021. 120 : p. 104035.

[20] Jiao, D., et al., Quantitative assessment of the influence of external magnetic field on clustering of nano-Fe_3O_4 particles in cementitious paste. *Cement and Concrete Research*, 2021. 142 : p. 106345.

[21] Jiao, D., et al., Structural evolution of cement paste with nano-Fe_3O_4 under magnetic field - Effect of concentration and particle size of nano-Fe_3O_4. *Cement and Concrete Composites*, 2021. 120 : p. 104036.

[22] Jiao, D., et al., Structural build-up of cementitious paste under external magnetic fields. In *Rheology and processing of construction materials.* 2020, Springer. p. 36–42.

[23] Jiao, D., et al., Structural build-up of cementitious paste with nano-Fe3O4 under time-varying magnetic fields. *Cement and Concrete Research*, 2019. 124 : p. 105857.

[24] Jiao, D., et al., Flow behavior of cementitious-like suspension with nano-Fe3O4 particles under external magnetic field. *Materials and Structures*, 2021. 54 : p. 209.

[25] Rich, J.P., P.S. Doyle, and G.H. McKinley, Magnetorheology in an aging, yield stress matrix fluid. *Rheologica Acta*, 2012. 51(7): p. 579–593.

[26] Chen, C., Y. Zhang, and C. Zhang, *Process for improving thixotropy of 3D printing cement-based materials.* 2019, China, patent application CN110092600A.

[27] Kokkinis, D., M. Schaffner, and A.R. Studart, Multimaterial magnetically assisted 3D printing of composite materials. *Nat Commun*, 2015. 6: p. 8643.

Chapter 8

Numerical simulation of active rheology control of cementitious materials

Robin De Schryver, Karel Lesage, and Geert De Schutter

CONTENTS

8.1	Constitutive ARC behaviour	233
	8.1.1 Constitutive magneto-rheological behaviour	235
	8.1.2 Rheological ARC quantification	238
	8.1.2.1 Relative magneto-rheological response	239
	8.1.2.2 Magneto-rheological structural build-up	241
	8.1.2.3 Magneto-rheological structural break-down	242
	8.1.2.4 Bingham parameters	242
	8.1.2.5 Application interpolation	243
8.2	Numerical methodology	244
8.3	Numerical validation	246
	8.3.1 Numerical reliability	246
	8.3.2 Experimental ARC Valorization	251
	8.3.2.1 Falsified alternative hypotheses	251
	8.3.2.2 Numerical ARC viability	254
8.4	Further research and outlook	258
	8.4.1 Projections of ARC modelling	258
	8.4.2 Generality of ARC model	260
	8.4.3 Future numerical ARC studies	260
8.5	Summary	261
References		263

8.1 CONSTITUTIVE ARC BEHAVIOUR

Two forms of active rheology control (ARC) were depicted in Chapter 3 and Chapter 4. The first one is ARC by addition of responsive minerals and the second one is by addition of responsive polymers. ARC by addition of responsive minerals in the form of magneto-rheological (MR) behaviour is used to highlight the capability of using numerical simulations for ARC. Although a constitutive behaviour model was derived based on the

DOI: 10.1201/9781003289463-8

philosophy to capture a MR response of cementitious suspensions as a particular form of ARC, its context may be extended to a broader range of ARC exhibited in other forms such as those elucidated in Chapter 3 and Chapter 4.

Particular to MR fluids, a magnetic response is fundamentally understood as the formation of magnetic spikes, clusters, networks or structures in general (cf. Chapter 3). Since thixotropy is generally understood as the formation of an internal structure [1], magneto-rheology could be considered as the actively controllable extension of passive, natural thixotropy. Similar to thixotropy, the kinetic process of an internal structure λ may be described by a structural build-up term driven by magnetization and a structural break-down term by demolition of magnetic clusters under shear action.

Nevertheless, the concept of describing the temporal evolution of an internal structure may be interpreted in a wider context of ARC. As such, ARC may generally be captured by the temporal evolution of a control state λ, whose control process may be described by structural build-up and break-down. The concept may be irrespective of whether a responsive polymer is added to the mixture or whether a responsive mineral is added. Moreover, it is irrespective of whether the active control would be enabled by an external magnetic field as a trigger or whether an external electrical field would be used as a trigger. Hence, whether it is a magnetic, chemical or physicochemical process that enables ARC, its behaviour description may be captured by the concept of a kinetic temporal evolution of an internal structural control state λ, whose development and demolition are described by structural build-up and break-down.

To do so, a kinetic equation can be used to express the temporal evolution of an internal control state parameter λ, which is actively controlled by a structural build-up term (R_{bu}) and a structural break-down term (R_{bd}). To link the control state to the rheological response, the control state parameter λ can be attributed to rheological properties, e.g. by a relative contribution χ to the yield stress τ_0 of the cementitious suspensions. As a result, the rheology is actively controlled by an internal control state parameter, which can be described by equation 8.1.

$$
\begin{cases}
\tau = \tau_0 \left(1 + \chi\lambda\right) + \mu\dot{\gamma} \\
\dfrac{D\lambda}{Dt} = R_{bu}\left(1 - \lambda\right) - R_{bd}\lambda
\end{cases}
\tag{8.1}
$$

In that manner, ARC can be time-dependently described by a control state parameter equal to 1, 0 or somewhere in between, whose response can be instantaneous (i.e. R_{bu} or R_{bd} equal to infinity), or more of transient character (i.e. finite values for R_{bu} or R_{bd}). In case of an instantaneous response, the control state is either in 'on-state' or 'off-state' ($\lambda = 1$ or $\lambda = 0$), which would result in an immediate, constant in- or decrement in yield stress. This can, for instance, be the case for MR or ferro-fluids. However, since cementitious suspensions as MR carrier fluid exhibit viscous or yielding behaviour, the response takes a finite time before full activation is established.

Since the context of this chapter relates to ARC established in the form of magneto-responsive behaviour, a constitutive MR behaviour model was further elaborated as such. In particular, it was developed for an MR fluid that was established by exploiting the magnetization capabilities of fly ash. A magnetic fly ash was introduced in a cementitious paste suspension as carrier fluid. In that way, a cementitious MR suspension was examined for its fundamental behaviour description in view of ARC. Based on the understanding of an MR response in Chapter 3 and previous works on MR fluids in the literature [2–28], a constitutive MR behaviour model as a particular form of ARC was elaborated.

Doing so, the first subsection elucidates a developed constitutive MR behaviour model. The second subsection elaborates upon developed rheological ARC quantification methodologies for the constitutive MR model.

8.1.1   Constitutive magneto-rheological behaviour

Since the literature attributes MR behaviour to an increased yield stress contribution in Bingham fluids [2–28], this approach is imposed. Due to the viscous and yielding character of cementitious suspensions as the MR carrier fluid, and as observed from rheological experiments (cf. Chapter 3 [13,15–20]), the MR response of cementitious suspensions exhibits time-dependent behaviour [1]. Therefore, a constitutive model based on equation 8.1 is proposed for the kinetic temporal evolution of an internal magnetic structure λ with as little parameters as possible, in a similar fashion as thixotropy models were proposed by Moore, Carleton et al. or by Houska [1, 29–34]. Especially because cementitious suspensions are generally perceived as Bingham fluids, the magnetic response state λ is attributed to the yield stress τ_0 of a Bingham fluid, in a similar fashion as done by Carleton et al. [30]. Hence, a constitutive magneto-structuration model for cementitious MR fluids is proposed in equation 8.2.

$$
\begin{cases}
\tau = \tau_0 \left(1 + \chi\lambda\right) + \mu\dot{\gamma} \\
\dfrac{D\lambda}{Dt} = R_{bu}\left(1 - \lambda\right) - R_{bd}\lambda
\end{cases}
\quad
\begin{cases}
\chi = \chi_\phi f_M \left(\dfrac{\left|\vec{H}\right|}{\left|\vec{H}_{sat}\right|}\right)^c \\
R_{bu} = \chi_\phi f_M \left(\dfrac{\left|\vec{H}\right|}{\left|\vec{H}_{sat}\right|}\right)^c \\
R_{bd} = \alpha\dot{\gamma}^a + \beta\left|\dfrac{\partial\vec{B}}{\partial t}\right|^b
\end{cases}
\qquad (8.2)
$$

The following parameters are attributed to equation 8.2. The temporal evolution of the internal magnetic structure λ is described by a structural build-up term and a structural break-down term. The rate of structural build-up R_{bu} (s^{-1}) is described by a magnetic growth rate δ_M (s^{-1}) and scaled by a magnetization function f_M in proportion to the magnetic field of saturation $\vec{H}_{sat}$ with a growth power d. The rate of structural break-down R_{bd} (s^{-1}) is described by a break-down due to shear action and a potential break-down due to micro-vibration induced by magnetic oscillations. Hence, similar to the model proposed by Roussel [35], a de-flocculation rate α is attributed and scaled proportionally with the shear rate $\dot{\gamma}$, with a power index a. Potentially induced micro-vibrations by magnetic oscillations may be described by a magnetic vibration disturbance rate β, which could proportionally be scaled by the temporal magnetic variation $\partial \vec{B} / \partial t$ with a power index b. Lastly, the magnetic response is incorporated into a relative yield stress scaling. A relative magnetic response χ is defined, which is scaled by the internal magnetic structure λ. The relative magnetic response is determined at a certain magnetic field near saturation ($\vec{H}_{sat}$) for a certain concentration ϕ. To simplify the complexity of the magnetic response as a function of the magnetic field $\vec{H}$ itself, the intermediate magnetization at magnetic fields $\vec{H}$ lower than the saturated magnetization is scaled by a magnetization function f_M, with a response power index c. Beyond the saturated magnetization ($|\vec{H}| \geq |\vec{H}_{sat}|$) it is assumed that no further increase in response occurs, i.e. $f_M \leq 1$. Alternatively, one could also express the magnetization scaling as a function of the corresponding magnetic field flux density $\vec{B}$ (in T), instead of the magnetic field intensity $\vec{H}$ (in kA / m). Nevertheless, both concepts are interchangeable.

It can be assumed that the MR response properties vary as a function (f_M) of the applied magnetic field $\vec{H}$ in proportion to the material's magnetization. This assumption is based on the fact that the magnetization of the MR cement paste is the driving force behind the overall MR response. To simplify the material's magnetization (cf. Chapter 3), it could be linearized between a zero magnetic field and a magnetic field of saturation $\vec{H}_{sat}$, after which no further response is expected. Therefore, the magnetization scaling function can be simplified to $f_M = |\vec{H}| / |\vec{H}_{sat}| \leq 1$. This (pseudo-plastic) linearization of the magnetization response is illustrated in Figure 8.1. Although a more accurate scaling than a linear (pseudo-plastic) approximation could be provided to mimic the material's magnetization response for fields below saturation, it could not be further verified and it is out of the scope of the proposed constitutive model.

Further simplifications can be done for the proposed model. For instance, it is known from the literature that the magnetic force between magnetic particles is proportional to the square of their magnetization [2–28]. The magnetic increase in yield stress τ_0 could be described by the

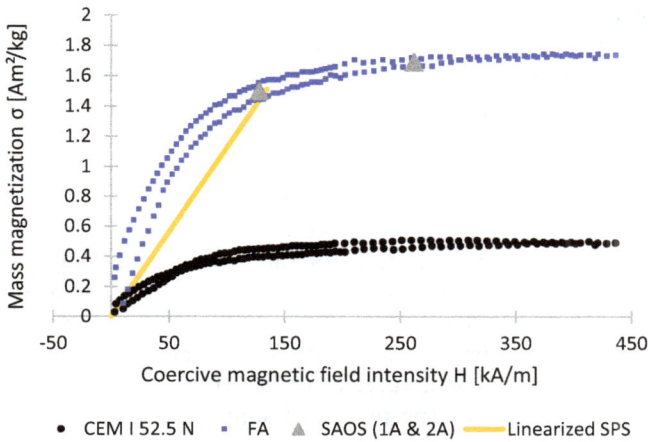

Figure 8.1 Magnetization of used cement and fly ash. The magnetic response is expressed as the mass magnetization σ (in emu / g in the centimetre-gram-second electromagnetic units (CGS-EMU) system or Am² / kg in SI units) as a function of the applied magnetic field intensity H(Oe in the CGS-EMU system or A / m in the SI system). Approximation of the fly ash magnetization response to a pseudo-plastic linearization, interpolated between the rheometer SAOS measurements (i.e. at 1 A and 2 A) to the magnetic field as used in the pumping experiment (i.e. B = 0.169 T). Source: originating from [1].

frictional force among magnetized particles. Hence, power indices c and d could be simplified to 2. Magnetic micro-vibrations were not explored in the numerical simulations, nor in validating experiments considered here. Hence, the second structural break-down term (scaled by β) can be omitted from equation 8.2. Similar to the thixotropic model proposed by Roussel [35], the shear action is more likely to be scaled by the shear rate $\dot{\gamma}$ with a power a equal to 1. The combination of these simplifications yields the proposed simplified magneto-structuration model given in equation 8.3.

$$\begin{cases} \tau = \tau_0\left(1+\chi\lambda\right)+\mu\dot{\gamma} \\ \dfrac{D\lambda}{Dt} = R_{bu}\left(1-\lambda\right)-R_{bd}\lambda \end{cases} \qquad \begin{cases} \chi = \chi_\phi f_M^2 \\ R_{bu} = \chi_\phi f_M^2 \\ R_{bd} = \alpha\dot{\gamma} \end{cases} \tag{8.3}$$

Although the model is proposed for magnetic fields perpendicular to the flow direction, it may still be applicable to magnetic fields aligned with the flow direction or a combination thereof. Due to the vectorial character of magnetic fields and a respective magnetic structure λ, more complicated anisotropic models may be defined. Furthermore, spatial magnetic gradients may also induce accelerations or decelerations, as supposed in 'magnetic guns or rivers' (cf. Chapter 4). However, instead of contributing to internal magnetic structuration, such magnetic gradients are rather expected to impose an additional external acceleration field ($\vec{a}$) induced by a magnetic driving force, similar to a gravity field ($\vec{g}$), but in the direction of the spatial magnetic gradient field. Nevertheless, more complicated anisotropy and spatial magnetic gradient fields were not explored. More complicated models, thus, remain an open path for future research. Magneto-rheology and thereby active stiffening control (ASC) and ARC were investigated from a preliminary but fundamental viewpoint – that is, verifying whether magneto-rheology can be described by a kinetic approach of the temporal evolution of an internal structure. In order to do so, a rheological quantification methodology for the proposed simplified magneto-structuration model (Eq. 8.3) is elaborated.

8.1.2   Rheological ARC quantification

Although some works exist in the literature of magneto-rheology for designing smart damping systems of structures and brakes, rheological quantification of cementitious MR fluids is still a missing link. Existing methods to quantify the rheological response are not applicable or contain gaps such as lack of time-dependency, which is especially the case for cementitious pastes as a viscous Bingham carrier fluid [2–28].

To overcome these issues and to bridge the gap towards rheologically quantifying ASC and ARC, invoked by magneto-rheology, a rheological quantification methodology was developed for the proposed simplified magneto-structuration model (Eq. 8.3).

To this end, a rotational parallel plate rheometer was used, in which test samples could be subjected to a magnetic field induced by a magnetic control unit. Since no reliable flow curve data could be obtained under magnetic fields, another approach was followed. As depicted in Chapter 2, a small amplitude oscillatory shear (SAOS) test was used under different magnetic fields. The relative magnetic response χ and magnetic growth rate δ_M were directly determined for the considered applied magnetic fields (i.e. 0.161 T and 0.329 T). Practically this means that no additional scaling was necessary to compensate for fields $\vec{H}$ different from saturation field $\vec{H}_{sat}$, since one can directly determine values for the relative magnetic response χ and magnetic growth rate δ_M, considered at a specific value of the magnetic field $\vec{H}$. Therefore, the proposed model (Eq. 8.3) can be formulated as a function of magnetic properties considered for a specific value of the magnetic field $\vec{H}$. Then, in order to be able to determine both the relative magnetic response ($\chi = \chi \mid^{\vec{H}}$) as well as the magnetic structural build-up rate ($\delta_M = \delta_M \mid^{\vec{H}}$), the rheological response can be decomposed into a reference without magnetic field and a magnetic response via equation 8.4.

$$\begin{cases} \tau = \underbrace{\tau_0 + \mu\dot{\gamma}}_{reference} + \underbrace{\tau_0\chi\lambda}_{magneto} \\[2ex] \dfrac{D\lambda}{Dt} = \delta_M \mid^{\vec{H}} (1-\lambda) - \alpha\dot{\gamma}\lambda \end{cases} \qquad (8.4)$$

Through decomposition of the Bingham reference and magneto-response, the magnetic response parameters can be determined. The Bingham reference follows from a reference SAOS test without magnetic field, and the response parameters from a SAOS test with an applied magnetic field of interest (e.g. 0.161 T or 0.329 T at 1 A or 2A). The complex valued linear elastic response in a SAOS test can be decomposed into a real-valued linear elastic (LE) response and a linear viscous (LV) response. Under the assumption that the magnetic response is attributed to the yield stress τ_0, it suffices to consider the real-valued LE response. Since a SAOS test is in the linear elastic region under static conditions, the shear rate $\dot{\gamma}$ is nearly zero ($\dot{\gamma} = 0$). In that way, the viscous contribution is nevertheless omitted, let alone that it would be significant for nearly zero shear rates. Irrespectively, the yield stress contributions of both the reference without magnetic field and with magnetic field are proportional ($LE \propto \tau$) – if not equal ($LE = \tau$) – to the LE SAOS response. More specifically, the LE response is given by the product of the storage modulus G' and the shear strain amplitude ($\gamma = 50 \mu rad$). In that way, one can make use of the LE response to determine the magnetic response parameters χ and δ_M. The two following paragraphs, therefore, outline the experimental methodology to quantify the relative magnetic response ($\chi = \chi \mid^{\vec{H}}$) and the magnetic structural build-up rate ($\delta_M = \delta_M \mid^{\vec{H}}$).

8.1.2.1   Relative magneto-rheological response

Knowing that the SAOS test without magnetic field serves as the LE SAOS reference response, the absolute magnetic response can be captured by taking the difference between an LE response from a SAOS test with and without magnetic field. After temporal equilibrium for a fully established response ($\lambda = 1$) has been achieved for the absolute magnetic response $\Delta\tau$, the relative magnetic response χ can be found from equation 8.5 after division by the LE SAOS reference response without magnetic field, i.e. τ_0.

$$\begin{cases} \Delta\tau = \tau_0 \chi \lambda \\ \dfrac{D\lambda}{Dt} = \delta_M (1-\lambda) \\ \lambda(t) = 1 - \exp(-\delta_M t) \approx 1 \end{cases} \qquad (8.5)$$

To practically do so, a thixotropic Bingham model is fitted first to the reference LE response without magnetic field in order to eliminate influences by noise disturbances, especially because of the sensitivity of SAOS measurements. Since thixotropic behaviour is inherently present in cementitious suspensions [35–38], the reference response also includes a thixotropic response under static conditions. Although the thixotropic growth is not significant for the considered mixture design (i.e. A_{thix} ca. 0.09 Pa / s based on the SAOS test), it is still inherently present in the LE response as a linear increase over time. Furthermore, during the first initial time period, a rapidly growing thixotropic structure may manifest in the material [35, 39, 40]. Hence, apart from an LE response with a constant slope, a non-linear exponential decay in accordance with [40] is also fitted to the reference LE response. This data curation allows to eliminate noise disturbances from the reference, which are subtracted from the magnetic response to obtain the absolute response as illustrated in Figure 8.2.

Furthermore, by taking the absolute difference between the SAOS test with and without magnetic field, the thixotropic contribution is also eliminated. The absolute magnetic response is then computed as the temporal average of the absolute difference (i.e. increment), after magnetic structuration equilibrium ($\lambda = 1$) has been achieved. Practically, this meant taking the

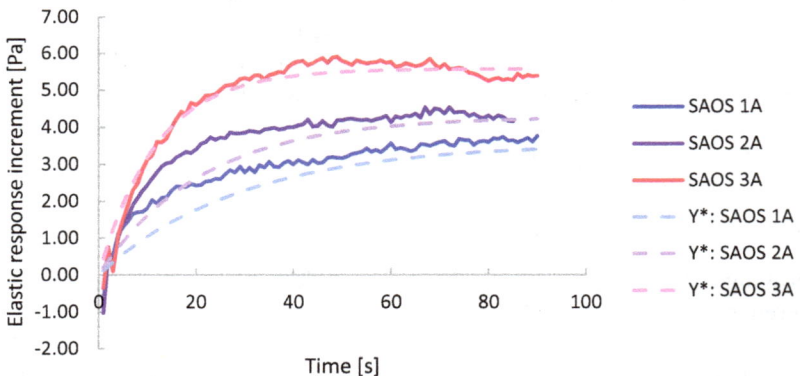

Figure 8.2 Illustration of temporal absolute difference of the linear elastic (LE) response between a SAOS test with magnetic field and without magnetic field as a reference (after reference model fitting).

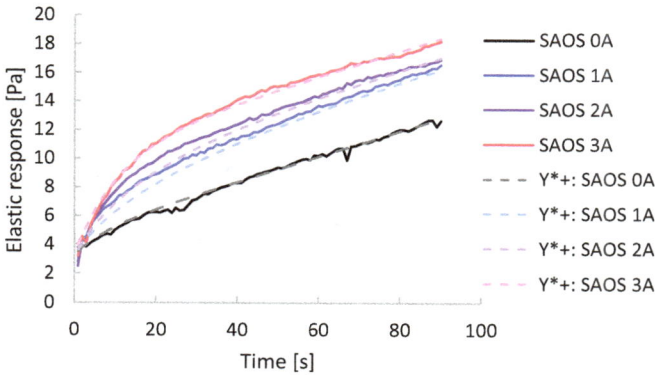

Figure 8.3 Illustration of temporal linear elastic (LE) response of a SAOS test with magnetic field and without magnetic field as a reference. The dashed lines indicate the regressions fitted by the depicted methodology.

temporal average between 30 s and 90 s (if no obvious noise disturbances were introduced in the measurement) of the absolute difference or increment $\Delta\tau$. To obtain the relative magnetic response, the absolute magnetic response is divided by the Bingham reference state of the cementitious pastes. The thixotropic linearity over time was eliminated for the Bingham reference state by considering the intercept of the thixotropically fitted LE response, as illustrated in Figure 8.3. In that way, a relative magnetic response χ could be obtained, which is relative to the intrinsic yield stress τ_0 of the suspension. Since the relative magnetic response χ is a proportional quantification, it is invariant for a potential bias in the rheology device.

8.1.2.2 Magneto-rheological structural build-up

To quantify the magnetic structural build-up δ_M, a time-dependent analysis should be performed. The temporal parameters can be determined from the transient stress decay or evolution of the absolute difference between the SAOS test with magnetic field and the reference SAOS test without magnetic field (Figure 8.2). The magnetic structural build-up rate δ_M is unravelled after taking the temporal derivative of the stress evolution. Since there is no shear action during a static SAOS test, the absolute difference and its temporal derivative simplify to equation 8.6 and equation 8.7, for which a logarithmic regression can be made from equation 8.8.

$$\Delta\tau = \tau_0\chi\left(1 - \exp\left(-\delta_M t\right)\right) \tag{8.6}$$

$$\frac{\partial \Delta \tau}{\partial t} = \tau_0 \chi \delta_M \exp(-\delta_M t) \qquad (8.7)$$

$$\ln\left(\frac{\partial \Delta \tau}{\partial t}\right) = \ln(\tau_0 \chi \delta_M) - \delta_M t \qquad (8.8)$$

Hence, one can obtain an estimation for the magnetic growth rate δ_M by a log-linear regression of equation 8.8. Practically, the temporal derivative was calculated by employing a noise smoothing differentiation algorithm. In that way, experimental noise which is potentially amplified after taking a derivative could be suppressed. The well-known Lanczos differentiation method was employed, since it is a commonly used and easily applicable noise smoothing algorithm for differentiation, suppressing higher temporal frequencies commonly attributed to noisiness [41–44]. For the low-noise Lanczos method, a weighted, smoothed interval of 11 points ($N = 11$) was chosen with second-order accuracy ($M = 2$), because longer smoothers are less subjected to wavy filter responses at high frequencies [1, 41–44]. Because of the simplicity of the algorithm it could easily be implemented in the data processing (in MS Excel), for which further details can be consulted in the literature [1, 41–44]. As such, the methodology is invariant for a linear transformation of the shear stress ($\tau = a + b\tau'$). Hence, even if the LE response was to be only proportional to the shear stress, instead of being equal to it, the result for the magnetic growth rate δ_M would be identical.

8.1.2.3   Magneto-rheological structural break-down

The MR structural break-down term could not be quantified due to shear-induced particle migration during flow curve tests under imposed magnetic fields. Hence, the de-flocculation rate α was calibrated for the conducted numerical simulations, based on the conducted experiments. If the de-flocculation rate is feasible as such, it at least means the proposed model is viable to conceptualize ARC exhibited in the form of magneto-rheology. Furthermore, one could presume a de-flocculation rate identical to the one considered in thixotropic studies of cementitious suspensions, since it concerns the same structural kinetics of break-down processes for cementitious suspensions. Hence, based on the work of Roussel [35], one could assume a de-flocculation rate $\alpha \approx 0.005$.

8.1.2.4   Bingham parameters

The Bingham yield stress τ_0 and plastic viscosity μ can be determined according to available methods in the literature. However, in order to be more closely related to the occurring situation and to impede a potential bias due to differences in rheological devices, the Bingham parameters were determined from the pumping experiment itself. For the conducted pumping experiment, this resulted in a yield stress τ_0 of 111.04 Pa and a plastic viscosity μ of 0.70 Pa.s.

8.1.2.5 Application interpolation

Following aforementioned rheological quantifications, one is able to determine all parameters of the proposed simplified magneto-structuration model. As such, the magneto-structuration properties (χ and δ_M) are determined for a given value of the imposed magnetic field in the rheometer set-up – that is, 0 T, 0.161 T and 0.329 T, which correspond to an electric current of 0 A, 1 A and 2 A in the magnetic induction coil unit of the rheometer set-up. However, since the pumping experiment practically applied ca. $\vec{H} \approx 0.169 T$, the magneto-structuration parameters were interpolated accordingly (cf. Figure 8.1).

Furthermore, instead of linearizing the magnetization in proportion to the field of saturation ($\vec{H}_{sat}$), the magnetization can be linearized in proportion to the field of application $\vec{H}_{app}$ for which the response parameters (χ and δ_M) were interpolated. In transient zones of the application, where the magnetic field is smaller than the field of the application ($|\vec{H}| \leq |\vec{H}_{app}|$), one may scale the magnetic properties accordingly, i.e. $f_M = |\vec{H}| / |\vec{H}_{app}| \leq 1$.

In that way, the magnetization is (pseudo-plastically) linearized in transient zones differing from the main application, as illustrated in Figure 8.1. As mentioned before, instead of linearizing the response to transient zones differing from the application, a more accurate magnetization response as a function of the magnetic field $\vec{H}$ could be refined. However, it is not the objective of this work to outline the exact relation of the magnetic response (characterized by χ and δ_M) as a function of a continuous spectrum of the magnetic field $\vec{H}$. On the one hand, this is because the experimental work would be too elaborate. Moreover, the response behaviour in transient zones is expected to be subordinate compared to the main application zone. On the other hand, apart from very small transient zones, only a single value of the magnetic field occurs (i.e. $\vec{H}_{app} = 134.5 \, kA/m$ or $\vec{B}_{app} = 0.169 T$ for the pumping experiment as depicted in Chapter 5). A response based on a single value determined for the field of application is thereby sufficient for the application. Moreover, the response behaviour in transient zones could not be further verified or validated. Hence, further refinement of the magnetization scaling (f_M) as a function of a continuous spectrum of magnetic fields $\vec{H}$ is out of the scope of this work.

8.2 NUMERICAL METHODOLOGY

In the literature, many numerical methods exist to model the fresh flow behaviour of cementitious materials, which can generally be referred to as CFD. Commonly used examples are the finite volume method (FVM), volume of fluid (VOF) method, discrete element method (DEM), Lattice Boltzmann method (LBM), etc. Numerical simulation techniques may be grouped into three categories, i.e. based on a continuous mesh-based approach, based on a discrete mesh-free approach and based on a combination thereof, known as hybrid approaches. Each technique has its advantages and disadvantages. Depending on the aimed outcome, such as microscopic, discrete particle or macroscopic, continuous phase modelling, a different technique can be selected. However, since behavioural aspects of ARC are missing in current techniques, a numerical simulation technique was developed to model ARC. This section therefore briefly depicts the context of the developed numerical simulation technique to model ARC.

In view of modelling the flow behaviour of ARC from a macroscopic viewpoint of, e.g., concrete pumping, formwork casting, three-dimensional (3D) printing or other applications, a continuous, mesh-based CFD approach was used in Open source Field Operation And Manipulation (OpenFOAM). More specifically, FVM and VOF methods were extended to incorporate non-Newtonian, time-dependent behaviour of ARC. To this end, transient FVM solver 'nonNewtonianIcoFoam' and transient VOF solver 'icoFoam' were adapted to include the constitutive ARC magneto-structuration behaviour model proposed in equation 8.3. Since a pumping experiment with a cementitious MR paste suspension in view of assessing ARC simulation was aimed for, only the adapted FVM solver was used in spite of further potential applications of the VOF solver.

To numerically simulate flow behaviour, physical conservation laws are often used in continuous CFD approaches. After imposing assumptions and simplifications, partial differential equations are typically obtained which need to be solved. More specifically, mass conservation and momentum conservation equations need to be solved, which are, respectively, known as the continuity and Navier–Stokes equations and depicted in equation 8.9 and equation 8.10 [45].

$$\nabla \circ U = 0 \tag{8.9}$$

$$\rho \frac{DU}{Dt} = \rho g - \nabla p + \nabla \circ (\nu \nabla U) + \nabla U \circ \nabla \nu \tag{8.10}$$

In fact, equation 8.9 and equation 8.10 describe a continuous domain of space and time, known as the continuum, with macroscopic properties such as a velocity field U, pressure field p, density ρ and Newtonian viscosity μ. In these equations, the material derivative is denoted as

$DX / Dt = \partial X / \partial t + U \circ \nabla X$ and the inner (tensor) product as $\circ$. To be able to solve these equations for fields U and p, they are discretized over a grid, known as a mesh cell, which in turn transforms the equations into a set of algebraic equations that can be computed numerically. The mesh cells are in fact small control volumes in case of FVM techniques, since the conservation equations are solved for each one of these cells. Hence, the technique is called the finite volume method. After numerical computation a solution is obtained for the numerical fields discretized across the mesh, which is an approximation constrained by the discretization settings, solver algorithm, solver settings and imposed assumptions [1, 46–49].

Furthermore, equation 8.9 and equation 8.10 are only valid for laminar incompressible flows with a constant density ρ and a Newtonian viscosity μ. As most fluids, cementitious suspensions are incompressible [50]. Although their flow is laminar, it is non-Newtonian, the more so in case of ARC [50]. To allow for non-Newtonian flow behaviour, the viscosity μ is therefore replaced with an apparent Newtonian viscosity v, which is equal to the non-Newtonian shear stress τ divided by the locally occurring shear rate ($v = \tau / \dot{\gamma}$). This approach is known as the generalized Newtonian approach (GNA) or the generalized Newtonian fluid (GNF) approach. Although it is a compromise, it allows for non-Newtonian flow behaviour description. However, since the apparent viscosity is thereby unbounded, numerical stability or accuracy issues may arise from the GNA of yield stress fluids, known as the unregularized viscoplastic problem. One way to cope with these issues is by regularizing the apparent viscosity and velocity under-relaxation to a limited extent. Hence, a bilinear regularization approach was applied with a maximum apparent viscosity ratio of 1000 compared to the plastic viscosity and a velocity under-relaxation of 0.95. More details on different forms of regularization and alternative coping techniques can be consulted in the literature [1].

Furthermore, the external magnetic field $\bar{B}$ as a trigger field enabling ARC was simulated and linked to the constitutive ARC magneto-structuration model proposed in equation 8.3. Although solvers were also developed to simulate a(n) (electro)magnetic field, the magnetic field of the active magnetic control unit was conceptually simulated. Practically, this meant that a constant external magnetic field was imposed with a value of $B = 0.169\,T$ in the vertical direction (y) as experimentally measured inside the magnetic pipe section with a Teslameter probe.

Making use of the adapted FVM solver, pipe flow of a cementitious paste suspension with ARC was simulated. More specifically, the magnetic set-up of the small pumping system (SPS) depicted in Chapter 5 was simulated. To verify the validity of the numerical simulations and to valorize the numerical ARC modelling, a sound numerical validation strategy was conducted.

8.3   NUMERICAL VALIDATION

This section elaborates upon two aspects of numerical validation. Firstly, the validity of the numerical simulations is verified in a section on numerical reliability. Secondly, in a section on experimental valorization, numerical ARC results are valorized based on the simulation of an MR pumping experiment.

8.3.1   Numerical reliability

To build confidence in the numerical simulations, the numerical accuracy or reliability was assessed based on a fivefold validation strategy. The fivefold strategy comprised a validation by qualitative concept, by method independence, by literature, by theory and lastly by experiments. The first four validation strategies are independent of the rheological reliability, while the fifth one directly depends on uncertainties involved with rheological input parameters. To do so, cementitious pipe flow simulations were assessed for their accuracy for each validation strategy. To not further complicate the conciseness of this chapter, this subsection outlines the numerical reliability only briefly for these five validation strategies. Further details on the numerical reliability can be consulted in earlier published works [1, 48].

Conceptually it was concluded that the numerical Bingham simulations of cementitious pipe flows correspond qualitatively well. This means that, though small, a Bingham plug was successfully simulated capturing a quasi-quadratic velocity profile as expected for Bingham pipe flow simulations. After the pipe flow was fully developed steady-state, a linear pressure loss evolution was obtained from the simulations. Also thixotropic behaviour was taken into account by the evolution of an internal structure in accordance with the cementitious thixotropy model proposed by Roussel [35], and it was concluded that also the temporal and spatial evolution of the internal structure λ corresponded logically well. Hence, it was qualitatively concluded that the numerical simulations were at least conceptually sound.

Method independence was also verified on a profound set of numerical pipe flow simulations. On the one hand, numerous (270) simulations were conducted with different cross-sectional, longitudinal mesh resolutions and cross-sectional aspect ratios. As such, a final mesh design was established as illustrated in Figure 8.4 for the ARC pipe flow simulations. On the other hand, numerous (240) simulations were conducted to justify appropriate use of a bi-linear regularization with a GNA viscosity ratio of 1000 and under-relaxation of the velocity limited to 0.95. Apart from a default set or CFD-advised solver settings, method independence was verified.

Simulations were also conducted and compared with the ones from the literature. To do so, the concrete pumping experiments simulated

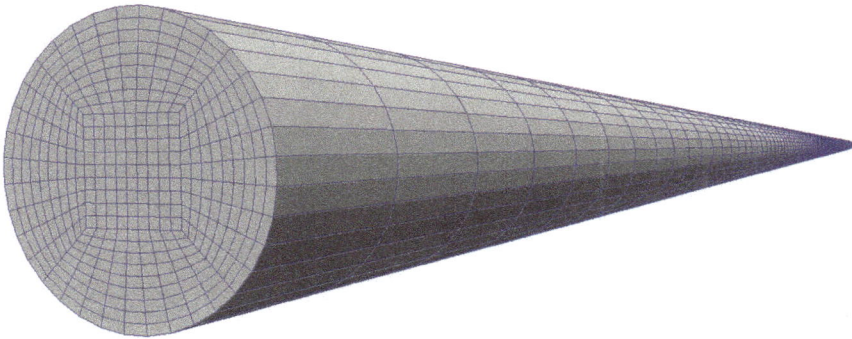

Figure 8.4 **3D rendered illustration of the final pipe mesh design after a profound mesh independence study.**

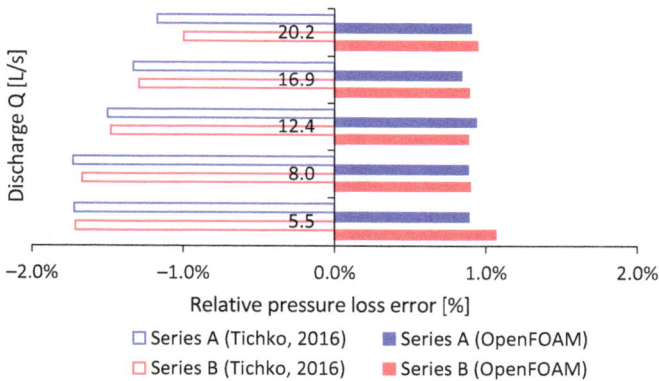

Figure 8.5 Literature comparison of the numerical concrete pipe flow simulations performed by Tichko, relative to the expected Buckingham–Reiner theory. The results indicate that OpenFOAM performs at least as well as, if not better than, commercial CFD software ANSYS FLUENT. Source: [51]; originating from [48].

in ANSYS FLEUNT with FVM by Tichko [51] were compared with simulations performed in the used FVM OpenFOAM solver. As observed from Figure 8.5 for this literature comparison, it was concluded that the OpenFOAM simulations perform at least as well as, if not better than, the ones from ANSYS FLEUNT.

In addition, an elaborate analysis was conducted by comparing numerical cementitious pipe flow simulations with the expected theory. To do so,

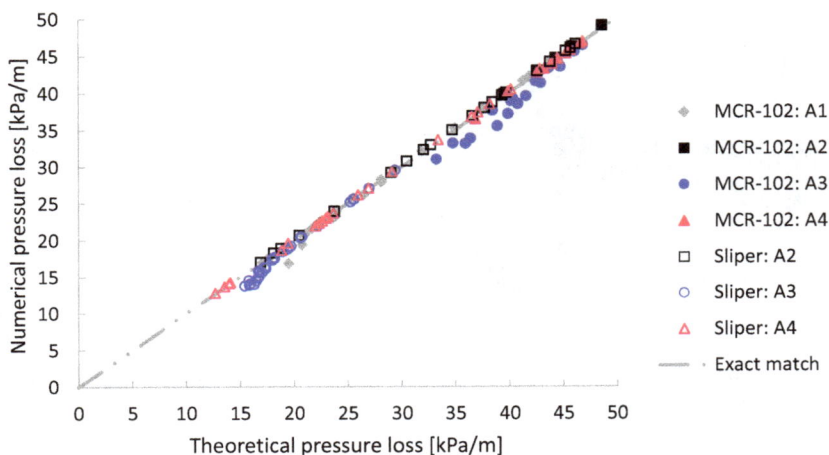

Figure 8.6 A good agreement is obtained between the simulated pressure loss and the expected theoretical Buckingham–Reiner solution. Source: [48].

pipe flow simulations were used, whose rheological input parameters originate from conducted pumping validation experiments. These simulations served a twofold purpose. On the on hand, they served to assess the numerical accuracy independent of the rheological uncertainty, by comparing the results with the theoretical solution, i.e. the Buckingham–Reiner solution. On the other hand, the simulations served to assess the numerical experimental reliability dependent on the rheology of different origins. In total, 213 simulations were conducted, which comprised a variety of pumping discharge measurements, for a range of four different Bingham paste suspensions (A1–A4). Pumping experiments were repeated at least three times (A1-1 to A1-3, etc.). Two different rheometers were emlployed to determine the rheological Bingham parameter as input for the numerical simulations. The first one was a parallel plate rheometer (MCR-102), while the second one was the sliding pipe rheometer (Sliper). Based on the numerically computed pressure loss in Figure 8.6, the simulations were in good agreement with the expected theoretical solutions of the Buckingham–Reiner equation.

Furthermore in Figure 8.7, the simulation accuracy is quantitatively expressed in non-dimensional form after transforming the pressure loss Δp as well as the pumping discharge Q via the pressure number $Pn = \Delta p R / \left(2L\tau_0\right)$ and dimensionless discharge $\hat{Q} = Q / \left(R^3 \tau_0 / \mu\right)$, after [1, 48, 52]. From Figure 8.7 it can be concluded that the numerical simulations are significantly accurate with theoretical errors below 1%, at least if the pressure number

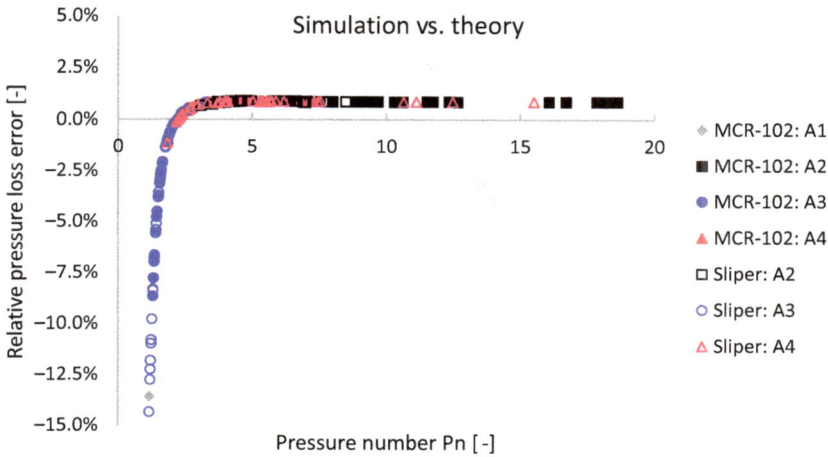

Figure 8.7 The relative simulation pressure loss error (compared to Buckingham–Reiner) as a function of the pressure number Qreveals the influence of the flow regime or unregularized viscoplastic problem. The accuracy decreases for lower pressure numbers or for more yield stress dominant flow regimes. Source: originating from [48].

is sufficiently high, i.e. $Pn > 5/4 = 1.25$ or more strict $Pn > 20/11 \approx 1.82$. This is equivalent with a dimensionless discharge of, respectively, $\hat{Q} > 0.069$ or more strictly $\hat{Q} > 0.424$ [1, 48]. For lower pressure or discharge numbers, the numerical accuracy diverges. From this analysis it was concluded that numerical inaccuracy or even instability arises from the lack of being able to properly simulate yielding behaviour as a compromise of a GNF regularization approach. Nevertheless, the pressure number Pn and/or the non-dimensional discharge $\hat{Q}$ indicate whether significantly accurate (laminar) numerical pipe flow simulations are obtained.

Lastly, the latter simulations were also assessed with regard to their reliability compared to the pumping experiments. In this case also the uncertain character of the rheological parameters is involved in the reliability analysis, which is clearly observed from Figure 8.8. The rheological parameters obtained from the sliding pipe rheometer (Sliper) perform significantly better than those obtained from the parallel plate rheometer (MCR-102). More specifically, Sliper-based predictions range from an exact match to a relative error of 100%, compared to a factor of 8 for MCR-based predictions (cf. [48] or [1] for more details).

The inaccuracy of the numerical simulations compared to experimental observations is attributed to several factors. The most predominant factor is

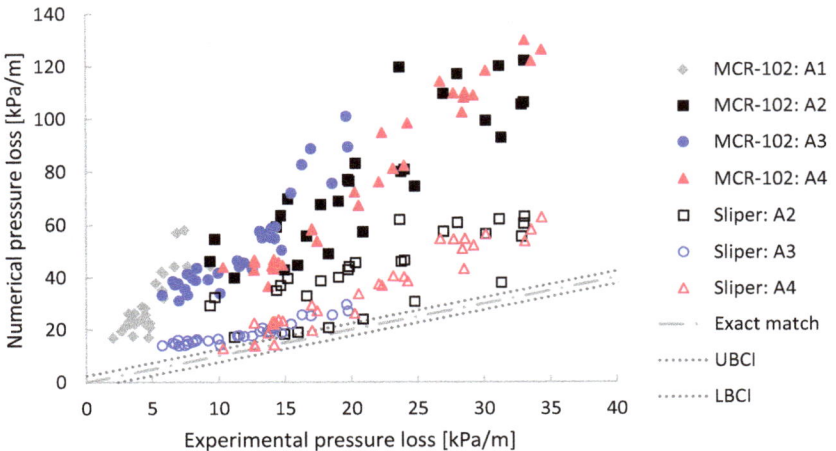

Figure 8.8 Comparison of numerically simulated pressure loss with experimentally obtained pumping results, for four different mixture designs (A1 to A4), based on rheological input from the MCR-102 rheometer and the Sliper. Pressure sensor uncertainty is indicated by the upper and lower boundary (UB and LB) confidence intervals (CI). Source: adapted from [48].

a bias involved in the rheological quantification. Indeed, significant disparities exist between different rheometers or rheometric geometries. A second, though less significant, source is a theoretical bias if a dominant physical phenomenon is overlooked. Even though the paste pumping experiments did not show any lubrication or slippage effects based on ultrasound velocity profile (UVP) measurements and even though most residual pre-usage cleaning water was pumped out of the pumping circuit, the rheology may have been influenced by intermixed water near the pipe wall surface to a limited extent. This may have resulted in lower experimental pressure loss due to small changes in composition. That is why rheological properties are better obtained from a device closely related to the aimed application. Indeed, Sliper-based predictions, which are more close the character of a pumping set-up, performed significantly better than the ones based on MCR. Of course, doing so, one should not overlook dominant phenomena such as lubrication flow, which is not present for paste suspensions, but can be the case for actual concrete suspensions. Additionally, this does not mean rheometers are not useful. The difficulty lies in a change in context. Lastly, experimental accuracy may have been an additional source of error to an unknown extent. For example, the density-based measured discharge may have been biased by errors in density and mass evolution measurement, a small paste membrane found in front of the inserted pressure sensors may have reduced measured pressure losses.

Hence, based on this fivefold validation strategy it is concluded that the used numerical framework is sound. Significantly accurate pipe flow simulations can be expected with theoretical errors below 1%, at least if proper meshing is applied, if numerical regularization is dealt with properly (i.e. $GNAR = 1000$), if the pressure number is higher than 1.25 or 1.82 more strictly. Bearing in mind that rheological properties are better obtained from a device more closely related to the aimed application without overlooking important physical phenomena such as lubrication, numerically reliable results can be obtained for numerical pipe flow simulations.

8.3.2 Experimental ARC Valorization

To valorize the simulation of ARC, this subsection elaborates upon the simulation of the MR ARC pumping experiment depicted in more detail in section 5.3, Chapter 5. To this end, two paragraphs are distinguished. First, three alternative hypotheses which could explain the MR response are evidence-based falsified. Second, the numerical viability is illustrated for its capacity to simulate the ARC pumping experiment, based on the magneto-structuration model proposed in equation 8.3.

8.3.2.1 Falsified alternative hypotheses

The main hypothesis to model ARC is based on the concept of the kinetic temporal evolution of an internal structure or control state λ, driven by a structural build-up term and diminished by a structural break-down term. However, other alternative hypotheses may also explain the exhibited response of the cementitious MR paste in the ARC pumping experiment. Three alternative plausible hypotheses are therefore evidence-based falsified. The first one is Magneto-HydroDynamics (MHD), the second one is a static magnetic response and the third one is flow confinement by magnetic agglomeration.

As a first alternative hypothesis, MHD originates from Lorentz force-induced magnetic flows. As any fluid, the paste suspension carries electrical charges, upon which a force is exerted when moving in a magnetic field. This force is known as the Lorentz force and is perpendicular to the fluid moving velocity and magnetic field. Due to the Lorentz force, flow paths deviate causing induced electromagnetism and second-order Lorentz forces resulting in magnetically induced eddies. The coupled effect of Lorentz forces on fluid flow is what is known as MHD. MHD finds its importance and applications in the field of fusion reactors, liquid metal flows, cosmic magnetic-gas-dynamics, astrophysical solar and planetary flows, plasma flows, magnetic fusion blankets and magneto-hydrodynamic pumps, with pioneering works by Hartmann [1, 55–59]. The significance of MHD can be assessed by dimensionless numbers, known as the Hartmann number M, the Stuart number N or alternatively the magnetic Reynolds number Re_m. Applied to the pumping experiment, these numbers are, respectively, estimated as $M \approx 4.10^{-2}$, $N \approx 1.7.10^{-4}$ and $Re_m \approx 5.10^{-4}$. For small values of the Hartmann and Stuart number ($N \ll 1$), the flow is unaffected by magnetic drag or effects due to the Lorentz forces and for low magnetic Reynolds numbers ($Re_m \ll 1$), the flow is self-inductionless [53–56, 58, 59]. Hence, the hypothesis that the ARC behaviour can be explained by MHD is rejected.

A static magnetic response, in which the cementitious MR suspension exhibits a constant increase in yield stress under influence of a magnetic field as external activator, is a second alternative hypothesis. Not only the fact that the MR pumping response was not constant across a range of pumping discharges, but also that the cementitious MR paste suspensions clearly showed time-dependent behaviour (cf. Chapter 3 and Figure 8.3), falsify the hypothesis of a static magnetic response.

A third explanation that was evidence-based rejected is that of flow confinement by magnetic agglomeration. In the pumping experiment, residual magnetic spikes and clusters could be visually observed, both after the magnetic pipe section was carefully flushed with water as cleaned with sponge balls, though only minor. The hypothesis is that a pure physical reduction in sectional area due to magnetic clusters and spikes attaching to the pipe walls could explain the response in pumping resistance. To verify the hypothesis of magnetic flow confinement, a series of numerical simulations were conducted, in which the aspect ratio of the reduction height h to pipe diameter D were varied. Although a reduction aspect ratio of $h / D \approx 0.554$ seems visually most likely (Figure 8.9), still a broad range of other possibilities were simulated to verify the significance of this hypothesis.

To do so, a mesh was designed in line with the same philosophy as outlined in section 8.3.1 on numerical validation. A longitudinal transition zone of 15 cm with a gradual transition from a circular to a stadium-like geometrical cross-section was also taken into account. Although a stadium-like geometry was an imposed assumption of the confined geometry, an alternative geometry would still result in results of the same order of magnitude, and thereby not further reject or assume the hypothesis. Based on results of the respective pipe section difference in Figure 8.10, it was concluded that magnetic pipe confinement is not the pre-dominant hypothesis.

That is because the simulation results are at least one order of magnitude higher than the experimentally measured pressure loss section difference. Only for very minor physical flow confinement, i.e. a reduction in the order of a few mm, the numerical simulations would have a similar order of magnitude. Moreover, the pressure loss section difference increases with an increasing pumping discharge, unlike the experimental observation

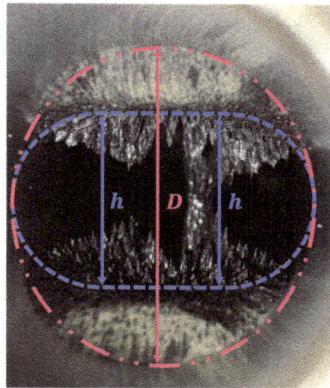

Figure 8.9 Illustration of expected flow confinement, with an estimated aspect ratio $h / D \approx 0.554$, assuming the dead zones on a qualitative basis, indicated by the blue stadium geometry.

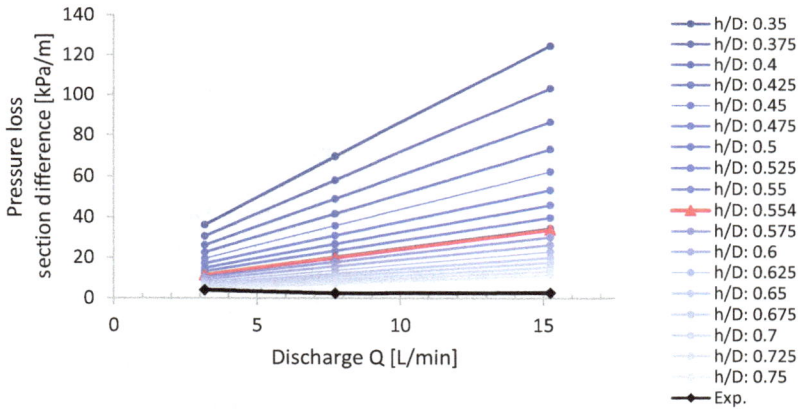

Figure 8.10   Simulated change in pressure loss section difference, between the magnetic section 2–3 and reference section 3–6 (kPa / m) as a function of the pipe flow confinement aspect ratio h / D. The global, simulated trend is increasing as a function of the discharge Q. The simulated results are significantly higher than observed from the experiment (Exp.), for an estimated aspect ratio $h / D \approx 0.554$, as well as for higher aspect ratios.

which first decreases for an intermediate discharge after which it slightly increases for the highest discharge. Hence, even though a few mm of fixed, residual magnetic structures could be present as visually observed after the pipe was cleaned with sponge balls, it is insufficient to explain to measured ARC response. Hence, magnetic flow confinement can only be plausible for a minor, insignificant contribution. As such, three alternative plausible hypotheses have been evidence-based falsified, which leaves the main hypothesis of dynamic magneto-structuration to simulate ARC.

8.3.2.2   Numerical ARC viability

To verify the viability of the proposed ARC magneto-structuration model (Eq. 8.3), numerical simulations were conducted based on the ARC pumping experiment as depicted in more detail in Chapter 5. To do so, the ARC parameters were experimentally determined as depicted in section 8.1.2. As such, stochastic expected values were computed for the magnetic field applied in the pumping experiment (i.e. a vertical magnetic field density of $B = 0.169\,\mathrm{T}$). This yielded a relative magnetic response $\chi = 0.63$ and a structural magnetic build-up rate $\delta_M = 0.14\,\mathrm{s^{-1}}$. The Bingham properties were determined from the pumping experiment itself from a reference section without magnetic field, via a reverse-engineering Buckingham–Reiner approach, resulting in a yield stress $\tau_0 = 111.04\,\mathrm{Pa}$ and a plastic viscosity $\mu = 0.70\,\mathrm{Pa.s}$. However, the de-flocculation rate α could not be determined. Therefore, a calibration set of parameter series was used to verify the viability of the model.

After computation, these simulations yielded the following results for the pipe section with the magnetic control unit as illustrated in Figure 8.11. On the one hand, in the zone where the magnetic field is applied ($B = 0.169\,\mathrm{T}$) as external trigger field, a respective magnetic structural build-up rate $R_{bu} = 0.14\,\mathrm{s^{-1}}$ is established. On the other hand, near the pipe wall boundaries high shear action occurs (i.e. a shear rate $\dot{\gamma}$ with values up to $93\,\mathrm{rad\,/\,s}$ or higher) which corresponds to a high structural break-down rate R_{bd} near the pipe wall boundary. There the internally built structures, networks or clusters are demolished, which results in a decrease in structural state. At each location in the pipe, a spatio-temporal equilibrium is established driving the evolution of an internal structure state λ. An internal structure of one corresponds to 100% or full structuration, while a value of zero corresponds to complete demolition or lack of structuration. As such, as soon as the active MR paste conveys through the magnetic control unit section, an internal magnetic structure λ starts to develop, while after it leaves the control unit, the magnetic structure remains unaltered apart from de-structuration in the vicinity of the pipe wall boundaries. That is why the internal structure remains unchanged in the central plug zone, since it is not affected by any shear action. Based on a simulation with a flow rate $Q = 3.17\,\mathrm{L\,/\,min}$, a most developed structure in the central plug zone of about 70% can be established, even for the considered, relatively short magnetic control length of 1.2 m.

Apart from these qualitative numerical results, further quantitative analysis was conducted. More specifically, Figure 8.12 depicts the difference in

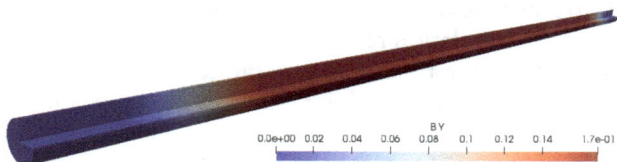

(a) Magnetic flux density field $\vec{B_y}$ in the y-direction (T).

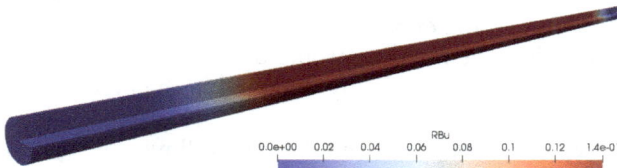

(b) Magnetic build-up rate field R_{bu} (s^{-1}).

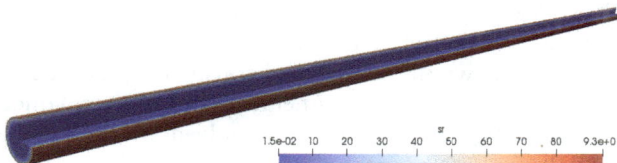

(c) Shear rate field $\dot{\gamma}$ (sr, rad/s).

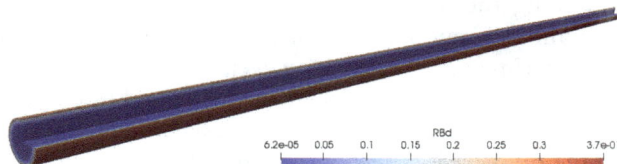

(d) Magnetic break-down rate field R_{bd} (s^{-1}).

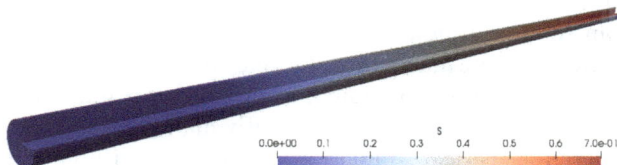

(e) Internal magnetic structure field λ (S).

Figure 8.11   Illustrations of numerical results of the proposed simplified magneto-structuration model after [1]. A close-up of magnetic section 2–3 of pipe flow ($D = 0.02664$ m) with a discharge $Q = 3.17$ L / min (inflowing from left) and a magnetic field $\bar{B} = 0.169$ T was considered with a relative magnetic response $\chi = 0.63$, a magnetic growth rate $\delta_M = 0.14$ s^{-1}, a de-flocculation rate of $\alpha = 4.10^{-3}$, a Bingham yield stress $\tau_0 = 111$ Pa and a plastic viscosity $\mu = 0.70$ Pa.s.

pressure loss between the magnetic pipe section and a non-magnetic reference section. Irrespective of the considered de-flocculation rate α, it is quite remarkable that the simulation results are in the same order of magnitude as the experimental pumping results. In general, the higher the de-flocculation rate, the lower the ARC response for a constant discharge Q. This is logical because a more structural break-down occurs and thereby less structuration or response can manifest for the same discharge. Nevertheless, it is even more remarkable that calibrated values of the de-flocculation rate in the order of magnitude of 2.10^{-3} to 4.10^{-3} resulted in a near-perfect match compared to the pumping experiment. That is exactly in the same order of magnitude as is the case for thixotropic de-flocculation rates α of cementitious suspensions as independently obtained in the literature ($\alpha \approx 5.10^{-3}$), cf. [35].

Furthermore, the numerical simulations follow the speculated ARC behaviour trend, i.e. a higher response for a lower discharge. However, the simulations cannot explain a slight increase or constant response going from an intermediate discharge to the highest discharge. On the one hand, this might be explained by the relatively high sensor uncertainty. On the other hand, if the same error would be assumed considering that it concerned the same sensors during the same experiment, it could be explained by minor effects due to magnetic flow confinement. Indeed, after cleaning the pipes with sponge balls, minor magnetic structures remained attached to the pipe walls to some extent due to strong magnetic gradients in the vicinity thereof. Combining the hypothesis of minor magnetic flow confinement with predominant dynamic magneto-structuration could perfectly explain the obtained experimental results. However, the degree of magnetic flow confinement due to residually attached magnetic spikes could not be further verified. Due to the additional, hypothetical contribution of minor flow confinement, the pressure loss response would slightly increase and thereby the calibrated de-flocculation rate α would result in a slightly higher value, e.g. ca. 5.10^{-3}.

Apart from relatively high sensor accuracy, the numerical results could also be subjected to some inaccuracy. As determined from the numerical reliability study in section 8.3.1, only significantly accurate numerical results are expected in case the flow regime is sufficiently high (i.e. $\hat{Q} > 0.424$). Applied to the considered numerical simulation, in the zone without magnetic field the non-dimensional discharge is $\hat{Q}_{\lambda=0} \approx 0.14$, while in the magnetic zone it could be estimated as $\hat{Q}_{\lambda=0.7} \approx 0.092$ (based on a yield stress increase for $\lambda = 0.7$). Therefore, the numerical results do not fulfil the prerequisites to be significantly accurate. Nevertheless, based on a more refined error analysis with corresponding theoretical errors, it can be expected that the simulated pressure loss should be approximately 13% higher or roughly 0.5 kPa / m. This in turn would result in a slightly higher calibrated de-flocculation rate α of ca. 5.10^{-3}.

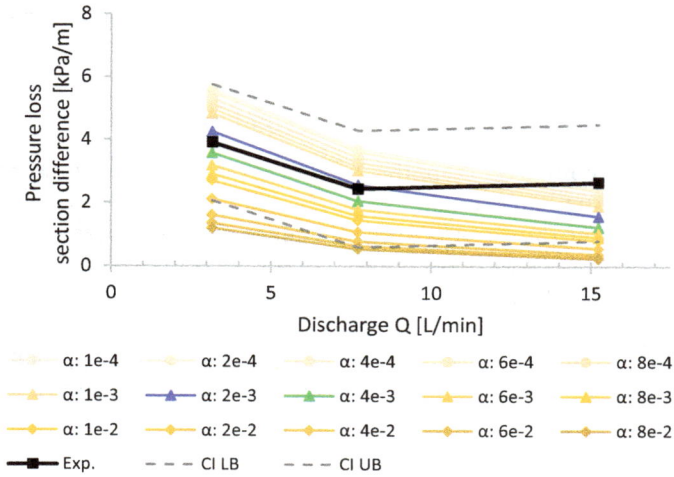

Figure 8.12 Numerical simulation results of the change in pressure loss section difference, between magnetic section 2–3 and reference section 3–6, before and after placing the magnetic control unit. The results are shown as a function of the discharge Q, as well as the considered de-flocculation rate α for a statistically expected relative magnetic response $\chi = 0.63$ and magnetic growth rate $\delta_M = 0.14\,\text{s}^{-1}$. The experimental results (Exp.) and their confidence intervals (CI, lower and upper bound (LB and UB)) are depicted for comparison. Source: originating from [1].

Combining the two artefacts – due to numerical accuracy and omitted plausible minor magnetic flow confinement – would result in a slight increase in pressure loss response and thereby a slight increase in viable values for the de-flocculation rate, e.g. ca. 5.10^{-3} to 6.10^{-3}. This is still in the exact same order of magnitude as is the case for thixotropic cementitious suspensions (cf. [35]). Hence, the same conclusions hold true.

Lastly, the quantified MR properties were subjected to some statistical scatter. Nevertheless, the average value was used, which is an unbiased stochastic estimator. The used rheological properties are therefore representative for the overall averaged bulk quantity and thereby to determine the overall macroscopic outcome. Even if the rheological properties were slightly shifted (due to a rheological bias in spite of its relative, proportional quantification), it would still result in similar numerical simulations [1].

Therefore, the hypothesis is accepted that ARC can be captured both qualitatively and quantitatively by the concept of the kinetic temporal evolution of an internal structure λ driven by structural build-up and break-down. Furthermore, for the depicted magnetic ARC response, the combination with subordinate magnetic flow confinement due to attached magnetic structures near the pipe wall seems plausible. Hence, the proposed ARC model is numerically viable.

8.4   FURTHER RESEARCH AND OUTLOOK

Although the combined hypothesis of predominant dynamic magneto-structuration and subordinate magnetic flow confinement is accepted, it does not mean that it is proven. It only means it is viable or feasible. Future results should further confirm, if not reject, this combined hypothesis. The scope of ARC is much wider and deeper than magnetic materials and responses. This section therefore addresses three future perspectives, i.e. first projections based on the considered ARC modelling, then the generality of the proposed ARC models and lastly potential future numerical ARC studies.

8.4.1   Projections of ARC modelling

Even though the proposed ARC model was employed to simulate active stiffening control applied to pumping of smart cementitious suspensions, it has direct applications and thereby projections. Considering the fact that an internal magnetic structure can be developed over a relatively short magnetic application distance (i.e. ca. 1 m), in combination with a discharge that is within practical range for 3D printing of cementitious materials, important projections can be made.

One projection that follows from the presented results is that if a constant magnetic field would be applied to the 3D printing offset area, the yield stress τ_0 of the printed material could be increased by ca. 44% (i.e. the relative magnetic response multiplied by the structuration state $\chi\lambda$), assuming that most of the printed material would be deposited from the central plug zone without any further disturbance. If not, it is expected that additional disturbance could be coped with through extending the printing nozzle or a special reducer for a seamless transition towards the printing nozzle. This at least holds true if the plug velocity is slow enough to allow for sufficient structural build-up depending on the structural build-up rate δ_M. Indeed, the most developed structure λ can be estimated from simple analytical equation 8.11 solved from equation 8.3 for the plug zone where no shear action takes place ($\dot{\gamma} \approx 0$), and where the application time t_{app} can be conservatively estimated by a parabolic Newtonian pipe velocity profile with average velocity U flowing through an activation length L as given in equation 8.12.

$$\lambda\left(t_{app}\right) = 1 - \exp\left(-\delta_M t_{app}\right) \tag{8.11}$$

$$t_{app} \approx \frac{2}{3}\frac{L}{U} - \frac{2}{3}\frac{\pi R^2 L}{Q} \tag{8.12}$$

Hence, the total increase in yield stress can be determined from knowing the relative response of the material χ and the developed internal structure λ. In its turn, the most developed structure depends on the magnetic growth rate δ_M and time of application t_{app}. The maximum acquired internal structure λ in the plug can be increased by either reducing the flow rate Q, increasing the pipe radius R or increasing the magnetic application length L. Alternatively, one may also design a certain mixture or section in order to achieve a certain structuration state λ and vice versa. The increase in yield stress remains active as long as an external magnetic field would be applied. If no magnetic field would be applied in the printing area, one could still account for an increased yield stress by residual magnetic remanence. Magnetic remanence can be significant, e.g. the residual magnetization of magnetite ($\sigma_r \approx 14.8$ Am^2 / kg in Table 3.2) can be 8.7 times higher than the saturated magnetization of a magnetic fly ash ($\sigma \approx 1.7$ Am^2 / kg in Table 3.3) [1, 18, 19]. Indeed, Chapter 7 already illustrated promising results in view of actively controlled 3D printing in which the material was shortly excited and stiffening was achieved by residual magnetic remanence. Both projections result in a magneto-rheologically increased yield stress and thereby increased buildability of the 3D printing process. Hence, the overall 3D printing performance could be improved by exciting an MR printing material momentarily before the printing nozzle. Even though magnetic remanence may not necessarily be significant, it might still be of practical value as the response is proportional to a rather high yield stress typically involved in 3D printing.

However, one drawback that was not considered in these projections is the feasibility of the considered rheological scope, especially since 3D printing applications typically involve mixtures with a rather high yield stress. Indeed, as indicated by the Mason number Mn or magnetic yield parameter Y_M (Eq. 3.3), the relatively stiff rheology of cementitious suspensions as MR carrier fluid can reduce the magnetic response and thereby performance [13, 18–20]. Nevertheless, a qualitative MR response can still be expected which is beneficial for the 3D printing process. Whether the magneto-response is accounted for by active coercion or by passive magnetic remanence, this illustrates that MR materials have a high potential to be used in formwork leakage control, cementitious 3D printing or other applications [18,19,21]. These projections already show that the proposed magneto-structuration model can be used to investigate the influence of the excitation field, activating nozzle or pipe design, MR design and the tailoring thereof.

8.4.2 Generality of ARC model

Although a simplified model was applied to the context of active stiffening control in the form of magneto-rheology during pumping, it has a wider, more general scope. A simple model implies that the model can be further improved. The proposed model could, for instance, include effects of a more precise magnetization scaling function, a scaling as a function of the magnetic (mass) concentration of the mixture design, magnetic remanence, magnetic micro-vibration, magnetic acceleration and even magnetic anisotropy. A more general ARC model (Eq. 8.2) was proposed in which a more complicated magnetic scaling function f_M can be accounted for and in which a second structural break-down term was foreseen to account for effects due to changing or oscillating magnetic fields imposing magnetic microvibration effects. Apart from that, the simulation model may be extended to incorporate effects due to magnetic force fields, e.g. in the form of an additional acceleration field based on magnetic forces originating from magnetic gradients.

Furthermore, even though the MR response was attributed to the yield stress in the proposed MR model similar to most literature works, it could still be attributed to both the yield stress and viscosity.

Last but not least, the scope of ARC is wider than magnetic materials. For example, as illustrated in Chapter 4, ARC may be established by electrochemical polymers. Nevertheless, the more generic model given in equation 8.1 can still serve to model ARC in general by the concept of a control state parameter λ. By means of its general structural build-up term R_{bu} and structural break-down term $R_{bd,}$ it is able to capture a wide scope of both time-dependent and time-independent (instantaneous) active rheologically controlled processes. Even though the proposed more generic models have the potential to be employed or further developed for a wider scope and more versatile materials, this potential was not explored in this chapter. Nevertheless, their generic character is highlighted here.

8.4.3 Future numerical ARC studies

Apart from ARC applied to pumping and by means of a VFM, other numerical cases studies and techniques could still be explored. However, in order to do so, a wider scope of material responses needs to be examined. For instance, in order to be able to design an actively controlled application, it would be useful to be able to describe the material response as a function of its trigger signal and/or active component concentration. On the one hand, a wide variety of materials exist that, for instance, exhibit a magnetization response. On the other hand, as the end goal is mostly mortar and/or concrete applications, the response as a function of the active component concentration in proportion to the rheology of the carrier fluid is also important to describe the behaviour. In the end, if one would be able to describe a full spectrum of constitutive responsive behaviour – as a function of its active component concentration, activation trigger field intensity and/or operation mode and in proportion to its unmodified reference mixture – numerical simulations are a useful means to assess or design ARC applications. Nevertheless, even with the limited behavioural spectrum of the material and signal scope examined in this chapter, some future numerical studies could be conducted.

First of all, future numerical ARC modelling studies could already include the design of activating 3D printing nozzles or pipe sections, whose perspective is in line with the results provided in Chapter 7. In that way, the efficiency of such activating control units could be examined and optimized for the considered material and vice versa. Alternatively, active stiffening control of formwork leakage control could be studied numerically, in which the reduction or control of leakage through a formwork slit could be numerically investigated (cf. Chapter 6). Mixture shape retention may also be studied, e.g. for shotcrete applications. Pumping itself could also be further investigated with respect to up-scaling to concrete level, since currently only ARC of a cement paste was simulated. Therefore, ARC simulation in combination with lubrication or slippage flow would be the next important step of up-scaling. As such, the proposed simplified magneto-structuration model could still be employed for both the lubrication or slippage layer and for the concrete bulk after respective response parameter values would have been determined experimentally. Apart from FVM, alternative numerical techniques could be investigated, although their complexity and validation may be somewhat cumbersome. For instance, one example would be the combination of a coupled computational fluid dynamics-discrete element method (CFD-DEM) technique, in which the response of a material would be simulated on a particle basis consisting of attracting or repulsion forces.

Nonetheless, numerical case studies of ARC have direct or indirect perspectives, as long as the necessary spectrum of material responsive behaviour is uncovered since it is a prerequisite to serve as input for the numerical simulation. Even though part of the material response spectrum was touched upon in this chapter, as stressed, the scope of ARC is much wider both in terms of material variety as in trigger fields and operation modes.

8.5 SUMMARY

In summary of numerically modelling ARC, it can be concluded that a first approach was developed that allows to conceptualize the results of ARC both qualitatively and quantitatively. To do so, more generic models in equation 8.1 and equation 8.2 were proposed to conceptualize the constitutive ARC behaviour by the concept of the temporal kinetic evolution of an internal structure or control state λ, driven by a structural build-up and break-down term. Applied to the context of MR pumping, a simplified magneto-structuration model (Eq. 8.3) was proposed based on the underlying mechanisms of magnetic clustering. A corresponding rheological quantification methodology was outlined, which is a powerful means since it is based on relative proportions of the response and intrinsic temporal process parameters. As such the rheological methodology is independent of or insignificantly dependent on rheological biases. Other rheological parameters were determined from the pumping experiment itself as a reference without active control.

Even though many CFD techniques exist, one methodology, i.e. a VFM in OpenFOAM, was employed due to its practical relevance to simulate fresh cementitious suspension flows from a macroscopic view. The methodology to do so was briefly outlined. To allow for flows incorporating external trigger fields and coupled actively controlled constitutive material behaviour, an existing OpenFOAM FVM solver was adapted and the proposed constitutive ARC model was implemented and coupled with the external magnetic trigger field via a generalized Newtonian approach.

To build confidence in the developed numerical framework, a sound fivefold validation strategy was conducted. It can be concluded that the employed numerical methodology is significantly accurate with theoretical errors below 1%, at least if proper numerical meshing, settings, GNA regularization and flow regimes are considered. However, considering the validation by experiments, rheological input for numerical simulations should be handled with care. To avoid biases from rheology, it is better to stay within the same rheological context or to determine rheological properties from a context as close as possible to the aimed application, at least if no important physical phenomena are overlooked. Furthermore, the viability of the numerical ARC simulation methodology was valorized based on the ARC pumping experiment depicted in Chapter 5. Three plausible alternative hypotheses (i.e. magneto-hydrodynamics, static magnetic response and magnetic flow confinement or restriction) were evidence-based falsified and rejected. Instead, based on the numerical simulation of the proposed simplified magneto-structuration model (Eq. 8.3), the hypothesis was evidence-based accepted that ARC can be captured by the concept of the temporal kinetic evolution of an internal structure or control state λ, driven by a structural build-up and break-down term. This does not mean that the hypothesis and corresponding model are proven. It only means it is viable. Future results should further confirm, if not reject, this hypothesis.

In perspective, the future holds a high potential for practical applications and future numerical studies. On the one hand, direct applications of active stiffening control could be further investigated, e.g. based on the proposed simplified magneto-structuration model. Preliminary projections based on the conducted ARC simulations showed that a better buildability or performance for 3D printing or formwork leakage control is within practical reach. On the other hand, the work of numerical ARC modelling just started. Not only is the accepted hypothesis to be further confirmed or investigated, also more complicated response phenomena could be incorporated or further investigated, such as, e.g., a wider scope of magnetic materials, a more appropriate magnetization scaling, magnetic microvibration, magnetic acceleration fields, magnetic anisotropy, etc. Moreover, the scope of ARC is wider than MR materials and responses, and therefore the more generic models (Eq. 8.1 and Eq. 8.2) may serve as building blocks for future developments or improvements. Even though the scope of ARC is wide, as long as the necessary spectrum of material responsive behaviour is uncovered, it can serve as input for a numerical simulation.

REFERENCES

[1] R. De Schryver, Active rheology control of cementitious materials: Numerical and experimental pumping investigation, Ghent, Belgium: PhD thesis, Ghent University, 2022.

[2] M. Ashtiani, S.H. Hashemabadi, and A. Ghaffari, A review on the magnetorheological fluid preparation and stabilization, *J. Magn. Magn. Mater.* 374 (2015): 716–730. doi:10.1016/j.jmmm.2014.09.020

[3] I. Bica, Y.D. Liu, and H.J. Choi, Physical characteristics of magnetorheological suspensions and their applications, *J. Ind. Eng. Chem.* 19 (2013): 394–406. doi:10.1016/j.jiec.2012.10.008

[4] M.R. Jolly, J.W. Bender, and J.D. Carlson, Properties and applications of commercial magnetorheological fluids, *J. Intell. Mater. Syst. Struct.* 10 (1999): 5–13. doi:10.1177/1045389X9901000102

[5] E.J. Park, L.F. da Luz, and A. Suleman, Multidisciplinary design optimization of an automotive magnetorheological brake design, *Comput. Struct.* 86 (2008): 207–216. doi:10.1016/j.compstruc.2007.01.035

[6] R. Patel, Mechanism of chain formation in nanofluid based MR fluids, *J. Magn. Magn. Mater.* 323 (2011): 1360–1363. doi:10.1016/j.jmmm.2010.11.046

[7] P.P. Phulé, Magnetorheological (MR) fluids: Principles and applications, *Smart Mater. Bull.* 2001 (2001): 7–10. doi:10.1016/S1471-3918(01)80040-X

[8] M. Pourjafar, F. Malmir, S. Bazargan, and K. Sadeghy, Magnetohydrodynamic flow of Bingham fluids in a plane channel: A theoretical study, *J. Nonnewton. Fluid Mech.* 264 (2019): 1–18. doi:10.1016/j.jnnfm.2018.12.005

[9] E.A. Schapova, A.N. Stashkov, S. V. Afanas'ev, and A.P. Nichipuruk, Magnetic properties and the structure of eutectoid steel with different pearlite morphology, *Phys. Met. Metallogr.* 122 (2021): 742–748. doi:10.1134/S0031918X21080135

[10] S.M. Thompson, and B.K. Tanner, The magnetic properties of pearlitic steels as a function of carbon content, *J. Magn. Magn. Mater.* 123 (1993): 283–298. doi:10.1016/0304-8853(93)90454-A

[11] J. Yang, H. Yan, Z. Hu, and D. Ding, Viscosity and sedimentation behaviors of the magnetorheological suspensions with oleic acid/dimer acid as surfactants, *J. Magn. Magn. Mater.* 417 (2016): 214–221. doi:10.1016/j.jmmm.2016.05.085

[12] M. Zubieta, S. Eceolaza, M.J. Elejabarrieta, and M.M. Bou-Ali, Magnetorheological fluids: Characterization and modeling of magnetization, *Smart Mater. Struct.* 18 (2009): 095019. doi:10.1088/0964-1726/18/9/095019

[13] D. Jiao, K. Lesage, M.Y. Yardimci, K. El Cheikh, C. Shi, and G. De Schutter, Rheological properties of cement paste with Nano-Fe3O4 under magnetic field: Flow curve and nanoparticle agglomeration, *Materials (Basel).* 13 (2020): 5164. doi:10.3390/ma13225164

[14] G. Bossis, O. Volkova, S. Lacis, and A. Meunier, Magnetorheology: Fluids, structures and rheology. In Odenbach, S. (Ed.) *Ferrofluids: Lecture notes in physics*, Berlin, Heidelberg: Springer, 2022, pp. 202–230. doi:10.1007/3-540-45646-5_11

[15] D. Jiao, K. El Cheikh, C. Shi, K. Lesage, G. De Schutter, Structural build-up of cementitious paste with nano-Fe3O4 under time-varying magnetic fields, *Cem. Concr. Res.* 124 (2019): 105857. doi:10.1016/j.cemconres.2019.105857

[16] D. Jiao, K. Lesage, M.Y. Yardimci, K. EL Cheikh, C. Shi, and G. De Schutter, Quantitative assessment of the influence of external magnetic field on clustering of nano-Fe3O4 particles in cementitious paste, *Cem. Concr. Res.* 142 (2021): 106345. doi:10.1016/j.cemconres.2020.106345

[17] D. Jiao, K. Lesage, M.Y. Yardimci, K. El Cheikh, C. Shi, and G. De Schutter, Rheological behavior of cement paste with nano-Fe3O4 under magnetic field: Magneto-rheological responses and conceptual calculations, *Cem. Concr. Compos.* 120 (2021): 104035. doi:10.1016/j.cemconcomp.2021.104035

[18] D. Jiao, K. Lesage, M.Y. Yardimci, K. El Cheikh, C. Shi, and G. De Schutter, Structural evolution of cement paste with nano-Fe3O4 under magnetic field - Effect of concentration and particle size of nano-Fe3O4, *Cem. Concr. Compos.* 120 (2021): 104036. doi:10.1016/j.cemconcomp.2021.104036

[19] D. Jiao, K. Lesage, M.Y. Yardimci, C. Shi, and G. De Schutter, Possibilities of fly ash as responsive additive in magneto-rheology control of cementitious materials, *Constr. Build. Mater.* 296 (2021): 123656. doi:10.1016/j.conbuildmat.2021.123656

[20] D. Jiao, *Active rheology control of cementitious materials using magnetic field*, Ghent University, 2021. http://hdl.handle.net/1854/LU-8688855.

[21] C. Chibulu, M.Y. Yardimci, D. Jiao, R. De Schryver, K. Lesage, and G. De Schutter, Active stiffening control by magnetically induced blocking in confined flow of fly ash pastes, *Constr. Build. Mater.* 313 (2021) 125485. doi:10.1016/j.conbuildmat.2021.125485

[22] W.W. Chooi, and S.O. Oyadiji, Design, modelling and testing of magnetorheological (MR) dampers using analytical flow solutions, *Comput. Struct.* 86 (2008): 473–482. doi:10.1016/j.compstruc.2007.02.002

[23] S. Genç, and P.P. Phulé, Rheological properties of magnetorheological fluids, *Smart Mater. Struct.* 11 (2002): 140–146. doi:10.1088/0964-1726/ 11/1/316

[24] J.M. Ginder, L.C. Davis, L.D. Elie, Rheology of magnetorheological fluids: Models and measurements, *Int. J. Mod. Phys. B.* 10 (1996): 3293– 3303. doi:10.1142/S0217979296001744

[25] R.Y. Hong, Z.Q. Ren, Y.P. Han, H.Z. Li, Y. Zheng, and J. Ding, Rheological properties of water-based Fe3 O4 ferrofluids, *Chem. Eng. Sci.* 62 (2007): 5912–5924. doi:10.1016/j.ces.2007.06.010

[26] W. Horak, Modeling of magnetorheological fl uid in quasi-static squeeze flow mode (2018). doi:10.1088/1361-665x/aab7c7

[27] H. Hu, Y. Yuan, S. Lim, and C.H. Wang, Phase structure dependence of magnetic behaviour in iron oxide nanorods, *Mater. Des.* 185 (2020): 108241. doi:10.1016/j.matdes.2019.108241

[28] W. Jiang, Y. Zhang, S. Xuan, C. Guo, and X. Gong, Dimorphic magnetorheological fluid with improved rheological properties, *J. Magn. Magn. Mater.* 323 (2011): 3246–3250. doi:10.1016/j.jmmm.2011.07.024

[29] M. Houska, *Engineering aspects of the rheology of thixotropic liquids*, Czech Technical University of Prague, 1981.

[30] J. Billingham, and J.W.J. Ferguson, Laminar, unidirectional flow of a thixotropic fluid in a circular pipe, *J. Nonnewton. Fluid Mech.* 47 (1993): 21– 55. doi:10.1016/0377-0257(93)80043-B

[31] A. Mujumdar, A.N. Beris, and A.B. Metzner, Transient phenomena in thixotropic systems, *J. Nonnewton. Fluid Mech.* 102 (2002) 157–178. doi:10.1016/S0377-0257(01)00176-8

[32] A. Wachs, G. Vinay, and I. Frigaard, A 1.5D numerical model for the start up of weakly compressible flow of a viscoplastic and thixotropic fluid in pipelines, *J. Nonnewton. Fluid Mech.* 159 (2009): 81–94. doi:10.1016/ j.jnnfm.2009.02.002

[33] A. Ahmadpour, and K. Sadeghy, An exact solution for laminar, unidirectional flow of Houska thixotropic fluids in a circular pipe, *J. Nonnewton. Fluid Mech.* 194 (2013) 23–31. doi:10.1016/j.jnnfm.2012.11.010

[34] L. Guo, X. Chen, S. Shi, Y. Wang, and X. Yu, Evaluation of thixotropic models for waxy crudes, *Open J. Appl. Sci.* 05 (2015): 304–312. doi:10.4236/ojapps.2015.56031

[35] N. Roussel, A thixotropy model for fresh fluid concretes: Theory, validation and applications, *Cem. Concr. Res.* 36 (2006): 1797–1806. doi:10.1016/ j.cemconres.2006.05.025

[36] J.E. Wallevik, Microstructure-rheology: Thixotropy and workability loss, *Nord. Concr. Res.* 31 (2004): 16. www.nvtf.org/ikbViewer/Content/224 632/I09 - Wallevik 2004-06-07.pdf.

[37] N. Roussel, and F. Cussigh, Distinct-layer casting of SCC: The mechanical consequences of thixotropy, *Cem. Concr. Res.* 38 (2008): 624–632. doi:10.1016/j.cemconres.2007.09.023

[38] D. Jiao, R. De Schryver, C. Shi, and G. De Schutter, Thixotropic structural build-up of cement-based materials: A state-of-the-art review, *Cem. Concr. Compos.* 122 (2021): 104152. doi:10.1016/j.cemconcomp.2021.104152

[39] N. Roussel, G. Ovarlez, S. Garrault, and C. Brumaud, The origins of thixotropy of fresh cement pastes, *Cem. Concr. Res.* 42 (2012): 148–157. doi:10.1016/j.cemconres.2011.09.004

[40] S. Ma, Y. Qian,and S. Kawashima, Experimental and modeling study on the non-linear structural build-up of fresh cement pastes incorporating viscosity modifying admixtures, *Cem. Concr. Res.* 108 (2018): 1–9. doi:10.1016/j.cemconres.2018.02.022

[41] P. Holoborodko, Noise robust smoothing filter, *Online Artic.* (2008). www.holoborodko.com/pavel/numerical-methods/noise-robust-smoothing-filter/.

[42] P. Holoborodko, Central differences, *Online Artic.* (2008). www.holoborodko.com/pavel/numerical-methods/numerical-derivative/central-differences/.

[43] P. Holoborodko, Smooth noise robust differentiators, *Online Artic.* (2008). www.holoborodko.com/pavel/numerical-methods/numerical-derivative/smooth-low-noise-differentiators/.

[44] P. Holoborodko, Low-noise Lanczos differentiators, *Online Artic.* (2008). www.holoborodko.com/pavel/numerical-methods/numerical-derivative/lanczos-low-noise-differentiators/.

[45] J.R. Welty, C.E. Wicks, R.E. Wilson, and G.L. Rorrer, Fundamentals of momentum, heat and mass transfer, *Int. J. Heat Mass Transf.* 13 (1970): 1641. doi:10.1016/0017-9310(70)90063-3

[46] R. De Schryver, K. El Cheikh, K. Lesage, and G. De Schutter, CFD implementation of time-dependent behaviour: application for concrete pumping. In E. Schlangen, G. De Schutter, B. Savija, H. Zhang, C. Romero Rodriguez (Eds.), *Proc. Symp. Concr. Model. CONMOD2018, RILEM*, 2018: pp. 122–130. http://hdl.handle.net/1854/LU-8598927.

[47] R. De Schryver, K. El Cheikh, M.Y. Yardimci, K. Lesage, and G. De Schutter, Fresh concrete pumping arrest investigation for thixotropy by a CFD modelling approach, in: V. Mechtcherine, K. Khayat,and E. Secrieru (Eds.), Rheol. Process. Constr. Mater. RheoCon 2019, SCC9, Springer, Dresden, 2019: pp. 580–587. https://link.springer.com/chapter/10.1007/978-3-030-22566-7_67.

[48] R. De Schryver, K. El Cheikh, K. Lesage, M.Y. Yardimci, G. De Schutter, Numerical Reliability Study Based on Rheological Input for Bingham Paste Pumping Using a Finite Volume Approach in OpenFOAM, Materials (Basel). 14 (2021) 5011. doi:10.3390/ma14175011

[49] N. Roussel, M.R. Geiker, F. Dufour, L.N. Thrane, P. Szabo, Computational modeling of concrete flow: General overview, Cem. Concr. Res. 37 (2007) 1298–1307. doi:10.1016/j.cemconres.2007.06.007

[50] D. Feys, Interactions between rheological properties and pumping of self-compacting concrete, Ghent University, 2009. http://hdl.handle.net/1854/LU-948561.

[51] S. Tichko, Hydrodynamic Modelling of the Flow of Self-Compacting Concrete in Formworks, 2016. http://hdl.handle.net/1854/LU-8071379.

[52] R. De Schryver, G. De Schutter, Insights in thixotropic concrete pumping by a Poiseuille flow extension, Appl. Rheol. 30 (2020) 77–101. doi:10.1515/arh-2020-0103

[53] A. Tassone, M. Setareh, H. Nilsson, Magnetic induction and electric potential solvers for incompressible MHD flow, in: CFD with OpenSource Softw., 2017: pp. 1–60. www.tfd.chalmers.se/~hani/kurser/OS_CFD_2016.

[54] L. Marioni, Computational Modelling and Electromagnetic-CFD Coupling in Casting Processes, Paris, France: PhD thesis, Université Paris Sciences et Lettres, 2018.

[55] F. Ferroni, Magneto-hydrodynamic Simulations of Liquid Metal Flows in Fusion Reactors (2012) 58. doi:10.13140/RG.2.1.4279.7602

[56] M.J. Wolfendale, M.J. Bluck, A coupled systems code-CFD MHD solver for fusion blanket design, Fusion Eng. Des. 98–99 (2015) 1902–1906. doi:10.1016/j.fusengdes.2015.04.025

[57] Fluent Inc., Magnetohydrodynamics (MHD) Module Manual (2005). www.afs.enea.it/project/neptunius/docs/fluent/html/mhd/main_pre.htm.

[58] ANSYS Inc., ANSYS FLUENT Magnetohydrodynamics (MHD) Module manual, Ansys. 15 (2013) 724–746.

[59] O.M. Al-Habahbeh, M. Al-Saqqa, M. Safi, T. Abo Khater, Review of magnetohydrodynamic pump applications, Alexandria Eng. J. 55 (2016) 1347–1358. doi:10.1016/j.aej.2016.03.001

Chapter 9

Overview and remaining challenges of active rheology control of cementitious materials

Karel Lesage and Geert De Schutter

CONTENTS

9.1   Introduction   269
9.2   Responsive mineral additions   270
9.3   Switchable admixtures   271
9.4   Challenges   273
9.5   Conclusions   274
References   274

9.1   INTRODUCTION

Workability of concrete is of significant importance during processing, requiring the appropriate flowability for a specific application. The target workability for a specific concrete mixture is normally obtained by means of an appropriate concrete mix design taking into account workability requirements in parallel to strength and durability requirements. Reaching a target workability often includes the use of superplasticizers. In spite of a correct mix design, the workability achieved on the construction site might deviate somewhat from the target, without further options for adjustment during processing, post-mixing. The deviating workability might be linked to lack of robustness of the concrete mixture, leading to unwanted levels of variation in flowability due to small changes in water content, e.g. due to insufficient precision while monitoring water content of aggregates before mixing [1].

Another workability challenge is the sometimes contradicting requirements in different steps of the entire processing and casting operation. While pumping, the fresh concrete should have an adapted workability in order to reduce pressure losses. The pumping process itself can also influence the rheological properties, which is an extra variable to the system [2]. After pumping, the fresh concrete should still be workable enough to ease formwork casting. However, after filling the forms, the concrete should stiffen fast to reduce formwork pressure rapidly.

DOI: 10.1201/9781003289463-9

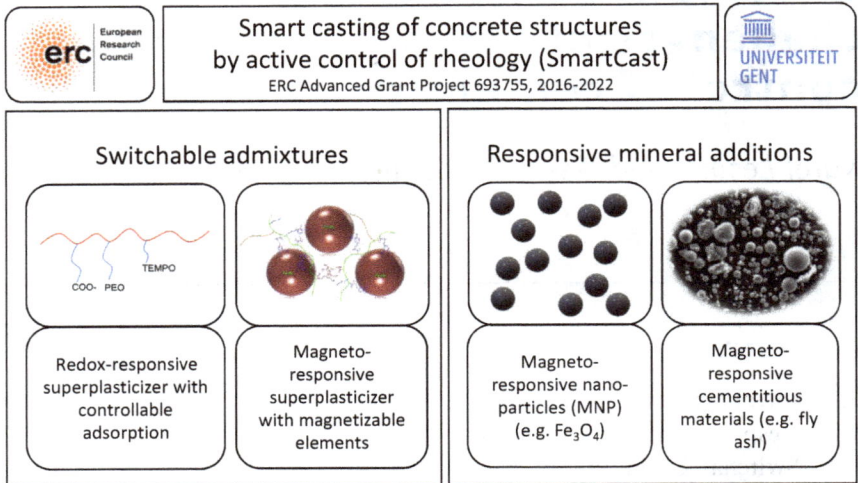

Figure 9.1 Overview of different routes to achieve active rheology control, as studied in the ERC Advanced Grant 'SmartCast' project.

In view of extrusion-based three-dimensional (3D) printing of concrete [3], the required change in rheological properties is even more drastic and rapid. While pumping the fresh concrete or mortar to the nozzle, it should be sufficiently flowable. Within the nozzle, the material should be extrudable. Immediately after extruding, the fresh concrete layers should stiffen fast to reach a good buildability. However, it should also not stiffen too fast, in view of still obtaining appropriate bond properties between the subsequent layers.

With current technologies, we can only rely in a rather passive way on the evolving rheological response of the fresh concrete within the different processing steps. A new methodology, currently under development at Ghent University, envisages active control methods to adjust the rheology of the fresh concrete while processing (i.e. pumping, formwork casting, 3D printing, etc.), and is named active rheology control (ARC), also including active stiffening control (ASC) [4]. This chapter briefly summarizes two different routes to achieve ARC (schematically represented in Figure 9.1), as explained in more detail in previous chapters.

9.2   RESPONSIVE MINERAL ADDITIONS

The first route to obtain ARC is based on the concept of magneto-rheological fluids (MRF). Magnetizable solid particles are added to the cementitious material, e.g. magnetizable nanoparticles (MNPs) consisting of Fe_3O_4 or more traditional cement-replacing materials like fly ash with some magnetic response. Upon applying a magnetic field, the magnetizable particles can

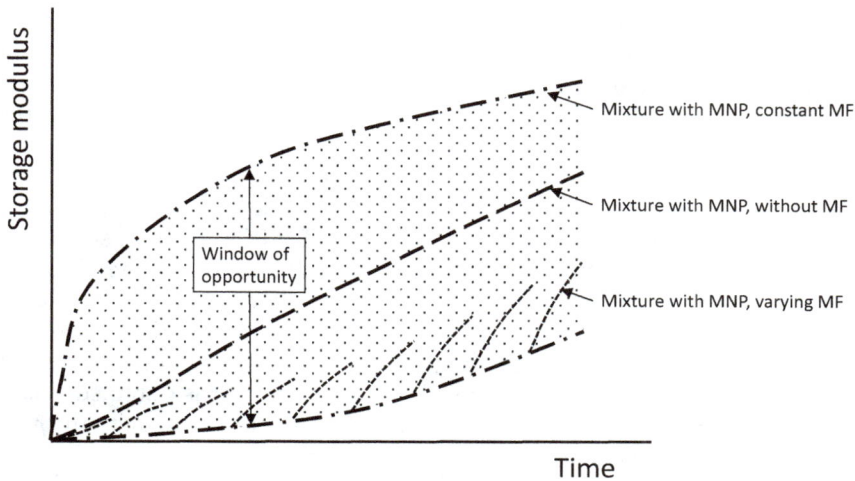

Figure 9.2 Schematic window of opportunity for active magneto-rheological control of storage modulus of cement paste containing MNP.

form clusters, inducing stiffening on demand. Alternatively, the acceleration of the magnetizable particles due to the electromagnetic forces temporarily gives the cementitious material some more liquid-like properties. Variable magnetic fields can be applied to actively control rheology, using the magnetizable particles as internal actuators.

The magnetic response of a cementitious material containing magnetizable solid particles depends on several factors, as studied in detail by Jiao [5] (see also Chapter 5). As a conceptual illustration, Figure 9.2 shows the window of opportunity for active magneto-rheological control of a cement paste containing MNPs. The evolution of the storage modulus (indicative for the elastic properties or stiffness of the material) in absence of magnetic field is shown as the central curve. This curve depends on different parameters, including water/cement ratio and MNP concentration. When a constant magnetic field is applied, a controlled faster stiffening is obtained as shown by the top curve. In case of a varying magnetic field, a controlled reduced stiffening is obtained as shown by the staggered lower curve. In principle, the area in between these two extreme curves, shown as the dotted area, is the potential field of application for active rheology control for the considered cement paste with MNP.

9.3 SWITCHABLE ADMIXTURES

A highly innovative and groundbreaking route is the development of switchable superplasticizers to enable ARC. This technology further builds on the working principle of the latest generation of superplasticizers, which are comb copolymers consisting of a backbone with negative charges in view of adsorption onto the surface of cement grains and of side chains in view of providing steric hindrance to disperse cement particles. Instead of using classical polycarboxylate ethers (PCE), additional functionalities are given to the superplasticizers by incorporating responsive elements into the polymer architecture.

One example is a comb copolymer consisting of a classical backbone with negative charges (methacrylic acid) and of side chains of a new type, namely 2,2,6,6-tetramethyl-1-piperidinyloxy (TEMPO) [6]. This side chain on the one hand can provide steric hindrance, and on the other hand can be triggered by an electric signal to become positively charged. Upon electric triggering, the positive charges of TEMPO+ will counterbalance negative charges form the backbone, reducing the adsorption capacity to the cement grain surface. In this way, a controllable superplasticizer is obtained, enabling active rheology control by moderating the adsorption capacity through external electric signals. In principle, this control mechanism is reversible. The responsiveness and controlled adsorption of the newly developed polymer is discussed in more detail in Chapter 4. The potential of the newly developed switchable superplasticizer is conceptually illustrated in Figure 9.3. Implementing the newly developed admixture, initially without external control signal, leads to a reduced flow curve (lower shear stress as a function of shear rate) compared to the reference paste, showing the plasticizing effect. Upon application of a redox signal, pTEMPO is given a positive

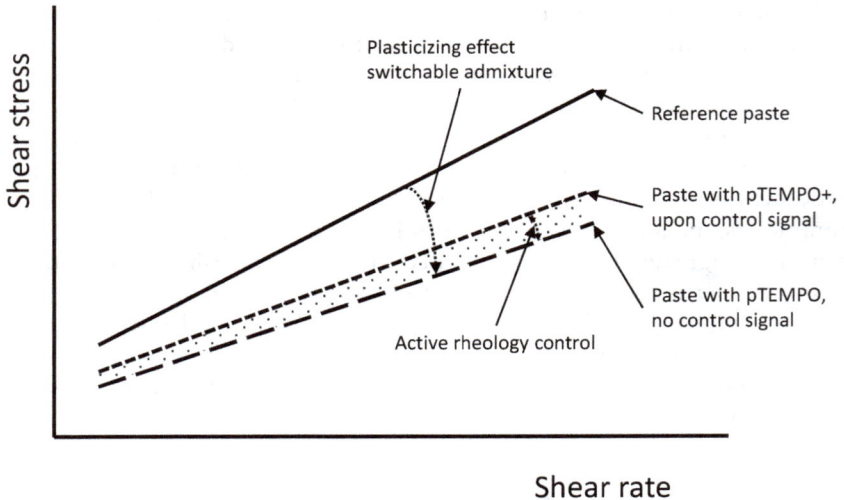

Figure 9.3 Conceptual illustration of the potential of active rheology control by means of redox-switchable superplasticizers.

charge, resulting in pTEMPO+. This is influencing (reducing) the adsorption to the cement grains, and thus the plasticizing effect, as illustrated by the higher flow curve. The dotted area conceptually shows a potential window of operation for the ARC by means of the redox-switchable plasticizer. In a similar way, an active stiffening control can also be obtained, adjusting the storage modulus on demand by means of a redox signal, as shown in more detail in Chapter 4.

Another example of a switchable superplasticizer is a magnetizable polymer composite consisting of polymer chains linking MNP (by means of a chelator) to the surface of a cement grain (by means of electrostatic adsorption), as discussed in more detail in Chapter 4. These polymers act as a superplasticizer due to adsorption on the cement surface and steric hindrance by the side chains. By means of an externally applied magnetic field, the MNP will try to cluster in a similar way as the previously mentioned responsive mineral additions, now also involving the cement particles to which the polymers are electrostatically adsorbed. This can provide a strong active stiffening effect.

9.4   CHALLENGES

At this moment, a laboratory proof of concept has been successfully obtained at paste level, both for the approach using responsive mineral additions and the approach applying switchable responsive plasticizers. This shows the great potential of ARC and ASC of cementitious materials. Nevertheless, it has to be mentioned that more in-depth scientific validation remains to be obtained regarding the fundamental mechanisms behind the newly developed technology, especially in the case of the redox-switchable superplasticizers. At this moment, sound scientific experimental characterization results are available regarding the obtained architecture of the synthesized polymers, showing that the intended polymer structure can be obtained. Fundamental experiments at polymer level clearly show their responsiveness, as intended. Indirect information obtained by means of zeta potential supports the hypothesis regarding the controllable adsorption to cement grains in case an electric (redox) signal is applied. The resulting active rheology and stiffening control has been illustrated by means of fundamental experiments, applying screen-printed electrodes giving the signal to small paste samples. All elements contribute to a validation of the hypothesis regarding underlying mechanisms, however, without more direct and more formal proof. Further fundamental research is needed to validate or falsify the working mechanisms as hypothesized, e.g. by means of molecular dynamics simulations based on first principles. This was beyond the scope of the SmartCast project.

Besides the further in-depth studies, several challenges remain in view of upscaling the newly developed technology to concrete level in industrial environment. A main challenge is the mode of application of the external signals, in a safe way, and with sufficient effect inside a larger material volume. In this respect, application of a magnetic signal will be more straightforward than application of an electric (redox) signal. Nevertheless, all external signals need to be studied in terms of signal strength (which level?), signal mode (constant or variable?), signal location (locally or more extended?) and signal duration (when and how long?). These questions further need to be combined with safety regulations related to the impact of electromagnetic or other fields on humans, animals and environment. And finally, the extra cost for implementation of ARC and/or ASC (materials, equipment and qualified staff) needs to be considered in relation to potential savings due to more reliable processes, improved quality and less processing failures.

Optimization of polymer architecture and plasticizer dosage also need further investigation. First applications are focussing on pumping lines and 3D printing nozzles. The scientific breakthrough regarding ARC of concrete in general, and regarding switchable superplasticizers more specifically, has been successfully reached. Nevertheless, further technological and valorization studies are still needed (and have been initiated) to bring the new technology to the market.

9.5 CONCLUSIONS

At a fundamental materials science level, the concept of ARC, including ASC, has been shown at paste level following two different routes.

The first route is based on the technology of magneto-rheological fluids, incorporating magnetizable solid particles within the cementitious materials that can be triggered by an external magnetic field.

The second route is based on the development of a new category of switchable superplasticizers that can be triggered by an external signal.

While a laboratory proof of concept has been obtained at paste level, further upscaling and valorization studies are ongoing in view of bringing the new technology to concrete industry.

REFERENCES

[1] Van Der Vurst F., Grünewald S., Feys D., Lesage K., Vandewalle L., Vantomme J., and De Schutter G. (2017) Effect of mix design on the robustness of fresh self-compacting concrete, *Cement and Concrete Composites*, 82, 190–201.

[2] Feys D., De Schutter G., Khayat K., and Verhoeven R. (2016) Changes in rheology of self-consolidating concrete induced by pumping, *Materials and Structures*, 49, 4657–4677.

[3]   De Schutter G., Lesage K., Mechtcherine V., Nerella V.N., Habert G., and Agusti-Juan I. (2018) Vision of 3D printing with concrete – technical, economic and environmental potentials, *Cement and Concrete Research*, 112, 25–36.

[4]   De Schutter G. (2017) *Smart casting of concrete structures by active control of rheology*, video presenting the ERC Advanced Grant Project 'SmartCast', YouTube, www.youtube.com/watch?v=ZgVwNkyvJFg&feature=youtu.be.

[5]   Jiao D. (2021) *Active rheology control of cementitious materials using magnetic field*, PhD dissertation, Ghent University, Belgium, 2021.

[6]   Ezzat M. (2022) *Concrete superplasticizers: From fundamental studies to the development of smart superplasticizers for rheology control*, PhD dissertation, Ghent University, Belgium, 2022.

Index

Note: The given page number refers to the start of the relevant section or chapter giving information regarding the indexed keyword.

accelerator 13, 66
activation energy 13
additive 50, 57, 69, 82, 112, 119, 126, 184, 191, 193, 199, 214, 216, 221, 230
additive manufacturing 66, 221, 229; *see also* printing
adhesion 19, 160, 161
admixture 2, 9, 13, 38, 41–3, 53, 119, 138, 221, 271
adsorption 3, 50, 141, 146, 149, 154, 161, 271, 274
agglomeration 3, 50, 82, 102, 111, 126, 141, 152, 203, 216, 224, 251
align(ment) 22, 86, 120, 158, 161, 163, 189, 201
anisotropic-isotropic transition 20, 21
anisotropy 20, 21, 86, 235, 260, 261
antiferrimagnetism 78
Arrhenius 13

Bagnold number 177
Bingham 22, 41, 47, 53, 57, 63, 82, 138, 141, 172, 175, 235, 238, 246, 254
Bingham number 172
biosensor 19, 139
bituminous materials 15, 86
blocking 3, 56, 57, 177, 184, 191, 195, 201, 203, 208, 209, 211, 213
Brownian motion 22, 223
Buckingham-Reiner 57, 172, 246, 254
buildability 66, 221, 229, 230, 258, 261, 269
buoyancy 22, 82

carbonyl iron powder (CIP) 15, 78, 112, 193
CFD 3, 244, 246, 260, 261
charge 21, 50, 86, 141, 146, 161, 178, 251, 271
clay 3, 20, 66
cleavage 21, 22
cluster 3, 22, 82, 95, 98, 104, 108, 111, 113, 117, 126, 203, 209, 210, 213, 214, 224, 270, 271
coercive field 78, 258
coercivity 93, 113, 122, 152
cold joint 62, 66, 69
collective magnetism 78
colloid 25, 43, 138
colloidal force 42, 138
colloidal interaction 22, 95, 102, 138
colloidal structure 111
concrete: concrete industry 149, 191, 274; concrete pH 21, 25; fibre reinforced concrete 15, 78, 120; self-compacting concrete (SCC) 2, 9, 53, 57, 62, 63, 189, 210, 216; sprayed concrete 13
conductivity 7, 21, 26, 93, 139
consistency factor 41
construction industry 7, 8, 26, 63
control: chemical intervention 13, 66, 95, 139, 184, 221, 229; electric intervention 17, 95, 139, 139, 141, 161, 271, 273; electro-rheological control 22, 69, 139, 146, 163; hygral intervention 14; intervention by pressure 15; magnetic intervention 15, 77, 139, 140, 149, 163, 181,

184, 189, 221, 233, 244, 246, 270, 271; magneto-rheological control 22, 69, 82, 86, 122, 140, 149, 163, 177, 181, 184, 189, 221, 233, 244, 246; mechanical intervention 11, 86; mechanical vibration 11, 86, 177, 184, 221; microwaves 18, 78, 221; thermal intervention 13
cooling 13, 21
corrosion 17
Coulomb friction 3, 56, 177
crack 13, 14, 78, 86, 120, 191
critical shear rate 172
critical shear strain 50, 66, 156, 157
critical temperature 78
crosslinking 20, 21
crystal 86, 93, 193
crystalline 86, 93, 113, 118, 119
crystallization 21
Curie temperature 78
curing 9, 13–15, 17
cyclic voltammetry (CV) 141

damage 8, 191
damper 20, 82
damping 9
demoulding 17
desiccation 14, 18
diamagnetic material 78
diamagnetism 78
dielectric 18, 26, 120, 139
diffusion 13, 14
diffusivity 13
digital design 66
digital fabrication 26
discharge 3, 9, 56, 57, 62, 69, 172, 175, 178, 181, 246, 251, 254, 258
dissipation 22, 95
Doppler effect 178
drag 82, 86, 122
drug delivery, drug release 19, 21, 22, 95, 139, 140
dry friction flow 171, 177, 184
durability 3, 9, 15, 50, 54, 66, 214, 269
dynamic light scattering 141

Edison 3
electrochemical 21, 45, 141, 184, 260
electrode 44, 141, 163, 273
electromagnet 86, 181, 184, 229
electron paramagnetic resonance (EPR) 141

electro-osmosis 3
electrostatic attraction 146, 271
electrostatic force 141
electrostatic interaction 138
electrostatic potential 141
electrostatic repulsion 22, 25, 43, 50
ERF 9, 20, 22
extrudability 63, 66, 221, 224, 269
extruder 63, 229
extrusion 63, 66, 69, 184, 221, 224, 230, 269

factory 3, 9
ferrimagnetic material 78, 122, 152
ferrimagnetism 78
ferromagnetic material 78, 86, 93, 112, 113, 119, 120, 122
ferromagnetism 78
fibre 15, 20, 78, 95, 120
filling 3, 53, 62, 178, 189, 195, 210, 213, 216, 269
filling ability 53
filling rate 56, 62
filtration 191, 208, 211, 214
finite volume method (FVM) 244
flocculation 42, 57, 66, 95, 102, 104, 158, 235, 242, 254
flow 3, 38, 43, 50, 53, 57, 93, 111, 120, 138, 141, 154, 171, 189, 221, 224, 233
flow area 203
flow curve 40, 46, 141, 154, 238, 242, 271
flow index 41
flow meter 178
fluid: electrorheological fluid (ERF) 9, 20, 22; magnetorheological fluid (MRF) 9, 20, 22, 82, 86, 93, 149, 233, 235, 238, 258, 260, 270, 274; Newtonian 22, 41, 82, 172, 244
fly ash 15, 17, 112, 184, 193, 199, 209, 210, 213, 216, 233, 258, 270
foaming 9
formwork: bottom-up pumping 3, 56, 62, 210; casting, filling 3, 56, 62, 111, 126, 138, 189, 195, 210, 213, 216, 244, 269; filling rate 62; joint 189, 191, 195, 216, 260; leakage 3, 56, 69, 189, 258, 260, 261; pressure 3, 56, 62, 69, 138, 189, 269; tightness 3, 56, 189, 216; top-down casting 3, 56, 62

freeform construction 66
frequency 11, 15, 18, 26, 42, 46, 47, 178, 224, 229, 241

gel 20, 21, 78, 139
generalized Newtonian fluid (GNF) 244

Hagen 172
hardening 13, 15, 17, 18
healing 15, 25
heat 13, 15, 17, 18, 86, 126, 153, 172, 229
heating 17, 21, 86, 221, 229
Herschel-Bulkley 41, 47, 63
hydration 3, 13–15, 18, 42, 50, 57, 66, 69, 78, 95, 108, 117, 138, 141, 153, 156, 158, 181
hydrodynamic 86, 111
hydrogel 14, 19–22, 25, 26
hydrophilic 19, 21, 150
hydrophobic 21
hydrophobicity 22

ion concentration 25
isothermal calorimetry 153
isotropy 20, 21, 120

Krieger-Dougherty 22

light 21, 26
limestone 111, 126, 223
Lorentz force 251
LS 50
lubrication 171, 172, 175, 184, 246, 260
lubrication flow 175
lubrication layer 3, 17, 26, 56, 57, 63, 66, 78, 171, 175, 260
LVER 42, 47, 138, 156, 224, 235

MAA 50, 141
macromolecule 139
magnetic acceleration 235, 260, 261, 270
magnetic chain 22, 82, 95, 224
magnetic cluster 82, 98, 104, 108, 111, 113, 117, 126, 203, 209, 210, 213, 214, 223, 224, 230, 233, 251, 254, 261, 270, 271
magnetic field 2, 3, 7, 9, 15, 17, 21, 22, 45, 69, 77, 115, 140, 149, 152, 156, 158, 160, 161, 163, 181, 184, 189, 221, 233, 270, 271, 273

magnetic force 82, 86, 95, 98, 126, 181, 210, 223, 235, 260, 270
magnetic saturation 78, 86, 93, 113, 115, 117, 118, 122, 126, 149, 209, 211, 213, 224, 235, 238
magnetic structuration 15, 86, 122, 181, 224, 233
magnetic yield parameter 82, 98, 102, 104, 126, 258
magnetism 78, 251
magnetization hysteresis 78, 122
magneto-hydrodynamics (MHD) 251, 261
manufacturing industry 66
masonry 3
membrane 21
memory 20
metal 66, 122
metal scrap 120
methacrylic acid (MAA) 50, 141
micelles 21
micellization 21, 22
micro-actuation 22
micro-actuators 19, 139, 270
micro-agitation 86, 122, 223
micro-movement 86, 160, 223
microspheres 20
micro-vibration 22, 86, 120, 184, 229, 235, 260, 261
micro-vibrator 22
mineral 2, 3, 7, 63, 77, 138, 271, 273
mixing 22, 66, 120, 138, 163, 199, 221, 224, 269
modulus: loss modulus 22, 42, 98, 102, 104, 122, 126, 156–8, 160, 161; storage modulus 22, 42, 47, 95, 98, 102, 104, 108, 115, 117, 122, 126, 138, 146, 156–8, 160, 161, 205, 209, 223, 224, 238, 270, 271
morphology 15, 78, 93, 117
MRF 9, 20, 22, 82, 86, 93, 149, 233, 235, 238, 258, 260, 270, 274

nanoclay 20
nano-Fe$_3$O$_4$ 20, 78, 82, 93, 140, 149, 163, 193, 199, 209, 216, 223, 224, 230
nanomotors 19
nanoparticles 20, 22, 78, 82, 93, 118, 126, 139, 141, 199, 209, 216, 224, 230, 270
nanotubes 20

Non Plus 3
nucleation 42, 138

opacity 26
opaque 26
orientation 15, 21, 22, 78, 120, 122
oxidation 113, 139, 141

PANI 139
paramagnetism 78, 93, 122, 152
particle: agglomeration 3, 50, 82, 102, 111, 126, 141, 152, 203, 216, 224, 251; cluster, clustering 1, 22, 82, 95, 98, 104, 108, 111, 113, 117, 126, 203, 209, 210, 213, 214, 224, 270, 271; concentration 22, 82, 98, 102, 104, 116, 126, 199, 205, 208, 209, 213, 216, 235, 260, 270; core-shell type particle 20; flow interaction 3, 95, 102, 104; friction 177; magnetic nanoparticle (MNP) 20, 77, 140, 152, 205, 209, 213, 224, 270; micro-vibration 22, 86, 120, 184, 229, 235, 260, 261; mineral particle 2, 3, 63, 77, 138, 271, 273; mineral powder 3, 7; responsive particle 22, 77, 233, 270, 273; settling 82; shape 86, 93, 117, 118, 189; size 22, 38, 43, 57, 66, 82, 86, 93, 104, 118, 122, 126, 149, 152, 189, 191, 193, 195, 203, 209; size ratio 203; translation 39, 86
passing ability 53, 54
PCE 50, 98, 126, 138, 271
PEGMA 149, 150, 153, 154, 160
permeability 21, 63, 78, 82, 86, 122
phase angle 42, 95, 102, 122, 156, 158, 160, 161
phase transition 139
Poiseuille 57, 171, 184
polarization 3, 17
polyaniline (PANI) 139
polymer: architecture 50; concentration 141, 154, 160; dispersity 150; electroactive (EAP) 139; electro-responsive 19, 21, 22, 139; entanglement 21; hydrogel 14, 19–22, 25, 26; hyperbranched 20; isomerization 21; redox-responsive 21, 141, 163, 271, 273; responsive 2, 3, 7, 19, 69, 137, 233, 271, 273; shape 139; shape memory 20; size 141; stimuli-responsive 19, 137, 271,

273; superabsorbing (SAP) 14, 20, 25; synthesis 141, 150; vitrimer 20
polythiophene (PT) 139
porosity 15, 18, 78
precast industry 2, 3, 7, 26, 53
pressure number 172, 175, 246
pressure sensor 177, 181, 184, 246
printing: 3D printing 2, 7, 26, 45, 63, 66, 69, 78, 138, 184, 221, 244, 258, 260, 261, 269, 273; 4D printing 26; head 26, 66, 221, 229, 230; nozzle 63, 66, 221, 230, 258, 260, 269, 273
processing: extrusion 63, 66, 69, 184, 221, 224, 230, 269; pumping 2, 9, 11, 13, 17, 25, 56, 57, 62, 63, 66, 69, 78, 126, 138, 171, 221, 224, 242–4, 246, 258, 261, 269, 273; see also pumping
PSS 139
pumpability 9, 57, 66, 69, 78, 111, 221, 230
pumping: design chart 57; lubrication layer 3, 17, 26, 56, 57, 63, 66, 78, 171, 175, 260; plug zone 56, 57, 63, 66, 120, 246, 254, 258; see also processing

RAFT 141
ready-mix industry 53
redox 21, 139, 141, 271, 273
remanence 122, 152, 258, 260
remanent magnetization 78, 86, 122, 152, 224, 230
response 21, 39, 42, 45, 47, 78, 93, 112, 119, 126, 138, 139, 141, 163, 223, 233, 269, 270
retardation 138, 141, 153
retarder 13
rheology: elastic response 22, 42, 95, 238, 239, 241; linear viscoelastic region (or regime) (LVER, LVR) 42, 47, 138, 156, 224, 235; numerical simulation 181, 233; small amplitude oscillatory shear (SAOS) 22, 47, 95, 122, 156, 158, 160, 163, 238, 239, 241; viscous response 22, 42, 47, 95
rheometer 40, 42, 43, 86, 122, 199, 205, 214, 224, 238, 243, 246; parallel plate 44, 111, 122, 193, 199, 224, 238, 246
rheometry 43, 146, 154, 224; creep test 47; flow curve 45, 141, 199, 205,

238, 242; inertia effects 46; linearity limit 47; oscillatory rheometry 22, 47, 95, 138, 156, 158, 160, 199, 224, 238; pre-shear 46, 224; protocol 45, 205, 224; steady flow 45, 86, 95, 102, 108, 111, 115, 117, 158, 160
robustness 3, 47, 56

safety 8, 27, 273
SAOS 22, 47, 95, 122, 156, 158, 160, 163, 238, 239, 241
SAP 14, 20, 25
saturation magnetization 78, 86, 93, 113, 115, 117, 118, 122, 126, 149, 209, 211, 213, 224, 235, 238
segregation 3, 9, 53, 54, 56, 57, 62, 66
self-assembly, self-assemble 21, 139
self-compacting concrete (SCC) 2, 9, 53, 57, 62, 63, 189, 210, 216
self-consolidating concrete see self-compacting concrete
separation 113
set-on-demand 9
setting 3, 13, 15, 18, 69
shape-morphing 21
shear rate 38–41, 44–6, 50, 86, 111, 126, 138, 205, 214, 224, 235, 244, 254, 271
shear strain rate see shear rate
shear stress 38–41, 44–6, 54, 57, 86, 111, 120, 138, 171, 172, 223, 224, 241, 244, 271
shear-thickening 3, 22, 41, 56, 172, 175
shear-thinning 22, 41
shrinkage 14, 78
signal 69, 221, 229, 230, 260, 271, 274; light 21, 26; redox 21, 139, 141, 271, 273; see also control
slag 122, 126
slipform 3
slippage 171, 175, 184, 246, 260
Smart Dynamic Casting (SDC) 3
smart windows 19
SMF 50
SNF 50
soft robotics 19, 139
solver 244, 246, 261
sorption 21
stability 3, 26, 50, 54, 57, 66, 78, 82, 244, 246

steric hindrance, steric effects 22, 25, 50, 138, 141, 146, 149, 152
stiffening see active stiffening control
strength 3, 13, 15, 17, 18, 45, 50, 54, 78, 86, 108, 111, 113, 120, 126, 158, 191, 193, 199, 211, 213, 214, 216, 221, 223, 224, 229, 269, 273
structural breakdown 42, 108, 126, 233
structural build-up 3, 7, 42, 66, 95, 102, 104, 108, 111, 115, 117, 122, 126, 138, 146, 156, 158, 160, 161, 163, 191, 203, 224, 229, 230, 233, 235, 238, 241, 251, 254, 258, 260, 261
structuration 15, 45, 66, 86, 122, 138, 213, 233
sulfonated polystyrene (PSS) 139
superconducting materials 78
superplasticizer 2, 3, 22, 50, 137; adsorption 3, 50, 141, 146, 149, 154, 161, 271, 273; lignosulfonates (LS) 50; polycarboxylate ethers (PCE) 50, 98, 126, 138, 271; polycondensates of sulfonated melamine formaldehyde (SMF) 50; polycondensates of sulfonated naphtalene formaldehyde (SNF) 50; polymer architecture 50; switchable superplasticizer 2, 137, 271, 273
suspension 22, 25, 44, 50, 57, 63, 78, 82, 86, 95, 98, 102, 104, 108, 111, 113, 117, 126, 141, 171, 177, 184
swelling 20–2, 25, 139

TEMPO 141
thixotropy 3, 22, 42, 46, 50, 56, 57, 69, 95, 158, 172, 177, 181, 189, 191, 210, 229, 233, 235, 239, 242, 246, 254
TMPMA 141
transmission 26, 152

Ultrasonic Velocity Profiler (UVP) 178

vacuum 15, 82, 86
valve 82, 193
Van der Waals forces 22, 43, 138
vibrating sample magnetometer (VSM) 122, 149
vibration 2, 3, 8, 9, 11, 15, 63, 86, 120, 184, 189, 221, 229, 235, 260, 261

viscoelasticity, visco-elasticity 42, 45, 47, 82, 98, 138, 146, 156, 157, 158, 160, 205, 223, 224
viscosity 3, 9, 22, 41, 42, 46, 53, 57, 63, 66, 82, 86, 98, 111, 126, 138, 141, 154, 158, 172, 175, 199, 205, 209, 210, 216, 223, 242, 244, 246, 254, 260
viscosity modifying admixture (VMA) 54
viscous constant 175
vitrimer 20
volume of fluid (VOF) method 244

workability 9, 15, 25, 38, 42, 43, 50, 54, 66, 111, 120, 138, 163, 195, 269

yield stress 3, 9, 11, 18, 22, 41, 42, 46, 47, 50, 53, 54, 56, 57, 63, 66, 82, 86, 98, 111, 120, 126, 138, 141, 149, 154, 172, 175, 193, 199, 201, 210, 216, 221, 233, 235, 238, 239, 242, 244, 251, 254, 258, 260; dynamic yield stress 47; static yield stress 47, 50, 66; yield descriptors 47

zeta potential 141

For Product Safety Concerns and Information please contact our EU
representative GPSR@taylorandfrancis.com
Taylor & Francis Verlag GmbH, Kaufingerstraße 24, 80331 München, Germany